T0130896

INTERNATIONAL
WILDLIFE MANAGEMENT

Wildlife Management and Conservation
Paul R. Krausman, Series Editor

INTERNATIONAL WILDLIFE MANAGEMENT

Conservation Challenges in a Changing World

EDITED BY
John L. Koprowski
and Paul R. Krausman

Published in Association with *THE WILDLIFE SOCIETY*

JOHNS HOPKINS UNIVERSITY PRESS | BALTIMORE

© 2019 Johns Hopkins University Press
All rights reserved. Published 2019
Printed in the United States of America on acid-free paper
9 8 7 6 5 4 3 2 1

Johns Hopkins University Press
2715 North Charles Street
Baltimore, Maryland 21218-4363
www.press.jhu.edu

Library of Congress Cataloging-in-Publication Data

Names: Koprowski, John L., editor. | Krausman,
    Paul R., editor.
Title: International wildlife management : conservation
    challenges in a changing world / edited by John L.
    Koprowski and Paul R. Krausman.
Description: Baltimore : Johns Hopkins University Press,
    2019. | Series: Wildlife management and conservation |
    Includes bibliographical references and index.
Identifiers: LCCN 2019004542 | ISBN 9781421432854
    (hardcover : alk. paper) | ISBN 1421432854 (hardcover :
    alk. paper) | ISBN 9781421432861 (electronic) |
    ISBN 1421432862 (electronic)
Subjects: LCSH: Wildlife conservation—Management. |
    Wildlife conservation—Government policy.
Classification: LCC QL82 .I59 2019 | DDC 333.95/4—dc23
LC record available at https://lccn.loc.gov/2019004542

A catalog record for this book is available from the British
Library.

Special discounts are available for bulk purchases of this
book. For more information, please contact Special Sales
at 410-516-6936 or specialsales@press.jhu.edu.

Johns Hopkins University Press uses environmentally
friendly book materials, including recycled text paper that is
composed of at least 30 percent post-consumer waste,
whenever possible.

To our students, who employ their passion,
creativity, and technical expertise to
manage and conserve wildlife around the globe

Contents

Contributors

Marissa C. G. Altmann
Wildlife Friendly Enterprise
Network
433 Sprout Path NW, Bainbridge
Island, WA 98110, USA

Walt Anderson
Environmental Studies and
Sustainability
Prescott College
220 Grove Avenue, Prescott,
AZ, 86301, USA

Victoria L. Atkin Dahm
Department of Biology
Southern Connecticut State
University
New Haven, CT, 06515, USA

Karen Bailey
Environmental Studies Program
University of Colorado
Sustainability, Energy, and
Environment Complex
4001 Discovery Drive, Boulder,
CO 80309-0397, USA

Gabrielle Beca
School of Biological Sciences
University of Western Australia

35 Stirling Highway, Crawley WA
6009, Australia

Delwin E. Benson
Professor and Extension Wildlife
Specialist
Department of Fish, Wildlife, and
Conservation Biology
Colorado State University,
Fort Collins, CO 80523, USA

Sandro Bertolino
Department of Life Sciences and
Systems Biology
Università degli Studi di Torino
Via Accademia Albertina 13,
I-10123 Torino, Italy

Hsiang Ling Chen
Department of Forestry
National Chung Hsing
University
145, Xingda Rd., Taichung 402,
Taiwan

Amar N. Choudhary
Traffic India
WWF India Secretariat
172 B, Lodi Estate, New Delhi
110003, India

David Christianson
School of Natural Resources and
the Environment
University of Arizona
Tucson, AZ 85721, USA

Thomas K. Frazer
School of Natural Resources and
Environment
103 Black Hall, PO Box 116450
Gainesville, FL 32611-0430,
USA

Mauro Galetti
Departamento de Ecologia
Universidade Estadual Paulista
(UNESP)
13506-900 Rio Claro, São Paulo,
Brazil

José F. González-Maya
Proyecto de Conservación de
Aguas y Tierras
ProCAT Colombia/Internacional,
Bogotá, Colombia

S. P. Goyal
Wildlife Institute of India
Chandrabani, Dehradun 248007,
India

Deborah M. Hahn
International Policy Director
Association of Fish and Wildlife
Agencies
1100 First Street, NE, Suite 825,
Washington, DC 20002, USA

Menna Jones
School of Biological Sciences
University of Tasmania
Hobart, Tasmania, Australia

Marta A. Jarzyna
Department of Evolution
Ecology and Organismal Biology
Ohio State University,
Columbus, OH 43210, USA

Joseph M. Kiesecker
Global Conservation Lands
Program
The Nature Conservancy
Fort Collins, CO 80524, USA

John L. Koprowski
School of Natural Resources and
the Environment
University of Arizona
Tucson, AZ 85721, USA

Paul R. Krausman
School of Natural Resources and
the Environment
University of Arizona
Tucson, AZ 85721, USA

Yves Lecocq
DVM
Secretary-General (retired) FACE
Brussels, Belgium

Shane P. Mahoney
President and CEO

Conservation Visions
PO Box 5489, Stn. C, 354 Water
Street, St. John's, NL, A1C 5W4,
Canada

Adriano Martinoli
Environment Analysis and
Management Unit, Guido Tosi
Research Group
Department of Theoretical and
Applied Sciences
Università degli Studi
dell'Insubria
Via J. H. Dunant, 3-I-21100
Varese, Italy

Robert A. McCleery
Department of Wildlife Ecology
and Conservation
110 Newins-Ziegler Hall, PO Box
110430, Gainesville, FL 32611-
0430, USA

Gonzalo Medina-Vogel
Centro de Investigación para la
Sustentabilidad
Universidad Andrés Bello
Santiago, Chile

Shekhar K. Niraj
Additional Principal Chief
Conservator of Forests,
and Director,
Advanced Institute for Wildlife
Conservation
Vandalur, Chennai 600048,
Tamil Nadu State, India

David E. Naugle
Wildlife Biology Program
University of Montana
Missoula, MT 59812, USA

Colman O Criodain
Policy Manager, Wildlife Practice
WWF International
PO Box 62440-00200, Nairobi,
Kenya

John F. Organ
US Geological Survey
Reston, VA, 20191 USA

Shambhu Paudel
School of Natural Resources and
the Environment
University of Arizona
Tucson, AZ 85721, USA

William F. Porter
Department of Fisheries and
Wildlife
Michigan State University
East Lansing, MI, 48823, USA

Ronald J. Regan
Executive Director
Association of Fish and Wildlife
Agencies
1100 First Street, NE, Suite 825,
Washington, DC 20002, USA

Carlos Ruiz-Miranda
Laboratório de Ciências Ambientais
Universidade Estadual do Norte
Fluminense
Campos dos Goytacazes, Rio de
Janeiro, Brazil

Andrea Santangeli
The Helsinki Lab of Ornithology,
Finnish Museum of Natural
History
University of Helsinki
PO Box 17, Helsinki, FI-00014,
Finland

Shreya Sethi
Indian Institute of Technology
Department of Humanities and
Social Science
Powai, Mumbai 400076, India

Julie T. Shapiro
School of Natural Resources and
Environment
Department of Wildlife Ecology
and Conservation
110 Newins-Ziegler Hall,
PO Box 110430, Gainesville,
FL 32611-0430, USA

Uday R. Sharma
Public Sector Specialist Forestry
IMC Worldwide
Kathmandu, Nepal

Craig Spencer
Transfrontier Africa
PO Box 1187, Hoedspruit, 1380,
South Africa

Ronald R. Swaisgood
Institute for Conservation Research
San Diego Zoo Global
15600 San Pasqual Valley Road,
Escondido, CA 92027-7000, USA

Chiachun Tsai
Department of Forestry and
Natural Resources
Purdue University
195 Marsteller Street, West
Lafayette, IN 47907, USA

Bob van den Brink
Boerema & van den Brink
B.V./Wildlife Europe
The Netherlands

Basile van Havre
Director General Domestic and
International Biodiversity Policy
Canadian Wildlife Service
Environment and Climate
Change Canada
351 St. Joseph, Gatineau, QC,
K1A 0H3, Canada

Wouter van Hoven
Faculty of Economic and
Management Sciences
North West University, South
Africa

Dun Wang
Department of Entomology
Northwest A&F University
Yangling, Shaanxi, 712100, P. R.
China

Lucas A. Wauters
Environment Analysis and
Management Unit, Guido Tosi
Research Group
Department of Theoretical and
Applied Sciences
Università degli Studi
dell'Insubria
Via J. H. Dunant, 3-I-21100
Varese, Italy

Samantha M. Wisely
Department of Wildlife Ecology
and Conservation
110 Newins-Ziegler Hall,
PO Box 110430, Gainesville,
FL 32611-0430, USA

Wei Hua Xu
Research Center for
Eco-Environmental Sciences

Chinese Academy of Sciences
18 Shuangqing Road, Haidian
District, Beijing, 100085, China

Tsuyoshi Yoshida
EnVision Conservation Office
Sapporo, Japan

Diego A. Zárrate-Charry
Forest Biodiversity Research
Network
Department of Forest Ecosystems
and Society
College of Forestry, Oregon State
University
Corvallis, OR 97331, USA

Jing Jing Zhang
Research Center for
Eco-Environmental Sciences
Chinese Academy of Sciences
18 Shuangqing Road, Haidian
District, Beijing, 100085,
China

Patrick Zollner
Department of Forestry and
Natural Resources
Purdue University
195 Marsteller Street, West
Lafayette, IN 47907, USA

Benjamin Zuckerberg
Department of Wildlife and
Forest Ecology
University of Wisconsin–Madison,
Madison, WI, 53706, USA

Preface

Like many in our profession, the authors of this volume were inquisitive about the natural world early in life, and that passion was fomented by experiences with family, friends, and teachers and by numerous excursions into deserts and forests. Many of us digested anything about wildlife we could find, in the wild or from outdoor and National Geographic magazines, Disney movies, and Jacques Cousteau television specials, and from Mutual of Omaha's *Wild Kingdom* (for the more senior members of our profession) to more contemporary daily doses of Animal Planet, the Discovery Channel, or online 24-hour zoo webcams. We learned about our local wildlife, and, growing up and living around the world, we were exposed to the diversity and challenges of wildlife in exotic locations. As professionals involved in wildlife management and conservation, most of us received broad training in our undergraduate career. Graduate training included increased specialization as we focused on disciplinary knowledge and skills to assist us in our research specialty. Careers progressed, but those early images of imperiled wildlife from distant lands continued to call us to seek involvement in international challenges but with little knowledge of how to do so. Our careers have enabled us to build international collaborations, host international students in our research groups, and facilitate our students' international experience. Hard work, passion, and understanding families have been a large part of this success, but good fortune has also played a significant role. We wanted to compile a volume that addressed the major challenges that we share in wildlife conservation around the world, written by a diverse group of international scientists with similar aspirations. Furthermore, we wanted to provide advice on how to get involved in international wildlife management early in a career, to minimize reliance on simple good fortune! *International Wildlife Management* includes the contributions of scientists from 17 countries on six continents in an effort to provide broad perspective as part of a true international collaboration. We hope that these efforts serve as an example and stimulate further cooperation to address the greatest challenges around the world.

Acknowledgments

Completing a volume coauthored by contributors from more than 20 countries across six continents requires considerable assistance. We thank our students, colleagues, and families for their understanding as we organized, wrote, and edited various components of this volume. Our authors showed similar patience while exercising diligence to provide well-polished manuscripts with excellent content. The University of Arizona and its School of Natural Resources and the Environment facilitated the effort through in-kind support. For their dedication in reviewing the chapters found herein, we thank J. A. Bissonette, V. C. Bleich, D. Decker, D. Dekelaita, J. Derbridge, M. Doring, V. Ezenwa, S. Fairbanks, A. Green, M. Hayward, D. Inky, A. J. T. Johnsingh, P. Lurz, J. Maerz, S. Mahoney, M. Mazzamuto, C. Mendes, M. Merrick, K. Nicholson, N. Ray, H. Sanderson, C. Spencer, S. Stevens, M. Stokes, N. Tamura, R. Valdez, D. Van Vuren, M. Vela-Vargas, S. Wells, and C. Williams. The concept behind this volume was developed with Johns Hopkins University Press and The Wildlife Society. Johns Hopkins University Press editors V. Burke and T. Gasbarrini were instrumental in bringing the work to completion. To all these individuals, we offer our sincere thanks.

INTERNATIONAL
WILDLIFE MANAGEMENT

1 International Wildlife
A Global Perspective

JOHN L. KOPROWSKI
PAUL R. KRAUSMAN
DUN WANG

Introduction

The challenges of managing and conserving global bio-diversity are grand and familiar. Dealing with increases in human population, temperatures, habitat fragmentation and conversion, biological invasions, and species extinctions requires professionals trained in the technical aspects of science and talented in coordinating management actions among diverse stakeholders (Lovejoy and Hannah 2006, Mawdsley et al. 2009). Wildlife conservation and management strategies are predicated on the sustainable use and preservation of biodiversity. Humans recognize that understanding wildlife means understanding scale; indigenous people from all parts of the globe have learned that a sustainable livelihood that incorporates wildlife is influenced by life beyond their population's borders. The aboriginal people of northern Australia capitalized on the seasonal migrations of sea turtles (Kennett et al. 2004). The Tohono O'odham people of the US Southwest and northwest Mexico recognized the return of pollinating hummingbirds and lesser long-nosed bats to their homelands (Pater and Siqueiros 2000). The Inuit people of the polar regions followed the movement of caribou (*Rangifer tarandus*), polar bears (*Ursus maritimus*), pinnipeds, and marine mammals in their annual harvests (Gilchrist et al. 2005).

The native people of the plains of East Africa prepared for the annual migrations of mixed herds of large ungulates (Spinage 2012). The Lakota and other native people of the plains of North America capitalized on the great migrations of bison (*Bos bison*: Zontek 2007). Understanding that wildlife depend on lands far from one's own has been part of human life.

Scale is critically important in the conservation and management of wildlife species, and the importance of including macroscale management in conservation efforts was advocated by none other than Aldo Leopold (1933). Managing waterfowl flyways for migratory ducks, geese, and swans began after the recognition of substantial declines in abundance caused by overharvesting and habitat conversion through the late 1800s and into the 1900s. The US Migratory Bird Treaty Act of 1918 protected game and nongame migratory birds in a large-scale approach to conservation and management (Nichols et al. 1995). Similar legislation exists on other continents, recognizing that conserving biodiversity includes protecting species that do not rely on a single nation to persist but require resources found among a coalition of nations. The Convention on the Conservation of Migratory Species of Wild Animals (CMS), or the Bonn Convention, is a United Nations Environment Programme treaty that includes more than 40 signatory nations that first signed in 1979. The CMS states that signatory nations are "Convinced that conservation and effective management of migratory species of wild animals require the concerted action of all States within the national jurisdictional boundaries of which such species spend any part of their life cycle" (https://www.cms.int/en/convention-text; all websites throughout this book were accessed on 7 January 2019). The treaty provides formal recognition of the importance of international collaboration in wildlife conservation and management. Between 16% and 26% of nearly 10,000 recognized bird species are migratory during some stage of their life history (Robinson et al. 2009, Cox 2010), and nearly 10% of mammals include large-scale migratory movements in their annual cycle (Wilcove 2008, Robinson et al. 2009). Migratory and other cross-border movements are of great importance in species management and conservation, as wildlife rarely recognize geopolitical borders except in rare cases where there are natural barriers such as rivers and mountain ranges or anthropogenic impediments including cities or fences. The CMS completed the progression from local to national regulation to international treaty, following similar United Nations Educational, Scientific and Cultural Organization (UNESCO) treaties on wetland conservation (the Ramsar Convention on Wetlands of International Importance especially as Waterfowl Habitat, in 1971) and the trade in threatened and endangered species (Convention on International Trade in Endangered Species of Wild Fauna and Flora, CITES, in 1973). Collaboration by local and national politicians, agency personnel, scientists, and the general populace is often required to ensure success (López-Hoffman et al. 2010); however, such interaction does not occur as often as desired (Dallimer and Strange 2015).

The Importance of International Collaboration

If collaboration is so widely understood to be of import to managing species, then why not work more collaboratively on a global scale? One purpose of this book is to provide an overview of the similar challenges that are faced around the world and to highlight the diversity of solutions that have been focused on international wildlife management. Science has become more collaborative over time, influenced in large part by scale and the technology that permits the incorporation of scale across geopolitical boundaries to address global challenges (Chester 2012).

The effect of incorporating international collaboration in science was first studied in detail in the 1970s (Davidson-Frame et al. 1977, Davidson-Frame and Carpenter 1979) and has been assessed in all decades since, during which time international collaboration has increased (Wagner et al. 2017). Papers with multinational authors now account for 25% of scientific publications (Wagner et al. 2015). Such assessments have identified several important patterns that facilitate international collaboration in wildlife management and conservation: applied research is less likely than basic research to include international collaboration; large national projects are less likely than projects coordinated by smaller research groups to be collaborative; and factors such as language, geography, and politics influence interaction in science (Davidson-Frame and Carpenter 1979). The barriers of distance have decreased with time and with the use of electronic and digital media (Frenken et al. 2009); however, the financial costs of international research need to be offset by considerable benefits to explain the growth of international collaboration (Barjak and Robinson 2008). The benefits of cross-nation collaboration in research include increased national involvement and coauthorship (Gazni et al. 2012), increased team size (Wuchty et al. 2007), enhanced citation rate (Glanzel and Schubert 2001, He 2009), and increased international mobility (Halevi et al. 2016). Increased mobility facilitates network formation (Franzoni et al. 2012) through face-to-face, in-person contact (Laudel 2002; Fig. 1.1). International collaboration leads to the creation of new

Figure 1.1   Chinese, Iranian, and US biologists, including two authors of this chapter (J. Koprowski, D. Wang) collaborate in biodiversity surveys in a montane meadow at 3,000 m near Kangding, Ganzi Prefecture, Sichuan Province, China. John Koprowski

connections in a network, to be capitalized on when future research needs become apparent.

International collaboration in science continues to grow and is particularly evident in the life sciences (Nature 2016): international authorship on published papers in high-impact journals in the life sciences increased from 50% in 2012 to 71% in 2015 (Nature 2016). The trend in international authorship across the sciences is one of exponential increase since 2000, and this pattern of network formation and growth is important to the internationalization of science (Ribeiro et al. 2018). Collaboration is most likely to occur between countries with trade relationships, similar socioeconomic status, similar scientific effect profiles, and in geograph-ical proximity; however, these patterns suggest that barriers are decreasing in strength, and ecologists will continue to have increasing opportunities for international collaboration (Parreira et al. 2017). The grand challenges and complex questions in ecology, such as global climate change, conservation of biodiversity at large geographical scales, habitat restoration, and pollution, require the expertise of scientists from different fields and countries and promote international scientific collaboration (Huang 2015). We need to prepare our wildlife conservation and management students to capitalize on professional opportunities in the future and to engage more routinely, as a discipline, with global challenges.

The Value of International Experience

International experience changes the perspectives of students who participate in such experiences (Potts 2015). Universities often lead the way in facilitating important international relationships that permit early career engagement and promote internationalization (Leask 2007). Students who have participated in a study abroad or other similar kind of international program are identified as demonstrating more intercultural sensitivity, respect, flexibility, tolerance, open-mindedness, sociability, critical analytical skills, and creativity than those lacking similar international experiences (Hodges and Burchell 2003, Quek 2005, Herfst et al. 2007). Recent assessments suggest that international experiences have a broader effect on participants than had been previously estimated (Potts 2015). Learning abroad contributes to the development of communication skills, teamwork skills, problem-solving skills, and self-management skills (Potts 2015). Skills promoted by engagement in international experiences and enhanced global perspectives are among the skill sets most frequently desired by employers of wildlife managers, conservation biologists, and natural resource professionals (Kessler 1995, Millenbah and Wolter 2009, Blickley et al. 2013, Lucas et al. 2017).

International experience increases the employability of students (Jones 2013). Students with international experience are viewed more positively by employers, who identify definitive connections between international experience and employability related to forging networks, learning through experience, acquiring language, and developing soft skills related to cultural understanding and diverse ways of thinking (Crossman and Clarke 2010). International experience early in one's career results in a demonstrable increase in career success, as measured by objective and subjective metrics such as salary and career satisfaction (Biemann and Braakmann 2013). Therefore, international experience can have lasting effects on the quality of life of practic-

ing professionals and positive effects of their conservation and management efforts.

Expanding Perspectives

Diversity needs to be enhanced in wildlife and fisheries conservation and management (Scott 2000, Black 2001, Allison and Hibbler 2004, Millenbah and Wolter 2009). Gender influences perception, and gender diversity results in a greater diversity of perspectives (Sanborn and Schmidt 1995). Women have been underrepresented but influential in wildlife management and conservation (Nicholson et al. 2008). Developing significant cultural intelligence in students is a heavily favored means of increasing diversity in the workplace, and international experience enhances the ability of new professionals to effectively recruit and manage diverse work teams (Ledwith and Seymour 2001). Capitalizing on diversity in our current institutions is a key step to successfully increasing inclusivity and diversity (Jones 2013).

Internationalization of organizations enhances multiculturalism, diversity, and inclusivity (Fitch and Desai 2012). Kessler et al. (1998) issued a clarion call to the profession of wildlife management for the internationalization of wildlife conservation curricula and experiences to enhance the quality and preparation of future wildlife professionals by training them to engage and make a difference in global conservation. Millenbah and Wolter (2008) further suggest that incoming students are interested in and expect a global perspective in their academic and professional development efforts. The global nature of our wildlife management and conservation challenges requires that we train our students in international collaborations and that scientists engage in and promote the internationalization of our field. The largest conservation- and management-focused professional societies, such as The Wildlife Society and the Society for Conservation Biology, can and should play a focal role with increased leadership. The emergence of bioinformatics, high-resolution remotely sensed imagery, and increasing resolution on the state of biodiversity make such global approaches more achievable and of greater importance than in the past. Current trends in climate and habitat change enhance the significance and increase the urgency. We hope that this book can be used as a guide to the continued internationalization of wildlife management and conservation to maximize our ability to address contemporary challenges and prepare students and professionals for a more collaborative future.

Summary

The scale of the challenges facing our land and biodiversity across the landscape requires that we continue to develop large-scale networks that facilitate the ability of professionals to engage in, address, and solve challenges across international boundaries. We contend that the benefits of such networks enrich many aspects of the wildlife management and conservation profession. Students must be trained to understand the importance of scale, develop skills for collaborative research, and value interaction in international networks. Forward-thinking professionals and administrators should provide this training and foster network growth. The diverse and international group of authors assembled for this volume present a current overview of the challenges that impede our ability to conserve and manage global wildlife populations. We begin with an examination of how wildlife is viewed and regulated around the world. Chapters then follow that deal with challenges and opportunity in management, including invasive species, private lands, habitat loss and fragmentation, climate change, energy development, disease, ecotourism, predators, reintroduction, illegal trade, and migration across borders. Finally, we conclude with a review of international organizations and programs, local and community-based approaches, and international careers that provide opportunities for engagement in international wildlife management and conservation. We hope the contents of this volume inspire enhanced international collaboration that increases our capacity to succeed in the conservation and sustainable management of wildlife and wildlands.

LITERATURE CITED

Allison, M.T., and D.K. Hibbler. 2004. Organizational barriers to inclusion: Perspectives from the recreation professional. Leisure Sciences 26:261–280.

Barjak, F., and S. Robinson. 2008. International collaboration, mobility and team diversity in the life sciences: Impact on research performance. Social Geography 3:23–36.

Biemann, T., and N. Braakmann. 2013. The impact of international experience on objective and subjective career success in early careers. International Journal of Human Resource Management 24:3438–3456.

Black, P. D. R. 2001. Women in natural resource management: Finding a more balanced perspective. Society and Natural Resources 14:645–656.

Blickley, J.L., K. Deiner, K. Garbach, I. Lacher, M.H. Meek, L.M. Porensky, M.L. Wilkerson, E.M. Winford, and M.W. Schwartz. 2013. Graduate student's guide to necessary skills for nonacademic conservation careers. Conservation Biology 27:24–34.

Chester, C.C. 2012. Conservation across borders: Biodiversity in an interdependent world. Island Press, Washington, DC, USA.

Cox, G.W. 2010. Bird migration and global change. Island Press, Washington, DC, USA.

Crossman, J.E., and M. Clarke. 2010. International experience and graduate employability: Stakeholder perceptions on the connection. Higher Education 59:599–613.

Dallimer, M., and N. Strange. 2015. Why socio-political borders and boundaries matter in conservation. Trends in Ecology and Evolution 30:132–139.

Davidson-Frame, J., and M.P. Carpenter. 1979. International research collaboration. Social Studies of Science 9:481–497.

Davidson-Frame, J., F. Narin, and M.P. Carpenter. 1977. The distribution of world science. Social Studies of Science 7:501–516.

Fitch, K., and R. Desai. 2012. Developing global practitioners. Journal of International Communication 18:63–78.

Franzoni, C., G. Scellato, and P. Stephan. 2012. Foreign-born scientists: Mobility patterns for 16 countries. Nature Biotechnology 30:1250–1253.

Frenken, K., J. Hoekman, S. Kok, R. Ponds, F. van Oort, and J. van Vliet. 2009. Death of distance in science? A gravity approach to research collaboration. Pages 43–57 in A. Pyka and A. Scharnhorst, editors. Innovation networks. Springer, Berlin, Germany.

Gazni, A., C.R. Sugimoto, and F. Didegah. 2012. Mapping world scientific collaboration: Authors, institutions, and countries. Journal of the American Society for Information Science and Technology 63:323–335.

Gilchrist, G., M. Mallory, and F. Merkel. 2005. Can local ecological knowledge contribute to wildlife management? Case studies of migratory birds. Ecology and Society 10:20.

Glanzel, W., and A. Schubert. 2001. Double effort = double impact? A critical view at international co-authorship in chemistry. Scientometrics 50:199–214.

Halevi, G., H.F. Moed, and J. Bar-Ilan. 2016. Researchers' mobility, productivity and impact: Case of top producing authors in seven disciplines. Publishing Research Quarterly 32:22–37.

He, T. 2009. International scientific collaboration of China with the G7 countries. Scientometrics 80:571–582.

Herfst, S., J. van Oudenhoven, and M. Timmerman. 2007. Intercultural effectiveness training in three western immigrant countries: A cross-cultural evaluation of critical incidents. International Journal of Intercultural Relations 32:67–80.

Hodges, D., and N. Burchell. 2003. Business graduate competencies: Employers' views on importance and performance. Asia Pacific Journal of Cooperative Education 4:16–22.

Huang, D.W. 2015. Temporal evolution of multi-author papers in basic sciences from 1960 to 2010. Scientometrics 105:2137–2147.

Jones, E. 2013. Internationalization and employability: The role of intercultural experiences in the development of transferable skills. Public Money and Management 33:95–104.

Kennett, R., N. Munungurritj, and D. Yunupingu. 2004. Migration patterns of marine turtles in the Gulf of Carpentaria, northern Australia: Implications for aboriginal management. Wildlife Research 31:241–248.

Kessler, W.B. 1995. Wanted: A new generation of environmental problem-solvers. Wildlife Society Bulletin 23:594–599.

Kessler, W.B., S. Csanyi, and R. Field. 1998. International trends in university education for wildlife conservation and management. Wildlife Society Bulletin 26:927–930.

Laudel, G. 2002. What do we measure by co-authorships? Research Evaluation 11:3–15.

Leask, B. 2007. Chapter 8: Internationalisation of the curriculum in an interconnected world. Pages 95–101 in G. Crosling, L. Thomas, and M. Heagney, editors. Improving student retention in higher education: The role of teaching and learning. Routledge, Abingdon, UK.

Ledwith, S., and D. Seymour. 2001. Home and away: Preparing students for multicultural management.

International Journal of Human Resource Management 12:1292–1312.

Leopold, A. 1933. Game management. Charles Scribner and Sons, New York, USA.

López-Hoffman, L., R.G. Varady, K.W. Flessa, and P. Balvanera. 2010. Ecosystem services across borders: A framework for transboundary conservation policy. Frontiers in Ecology and the Environment 8:84–91.

Lovejoy, T.E., and L. Hannah. 2006. Climate change and biodiversity. Yale University Press, New Haven, CT, USA.

Lucas, J., E. Gora, and A. Alonso. 2017. A view of the global conservation job market and how to succeed in it. Conservation Biology 31:1223–1231.

Mawdsley, J.R., R. O'Malley, and D.S. Ojima. 2009. A review of climate-change adaptation strategies for wildlife management and biodiversity conservation. Conservation Biology 23:1080–1089.

Millenbah, K.F., and B.H. Wolter. 2009. The changing face of natural resources students, education, and the profession. Journal of Wildlife Management 73:573–579.

Nature. 2016. International research collaborations on the rise. Nature index. http://www.natureindex.com/news-blog/international-research-collaborations-on-the-rise.

Nichols, J.D., F.A. Johnson, and B.K. Williams. 1995. Managing North American waterfowl in the face of uncertainty. Annual Review of Ecology and Systematics 26:177–199.

Nicholson, K.L., P.R. Krausman, and J.A. Merkle. 2008. Hypatia and the Leopold standard: Women in the wildlife profession 1937–2006. Wildlife Biology in Practice 4:57–72.

Parreira, M.R., K.B. Machado, R. Logares, J.A.F. Diniz-Filho, and J.C. Nabout. 2017. The roles of geographic distance and socioeconomic factors on international collaboration among ecologists. Scientometrics 113:1539–1550.

Pater, M.J., and B. Siquieros. 2000. Saguaro cactus: Cultural significance and propagation techniques in the Sonoran Desert. Native Plants Journal 1:90–94.

Potts, D. 2015. Understanding the early career benefits of learning abroad programs. Journal of Studies in International Education 19:441–459.

Quek, A. 2005. Learning for the workplace: A case study in graduate employees' generic competencies. Journal of Workplace Learning 17:231–242.

Ribeiro, L.C., M.S. Rapini, L.A. Silva, and E.M. Albuquerque. 2018. Growth patterns of the network of international collaboration in science. Scientometrics 114:159–179.

Robinson, R.A., H.Q. Crick, J.A. Learmonth, I.M. Maclean, C.D. Thomas, F. Bairlein, M.C. Forchhammer, et al. 2009. Travelling through a warming world: Climate change and migratory species. Endangered Species Research 7:87–99.

Sanborn, W.A., and R.H. Schmidt. 1995. Gender effects on views of wildlife professionals about wildlife management. Wildlife Society Bulletin 23:583–587.

Scott, D. 2000. Tic, toe, the game is locked and nobody else can play! Journal of Leisure Research 32:133–137.

Spinage, C.A. 2012. African ecology: Benchmarks and historical perspectives. Springer, Heidelberg, Germany.

Wagner, C.S., H.W. Park, L. Leydesdorff, and W. Glanzel. 2015. The continuing growth of global cooperation networks in research: A conundrum for national governments. PLOS One 10:e0131816.

Wagner, C.S., T.A. Whetsell, and L. Leydesdorff. 2017. Growth of international collaboration in science: Revisiting six specialties. Scientometrics 110:1633–1652.

Wilcove, D.S. 2008. No way home: The decline of the world's great animal migrations. Island Press, Washington, DC, USA.

Wuchty, S., B.F. Jones, and B. Uzzi. 2007. The increasing dominance of teams in production of knowledge. Science 316:1036–1039.

Zontek, K. 2007. Buffalo nation: American Indian efforts to restore the bison. University of Nebraska Press, Lincoln, USA.

2

RONALD J. REGAN
SHANE P. MAHONEY
BASILE VAN HAVRE
COLMAN O CRIODAIN
DEBORAH M. HAHN

Culture, Values, and Governance

Foundations to Systems of Global Wildlife Conservation

Introduction

Wildlife. Faune. Fauna. The very word "wildlife," in whatever language, evokes strong feelings and emotions of fear and joy, concern and hope across the globe. Indeed, the transcendent and emotive power of animals has extended over time and across human cultures, past and present. The cave art of Lascaux and Chauvet in France and Altamira in Spain, the earliest creation narratives, the archaeological digs of American Plains Indians, and the evidence from our first human experiences in East Africa all attest to the simple fact that wildlife has mattered and still does matter to humanity. While human cultural values and engagement with animals have changed over time, the current debates over the future of wildlife emphasize the ongoing relevance of wild animals to our lives.

Conserving wildlife has, nevertheless, proven a daunting challenge for human society. A significant part of this challenge involves the different worldviews associated with wildlife and what is deemed appropriate modern interaction with it. Yet a determination and a necessity remains to find successful models for conserving the diversity of nonhuman life and to ensure viable and sustainable human interaction with it. It is our contention that one can neither fully appreciate nor accurately evaluate the diverse tapestry of existing conservation initiatives without exploring the human cultures and values that have given rise to them. Only then can we make sense of the varied institutions, governance structures, paradigms of ownership, and funding constructs currently employed in conserving and sustainably managing wildlife populations and habitats around the globe.

We begin this chapter with an overview of two major conservation governance systems, those of Europe and North America, and apply the filters of ownership, hunting, and commercial trade across broader continental scales seasoned by historical, ecological, ethnographic, and geographic contexts. These cultural considerations form the basis for examining questions that consume modern-day wildlife managers and conservation institutions: How is nature best conserved? What does sustainable wildlife management look like, how is it governed, and how are the benefits of successful programs derived and conveyed, and to whom?

We transition next to illustrative narratives from two different parts of the world, focusing on very different, iconic wildlife species. The use, management, and conservation of these species takes place against a media-dominated worldview that does not always align with local cultural perspectives, expectations, or needs. We conclude with a few observations that knit together the themes of culture, values, and governance for durable, congruent, and relevant wildlife conservation programs of the future.

Wildlife Governance

Diverse cultures and histories have given rise to different relationships between people and wildlife in different parts of the world and have inevitably led to a variety of approaches to the conservation and management of wildlife and the ecosystems they depend on. Such approaches encompass a complex intersection of valuation and decision making relative to wildlife and to the local and wider human societies that live with and potentially benefit from wildlife resources. These approaches must be organized and administered to be effective, and the application of these frameworks must be governed by recognizable mechanisms of ownership, authority, and benefit-sharing.

Governance, in this context, refers to the instruments and mechanisms whereby governments, and sometimes other organizations or individuals, direct their programmatic activities toward the conservation, use, and management of wildlife. Governance includes the processes, laws, rules, and policies that help guide decision making; and governance approaches reflect how societies define goals and priorities in this endeavor. Governance structures are often complex and can include voluntary codes of conduct and diverse partnerships across political and social classes. Governance is often something governments do, yet quite often something for which governments are not solely responsible (Krausman and Cain 2013).

In many countries, living natural resources continue to be of great and direct importance to the livelihoods of millions of people; yet, these resources are often, legally, the property of the state. Depending on the region and political realities, this tension can pose challenges for effective governance overall, and the laws, rules, and regulations affecting wildlife management and use can often be restrictive, incomplete, vague, or discriminatory for stakeholders, sometimes leading to confusion, conflict, abuse of power, and discrimination. Frequently, stakeholders, including indigenous communities, do not have adequate opportunities to participate in decision-making processes related to the management and use of the natural resources on which they depend. Because those most directly dependent on natural resources are often the poorest citizens of the region, improving governance is an important factor in securing and improving livelihoods and promoting sustainable natural resource use and effective wildlife management and conservation. Thus, governance is a critical factor for enhancing and securing wildlife abundance and human opportunity.

Furthermore, sustainable wildlife management, conducted either by governments alone, at various levels, or in cooperation with nongovernmental organizations, occurs within existing and evolving governance structures, which are affected by international trends such as increased democratization and globalization of transport, trade, and information (Decker et al. 2012). These trends and, indeed, cultural shifts toward significant global connectedness have broad implications for how wildlife is viewed and valued and for how wildlife management may be governed. For example, the increasing human domination of ecosystems worldwide has sometimes necessitated the reallocation of certain responsibilities, previously within the scope of individual national or local authorities, to multiple sovereign states or agencies to achieve the best outcomes.

Yet, such distant governance may have increasing and deleterious effects on local users of the resource in question. This can set up a dichotomy of interest and of emphasis, involving possible benefit trade-offs between what is best for local and distant human cultures and communities and what is best for wildlife. Certainly, governance issues that affect, or are affected by, legitimate concerns for wildlife, and which also involve intersecting sovereign, subnational, and private interests, focus on complex issues that are difficult to resolve. Furthermore, the need for effective and collaborative governance is greatest when and where the stakes related to wildlife are greatest (Krausman and Cain 2013). These stakes are, in turn, strongly influenced by historical patterns of human engagement with wildlife. Thus the governance structures we observe today reflect a dynamic evolution over long periods of time or are the result of relatively rapid geopolitical shifts in authority over, attitudes toward, and ownership of wildlife.

Historically, large parts of Europe were governed by a feudal system of land allocation and use. Natural resources, including wildlife, formally belonged to elites entrusted by the monarch to conserve the resources and ensure their allocation to a peasantry who lived on and worked the land. For much of that time, lower human populations and lower-intensity agriculture and ranching techniques had more moderate effects on the landscape, and wildlife remained abundant. There was little need for complex regulations, and the system of governance was localized and relatively straightforward, if, manifestly, unfair.

As the population of Europe increased, however, so did the corresponding intensity of land-use activities. This led to natural resource declines and prompted landowners to implement more complex management approaches for natural resource use and access, which included the creation of laws and related prescriptions for their enforcement. Practically, this shift initially meant that landowners began employing game wardens to ensure the new regulations were followed. The duties of such wardens often combined aspects of conservation, habitat management, and regulation of hunting. At that point, the state played virtually no role in wildlife management. This approach to governance remained consistent in many European countries until well after World War II.

Gradually, European citizens and institutions became increasingly aware of the need for consistent regulatory and management regimes for wildlife across larger geographic scales. This awareness arose as human access to wider geographic areas increased and as technological innovation and advances in wildlife management occurred. These broad trends eventually exposed the ineffectiveness of previous approaches to conservation and the need for governance changes became apparent. Thus, in Europe, local wildlife management prescriptions eventually became ineffectual, in both their social acceptability and their capacity to ensure wildlife conservation. Over time, European governments acquired more responsibility for wildlife and natural resource management and responded by establishing national-level regulatory and enforcement regimes and by dramatically increasing wildlife management governance complexity and capacity.

Most modern Western democracies consider good wildlife governance, in the broad sense, to be participatory, transparent, and accountable (Krausman and Cain 2013). For example, the North American Model of Wildlife Conservation is one system that embodies these values (Geist 1995). The North American Model is a set of principles that have been applied collectively for more than a century, leading to the "form, function, and successes of wildlife conservation and management" in the United States and Canada (Organ et al. 2012: 123). The Model is described as North American in a conceptual, rather than actual, geographical context, because Mexico's conservation movement developed under significantly different social and political circumstances than that of Canada and the United States (Organ et al. 2012).

Unlike in Northern Europe, issues of wealth, status, and land ownership were not the historical bases for access to, or use of, wildlife in the United States and Canada. Rather, wildlife governance in these countries is most fundamentally based on the public trust doctrine, which legally enshrines public ownership and state responsibility for wildlife and ensures democratic access to wildlife resources. The well-established guiding principles for the Model include (1) wildlife are a public trust resource; (2) there is no large-scale marketing of wildlife; (3) appropriate allocation of wildlife use and access is prescribed by law; (4) wildlife can be killed only for a legitimate purpose; (5) wildlife is considered and managed as an international resource; (6) science is the proper tool to discharge wildlife policy; and (7) democracy of hunting is standard (Geist et al. 2001). The Association of Fish and Wildlife Agencies, representing national and international policy interests of state, provincial, and territorial fish and wildlife agencies, has formally adopted the Model as a construct to guide conservation policy of the future (Prukop and Regan 2005).

Over the course of the last century, the North American Model has achieved tremendous success in recovering many large vertebrate species severely depleted through commercialized and unregulated hunting during the first centuries of European settlement (Heffelfinger et al. 2013). One of the great achievements and surprising aspects of the North American conservation approach, especially given that excessive exploitation had led to serious declines and extinctions in wildlife populations and species (Krausman and Cain 2013), was the recognition that self-interest and consumptive use could play a central role in a long-term strategy for effective and sustainable wildlife management. Hunting emerged as a positive and constructive force in the development of the North American Model, with prominent recreational hunters committed to ending the overexploitation of wildlife by market hunters and driving species recovery for the preservation of wildlife populations and the continuation of their hunting traditions (Heffelfinger et al. 2013). It is interesting to note that, in Russia, the desire to hunt is also viewed as having been the major motivator for government-led efforts to restore and sustainably manage wildlife populations under Soviet rule (Baskin 1998). This approach is broadly similar to the South African model discussed elsewhere in this chapter.

Wildlife Ownership

As the two continental-scale management examples discussed above illustrate, varying approaches have emerged worldwide in which private and public ownership models of wildlife have been used, along with differing models of social organization and governance, to support conservation agendas. Ownership rights are a critically important dimension in these models, and may include rights to access, use, benefit from, and sell wildlife, and to derive benefits from hunting, tourism, and sale of wildlife products. Under various wildlife management systems, governments may exercise some of these rights, while private landowners or recreational users may exercise others. In the case of the North American Model, wildlife is owned by no single person, but is held in trust by the government for the benefit of all citizens; governments exercise most ownership rights, but strictly on behalf of the people and their interests (Mahoney and Jackson 2013).

Globally, government ownership of wildlife is, in fact, a fairly common approach. It remains the case across most African jurisdictions, for example, even where decades of emphasis on devolution of wildlife management to communities have taken place (Shyamsundar et al. 2005). It is specified as law in some countries, including Tajikistan, China, and Australia, and there is some evidence that the dominance of this system may be related to the transboundary nature of wildlife. Because wildlife cannot be expected to respect political or property boundaries, private ownership without elaborate fencing can be very problematic, especially in the case of migratory wildlife resources.

The privatized, fenced approach, however, has proven a viable alternative in some instances. For example, in South Africa, wildlife are formally res nullius, or owned by no one (Pack 2013). Yet, private landowners with fenced properties effectively exercise rights of ownership and are empowered to manage, use, control access to, and sell wildlife. Other approaches also exist. In the case of Namibia, private landowner and communal conservancies exercise most ownership rights regarding wildlife, though they are legally obligated to respect government restrictions on the use of certain species.

Further, despite much change in other aspects of governance, in many Western European countries today, ownership of wildlife remains vested in private landowners. This may automatically carry with it the right to hunt, as in the case of the United Kingdom, or it may not, as in the case of France. The continuing prevalence of landowner control of wildlife in Northern Europe reflects a long history of elite control of the use of and benefits from hunting of certain game species, and this contrasts greatly with the usual colonial and postcolonial model of centralized state control.

The importance of ownership rights in determining conservation outcomes is demonstrated by decades of more or less successful efforts in community-based wildlife conservation or natural resource management, primarily in Africa. Namibia provides what is perhaps the best example of strong private or community wildlife natural resource ownership rights and demonstrates how these effectively incentivize conservation. Under their system, private landowners or community conservancies may gain operative ownership of wildlife, so they are legally permitted to make most decisions about how it may be used and then may profit, retaining 100% of commercial benefits, from sustainable use (Weaver et al. 2011). The conservation benefits of this national approach have been widely recognized and include expansion of wildlife populations and land area dedicated to wildlife and reduced poaching and human-wildlife conflict. Namibia's system of wildlife management has been touted as so successful that many people now agree that the key problem undermining wildlife conservation efforts in many parts of the world is the reluctance of central governments to cede control of wildlife, and related benefits, including potential profits from sustainable-use endeavors, to private landowners or communities. The sparse population of Namibia is likely a facilitative factor in such success.

The Role of Hunting

The role hunting plays in relation to modern wildlife governance systems varies greatly across nations. In Australia, for example, there is no formal recognition in law or policy that hunting can be a mechanism for wildlife conservation, and hunting has instead been viewed as a detriment to management (Bauer and English 2011); in India, by contrast, almost all hunting is banned by law, an outcome in keeping with Hindu culture rather than of a pragmatic judgment about the utility or value of hunting (Bauer and Giles 2002). In South Africa, hunting is legally recognized as a fundamental component of wildlife management and conservation, because it provides incentives for wildlife as a land use, and in Tanzania, hunting is perceived as an important source of revenue, providing necessary monies to support the management of protected areas (Lindsey et al. 2007). In parts of Europe, the role of hunting relates more directly to the management of overabundant or otherwise problematic species than to providing incentives for conservation (Cedurland et al. 1998). In the United States and Canada, hunting and fishing have played direct, multiple, and significant roles in the management and recovery of wildlife species for more than a century (Mahoney 2013, Mahoney and Weir 2015). Furthermore, through license fees and excise taxes on hunting, angling, and recreational shooting products, as well as motor boat fuels, these traditional activities continue to be the most significant source of funding for wildlife management and conservation approaches, especially in the United States. Fundamental differences regarding the role of hunting in different locales reflect the variance in historical, cultural, ecological, and policy settings, and simple prescriptions advocating that what works in one setting should be applied in another are overly simplistic. Governance models for wildlife will vary across regions, and so will the application of hunting as a conservation mechanism.

Of course, conservation incentives can be monetary or not, and may involve consumptive or nonconsumptive use approaches. Nonhunting models of wildlife management certainly exist and can be effective, though these can be challenged by how to incentivize people to care for and value wildlife and its habitat, especially outside legally protected areas. Nevertheless, there exist excellent examples of nature-based tourism, the most common nonconsumptive approach, supporting wildlife conservation and improving local livelihoods, as in Costa Rica. This approach, however, is well suited only to specific areas that facilitate wildlife viewing, and its limited range hinders its capacity to contribute positively to conservation, which is a problem of increasing importance worldwide as human population and consumption pressures increase pressures on land in biodiversity-rich areas. Regardless of the role consumptive and nonconsumptive approaches may play, good governance of conservation and benefit-sharing remains critical. It is therefore not the form of wildlife use that is so critical to achieving wildlife conservation, but how such forms are governed and how transparent the rules of governance are.

Commercial Trade in Wildlife

Wildlife governance approaches also involve issues surrounding the commercial trade of wildlife and products made from dead wildlife, for example the sale of meat or skin from a hunted animal. The North American Model does not include trade, but many management and conservation approaches integrate aspects of commercial trade, though these are often accompanied by strict restrictions. For example, trade in bushmeat is illegal in numerous tropical countries, while some subsistence hunting is allowed (Collective Partnership on Sustainable Wildlife Management 2014). In Australia, indigenous people have rights to use wildlife for subsistence purposes, but not for trade (Cooney and Edwards 2009).

There are two concerns related to commercial trade: that commercial operations can mobilize social and economic pressures for larger harvests to meet market demand, potentially threatening sustainability, and that the existence of legal trade generally makes enforcement more difficult and complex. These concerns need to be balanced with positive aspects of sustainable commercial trade, including its capacity to remove incentive for illegal activities and its capacity to meet social and cultural needs.

International trade in at-risk species is governed by the Convention on International Trade in Endangered Species of Wild Fauna and Flora (CITES). Species listed on Appendix I of the convention cannot, in general, be traded commercially. Those listed on Appendix II may be traded commercially, subject to verification of the legality of acquisition and sustainability of harvest.

The different value sets regarding wildlife use and trade are seen in stark relief in Africa. On the one hand, countries like South Africa, Zimbabwe, and especially Namibia espouse a philosophy of sustainable wildlife use. They embrace hunting of wild game, including trophy hunting, and wild meat is widely available in shops and restaurants. Kenya, on the other hand, is ideologically opposed to such hunting. Wild meat is difficult to find, and trophy hunting is not allowed. These differences come to a head in relation to the international commercial trade of high-value wildlife commodities, such as ivory.

When it comes to ivory, southern African countries have made known a desire to sell the ivory they derive from natural elephant mortality and the killing of problem animals. Kenya and many other African countries oppose this on principle, and wish to maintain, or even reaffirm, the current global ivory trade ban. At the 2016 CITES Conference of the Parties, the African Elephant Coalition, a group of 29 countries including Kenya, tabled a proposal to formalize the ban by elevating the elephant populations of Botswana, Namibia, South Africa, and Zimbabwe from Appendix II to Appendix I. The current

Appendix II listing specifically excludes commercial trade in ivory, but an Appendix I listing would have precluded all commercial trade, including that involving live elephants and hides. The proposal was rejected by the meeting, but the de facto ban on ivory trade remains in place.

Botswana, which has the largest elephant population of any African country, is an interesting case in this regard. Until a few years ago, it too was a supporter of sustainable use of natural resources. It participated in ivory sales in 1999 and 2008, and it offered trophy hunting opportunities. Today, these policies have been reversed; the country has banned trophy hunting and supports the global ivory trade ban. Supporters of the Kenyan approach cite it as a model, but those who support the South African approach predict that reducing the economic value of wildlife, through banning of trophy hunting, will lead to reduced tolerance of human-wildlife conflict by local communities and, ultimately, negative conservation outcomes (C. L. Weaver, WWF Namibia, personal communication).

There certainly are examples of commercial trade increasing the value of wildlife to landowners, generating revenue for conservation, and decreasing illegal trade by supplying the market with a legal option. Several of Australia's kangaroo species are harvested at the scale of up to tens of millions per year, without causing significant declines in kangaroo numbers, and illegal trade remains insignificant (Cooney et al. 2012). In Europe, meat products from legally hunted wild boar (Sus scrofa) and other species may be legally sold in a number of countries, including Germany, where approximately 600,000 wild boar are harvested annually while populations continue to increase. In Namibia and South Africa, wildlife populations on private and communal lands are harvested for meat and skins, with revenues forming part of the economic rationale for maintaining wildlife-based land uses (Weaver et al. 2011), while commercial trade in vicuna (Vicuna vicuna) fiber contributes to indigenous and local community economies, leading to reduced domestic livestock levels and reduced vicuna poaching in Andean countries (Lichtenstein 2010). Despite these examples, however, the question of whether the existence of legal trade makes illegal trade more likely or threatens sustainability remains unanswered (Bulte et al. 2007).

The decisions made regarding wildlife resources directly affect large numbers of citizens globally. As such, it is unsurprising that wildlife management is one of many aspects of modern governance in which the demand for public participation is growing (Decker et al. 2012). There is no universally effective system of wildlife management, and there is no one-size-fits-all approach to governance. Decision makers and policy makers must work to identify best approaches, informed by science and traditional knowledge, at varying scales, given international and regional trends and cultural and geographical contexts. Two case studies will help flesh out these observations.

Contextual Narratives: Perspectives on Two Iconic Species

Polar Bear (*Ursus maritimus*)

The polar bear population is small (Wiig et al. 2015), at approximately 30,000 animals worldwide. Distributed across the circumpolar arctic region, polar bear habitat is well delineated and isolated from most of the world's human population, yet the polar bear is one of the world's most recognized wildlife species. Not surprisingly, given its size and coloration, its amphibious capacities and predatory lifestyle, images of the polar bear have been used by global brands like Coca-Cola for more than a century to promote notions of strength and vitality. An influential force in the ecology of arctic regions, the species has been of integral value and mythological importance to the human cultures that have most closely shared its range and presence. Yet, human valuations of the polar bear vary considerably depending on where one lives and whether polar

Figure 2.1 An unlikely pair—a polar bear and a domestic dog in Manitoba—at the crossroads of the north, the interface of wildness and cultural values. Lynn Bystrom

bears are a symbolic or real presence in one's life (Fig. 2.1).

For many, the polar bear is the very symbol of arctic ecosystems and a critical indicator of the health and resilience of such natural areas. With changing climate having a documented effect on arctic regions (Regher et al. 2017), a general expectation that the health and status of the polar bear should deteriorate in line with changes in climatic condition has emerged and been reinforced by a variety of media reports and more technical publications. Therefore, the polar bear, and particularly the perceived deterioration in its health and numbers, has been portrayed as a symbol for the need to address the cause and pace of climate change.

Although there is no doubt that the deterioration of habitat as a result of climate change is the primary threat to polar bears (Polar Bear Range States 2015), and that addressing its root cause is the most important action to improve the status of the polar bear, it has become increasingly evident that making a direct causal link between climate change and a population-level effect is not easy; short-term and local change of habitat characteristics have not, to date, resulted in corresponding changes in population numbers. In fact, organizations have stopped or decreased the use of polar bear images to promote action on climate change. If a link to wildlife is needed, other, less glamorous species can likely better illustrate the climate change–wildlife connection, such as barren-ground caribou (*Rangifer tarandus groenlandicus*).

With increasing urbanization, wildlife conservation—in particular, conservation of large mammals—is seen as valuable today. Yet most people will never experience the positive or the negative aspects of personally encountering a polar bear. A paradigm has been created in which the basic notion that nature has to be conserved in its entirety for people today and in the future has been extended to the notion that no harm should be done to any individual animal. According to this set of values, hunting should be prohibited and, consequently, trade banned. This is a set of values that must be treated with the same respect as the others discussed here.

Now let us turn our attention to the people who live within the home range of polar bear populations. These communities have depended on natural resources for their survival for centuries and have experienced the many challenges of living close to a top predator. Such challenges can range from destruction of meat caches to limitations on sending children to the local school because of fears of injury or loss of life and have led to the development of a value system that recognizes polar bears as a resource that supports human life and as an entity to respect and fear. Often, Inuit elders, for example, will be uncomfortable speaking about polar bears, because it is believed that they can hear and understand what is said about them.

To better understand this, it is useful to consider two concepts. First, the arctic economy, while changing rapidly, is still largely a subsistence-based economy. Today, local fauna and flora still provide most of the food to individuals and local communities, as the cost of transport for outside commodities is often prohibitive, and the local diet is founded on the local ingredients. As in any society, however, new needs that can be met only through a market-based economy have emerged. These include the need for modern transportation (snow machines) and access to electronic media. Often the only way for local communities to acquire the currency needed to obtain such goods is through the sale of wildlife products, such as polar bear hides.

Second, societies across the arctic are evolving very rapidly, yet the role of the hunter continues to be a keystone for families or community units. This role is defined by the capacity of individual males to provide for the needs of his dependents, and in return his dependents will turn to the hunters for advice and guidance. In a period of intense change, during which there are significant opportunities (through education) and threats (availability of alcohol and drugs), a tradition-based social structure is a critical element of social stability and growth. Depriving the hunter of the opportunity to fulfill his role in that society may consequently accelerate the diminution of a traditional structure with concomitant negative repercussions for a community.

The polar bear delineates the boundaries of a conservation spectrum: a no-harvest management paradigm, in keeping with a vision of a romanticized wild arctic landscape, contrasted with the need or desire for sustainable harvests in keeping with first principles of social stability and security in indigenous communities. Finding synergistic common ground to bridge these value distinctions is the common challenge of wildlife managers across the globe for a great many species.

African Elephant (*Loxodonta africana*)

Perhaps no species has divided the conservation community to the extent that elephants have. Richard Leakey, current and former head of the Kenya Wildlife Service, and Kenyan president Uhuru Kenyatta have, on repeated occasions, argued that "ivory is worthless unless it is on our elephants" (Makori 2016). And, indeed, to emphasize this point Kenya has publicly burned ivory stockpiles on four occasions. In fact, the first of these, in 1989 (Schiffman 2016) was instrumental in achieving an international ivory trade ban, which, despite two subsequent one-off ivory auctions, remains in place.

By contrast, the approach to elephant conservation in southern Africa, the source of the ivory stockpiles sold in the above-mentioned one-off sales, could not be more different. In March 1999, just prior to the first such sale, the British Broadcasting Corporation (BBC) ran a news story with the intriguing title "Shoot an Elephant, Save a Species" (Kirby 1999). The article, a report from a trophy-hunting expedition in Zimbabwe, included an account of the CAMPFIRE program, a community-based conservation program in that country that, inter alia, used trophy hunting to generate income for local communities that live in proximity to elephants. A representative of the program defended the approach, pointing out that it is only the incentive of such income that persuades rural communities to tolerate such dangerous and sometimes destructive animals.

As such, elephants are the best illustration of fundamentally different approaches to species conservation. One constituency argues that the only way to protect the species is to end the market in its products (Harvey 2016). The other argues that the solution lies in letting the market work, and that measures such as trade bans and stockpile destruction (Moyle and Stiles 2014) are counterproductive. In the fall of 2017, the United States government proposed lifting an importation ban on elephant ivory from Zimbabwe, resulting in a public backlash (Hincks 2017).

Another element to consider in elephant conservation is the cultural perceptions in China and neighboring countries. These perceptions, together with the spectacular growth of a wealthy middle class in the region, have resulted in a major shift in demand for wildlife products in recent decades. Historically, Europe was the destination for the bulk of the trade in wildlife products, including pelts and ivory (Parker 1936). This began to change in the nineteenth century, but changed decisively after World War II, partly because of the growing environmental movement, which placed a stigma on such products and lobbied successfully for restrictions on their trade, and partly because of increased wealth, first in North America and Japan, and then, later, China (Underwood et al. 2013). Ivory illustrates this dynamic very well. It is a commodity that is deeply embedded in Chinese culture, to the extent that demand for ivory contributed to the extinction of that country's elephants in historical times (Bishop 1921). However, it remained a commodity that only China's relatively small elite could afford. With the growth in China's economy in recent decades, that has changed, and today China is by far the biggest destination country for what is now illegal ivory, in addition to having partici-

pated in one legal sale in 2008 (Underwood et al. 2013).

One important cultural trait is that consumption of wildlife products in China and neighboring countries is not necessarily a private activity, but one that is often shared, through hospitality or gifting. This is linked to the concept of face, and the manner in which face is gained by sharing one's wealth with others. Conversely, failure to provide adequate gifts or hospitality on appropriate occasions constitutes loss of face.

Fundamentally different philosophical approaches about the most effective way to conserve wildlife, set against a cultural background of long-standing cultural aspirations that have become realizable with newfound wealth, have serious consequences for the often-distant communities that live in proximity to elephants (Fig. 2.2). As mentioned above, the southern African conservation model has been designed around sustainable use of wildlife for the benefit of local communities, and it has largely been successful. The historic growth of white rhinoceros (*Ceratotherium simum*) populations illustrates this (Adcock and Emslie 1994). But it remains to be seen whether this model can cope with the unsustainable level of demand for some wildlife products, including rhinoceros horn, in Asia. Meanwhile, in countries with weaker governance, where the model

Figure 2.2 Elephants are in the crosshairs of fundamentally different governance approaches to species conservation based on differing values and market paradigms. World Wildlife Fund

is not applied, or is more at risk from corruption, poaching levels of species such as elephants remain unsustainably high (CITES 2016).

Summary

The basic premise of this chapter has been the need for, or importance of, thinking about the intersection of culture, values, and governance in wildlife management systems globally. We illustrated that, starting from a common desire to conserve wildlife, various human communities, both within a species range and without, often have different sets of values and, therefore, come to different conclusions about conservation approaches and priorities. Both the polar bear and the African elephant are instructive case studies of the complex web of cultural and community imperatives, well-intended outside public interest, views about sustainable use (e.g., hunting), and the influence of international trade that surrounds such issues. Wildlife management across the world inspires passion and multifaceted convictions regarding desirable management outcomes. Consequently, amid such a milieu of often conflicting considerations, wildlife management is not for the unskilled or fainthearted. Nor can we forget the long shadow of history that often plays a significant role in how regional approaches to conservation have evolved.

Even though we cannot prescribe a universal model for success (what will work in one part of the world will not necessarily be effective in another), we contend that several principles of governance may offer enough elasticity for durable and relevant outcomes across a global spectrum of values or cultural constructs. We suggest that successful wildlife management systems, both those predicated by public trust principles and those predicated on private landowner economics, will be governed by science and traditional knowledge. In addition, they will be guided by a commitment to sustainability, with or without any harvest-based perturbations; will acknowledge the need for habitat conservation; and will offer some measure of transparency in the development of management priorities and decisions. To the extent that regulated hunting is contemplated or is a desired programmatic outcome, we suggest an explicit commitment to, or valuation of, the ethical approach to hunting, which will be desirable for broader social acceptance.

Wildlife has been and continues to be an integral part of human existence. Even with great passion for wildlife, humans continue to struggle with how best to successfully conserve it. How we interact with and manage wildlife continues to evolve as human society evolves and changes. One thing is certain: values and cultures across the globe will continue to influence the conservation and management of wildlife and influence its loss and imperilment.

LITERATURE CITED

Adcock, K., and R.H. Emslie. 1994. The role of trophy hunting in white rhino conservation, with special reference to BOP parks. Pages 35–41 *in* B.L. Penzhorn and N.P. Kriek. Proceedings of a symposium on rhinos as game ranch animals. Onderstepoort, Republic of South Africa.

Baskin, L. 1998. Hunting of game mammals in the Soviet Union. Pages 331–345 *in* E.J. Milner-Gulland and R. Mace, editors. Conservation of biological resources. Blackwell Science, Oxford, UK.

Bauer, J., and A. English. 2011. Conservation through hunting: An environmental paradigm change in NSW. Game Council of New South Wales, Orange, Australia.

Bauer, J.J., and J. Giles. 2002. Recreational hunting: An international perspective. CRC for Sustainable Tourism, Gold Coast, Australia.

Bishop, C.W. 1921. The elephant and its ivory in China. Journal of the American Oriental Society 41:290–306.

Brockington, D. 2008. Corruption, taxation and natural resource management in Tanzania. Journal of Development Studies 44:103–126.

Broussard, G. 2017. Building an effective criminal justice response to wildlife trafficking. Review of European, Comparative and International Environmental Law 26: 118–127.

Bulte, E.H., R. Damania, and G.C. Van Kooten. 2007. The effects of one-off ivory sales on elephant mortality. Journal of Wildlife Management 71:613–618.

Cedurland, G., J. Bergqvist, P. Kjellander, R. Gill, J.M. Gaillard, P. Duncan, P. Ballon, and B. Boisaubert. 1998. Managing roe deer and their impact on the environment:

Maximizing the net benefits to society. Pages 332–337 in R. Anderson, P. Duncan, and J.D.C. Linnell, editors. The European roe deer: The biology of success. Scandinavian University Press, Oslo, Norway.

CITES (Convention on Trade in Endangered Species of Flora and Fauna). 2016. Report on monitoring the illegal killing of elephants (MIKE). Seventeenth meeting of the Conference of the Parties. 24 September–5 October, Johannesburg, South Africa.

Collective Partnership on Sustainable Wildlife Management. 2014. Sustainable wildlife management and wild meat. United Nations Food and Agriculture Organization, Factsheet 2. http://www.fao.org/3/a-i5185c.pdf.

Cooney, R., and M. Edwards. 2009. Indigenous wildlife enterprise development: The regulation and policy context and challenges. Report to North Australian Indigenous Land and Sea Management Alliance (NAILSMA). https://www.researchgate.net/profile/Melanie_Edwards/publication/230559033_Indigenous_Wildlife_Enterprise_Development_Regulation_and_Policy_Context_and_Callenges/links/0fcfd501739987b85a000000/Indigenous-Wildlife-Enterprise-Development-Reguation-and-Policy-Context-and Challenges.pdf.

Cooney, R., M. Arche, A. Baumber, P. Ampt, G. Wilson, J. Smits, and G. Webb. 2012. THINKK again: Getting the facts straight on kangaroo harvesting and conservation. Pages 150–160 in P. Banks, D. Lunny, and C. Dickman, editors. Science under siege: Zoology under threat. Royal Zoological Society of New South Wales, Mosman, Australia.

Decker, D.J., S.J. Riley, and W.F. Siemer, editors. 2012. Human dimensions of wildlife management. Second edition. Johns Hopkins University Press, Baltimore, Maryland, USA.

Geist, V. 1995. North American policies of wildlife conservation. Pages 75–129 in V. Geist and I. McTaggart-Cowan, editors. Wildlife conservation policy. Brush Education, Calgary, Alberta, Canada.

Geist, V., S.P. Mahoney, and J.F. Organ. 2001. Why hunting has defined the North American model of wildlife conservation. Transactions of the North American Wildlife and Natural Resources Conference 66:175–185.

Harvey, R. 2016. Risks and fallacies associated with promoting a legalised trade in ivory. Politikon 43:215–229.

Heffelfinger, J.R., V. Geist, and W. Wishart. 2013. The role of hunting in North American wildlife conservation. International Journal of Environmental Studies 70:399–413.

Hincks, J. 2017. President Trump's controversial reversal of the elephant trophy ban, explained. Time Inc., 17 November. http://time.com/5028798/elephant-hunting-trophy-zimbabwe/.

Kirby, A. 1999. Shoot an elephant, save a species. BBC News, 18 March.

Krausman P.R., and J.W. Cain III, editors. 2013. Wildlife management and conservation: Contemporary principles and practices. Johns Hopkins University Press, Baltimore, Maryland, USA.

Lichtenstein, G. 2010. Vicuña conservation and poverty alleviation? Andean communities and international fibre markets. International Journal of the Commons 4:100–121.

Lindsey, P.A., P.A. Roulet, and S.S. Romanach. 2007. Economic and conservation significance of the trophy hunting industry in Sub-Saharan Africa. Biological Conservation 134:455–469.

Mahoney, S.P., editor. 2013. Monograph: Conservation and hunting in North America, I. Special Issue, International Journal of Environmental Studies 70 (3).

Mahoney, S.P., and J. J. Jackson. 2013. Enshrining hunting as a foundation for conservation: The North American Model. International Journal of Environmental Studies 70:448–459.

Mahoney, S.P., and J.N. Weir, editors. 2015. Monograph: Conservation and hunting in North America, II. Special Issue, International Journal of Environmental Studies 72 (5).

Makori, Ben. Kenya burns vast piles of elephant tusks as it seeks ban on trade. Reuters, 30 April. https://in.reuters.com/article/us-kenya-wildlife-idINKCN0XR0D5.

Milliken, T., and J. Shaw. 2012. The South Africa–Viet Nam rhino horn trade nexus: A deadly combination of institutional lapses, corrupt wildlife industry professionals and Asian crime syndicates. Trade Records Analysis of Flora and Fauna in Commerce (TRAFFIC), Johannesburg, South Africa.

Moyle, B., and D. Stiles. 2014. Destroying ivory may make illegal trade more lucrative. South China Morning Post, 3 February.

Organ, J.F., V. Geist, S.P. Mahoney, S. Williams, P.R. Krausman, G.R. Batcheller, T.A. Decker, et al. 2012. The North American Model of wildlife conservation. The Wildlife Society Technical Review 12-04. The Wildlife Society, Bethesda, Maryland, USA.

Pack, S., R. Golden, and A. Walker. 2013. Comparison of national wildlife management strategies: What works where and why? Heinz Center for Science, Economics, and Environment, Washington, DC, USA.

Parker, I.S. 1936. Ian Parker collection relating to East African wildlife conservation, Special and Area Studies Collections, George A. Smathers Libraries, University of Florida, Gainesville, Florida, USA.

Polar Bear Range States. 2015. Circumpolar action plan: Conservation strategy for polar bears. A product of the representatives of the parties to the 1973 Agreement on the Conservation of Polar Bears. https://polarbearagreement.org/circumpolar-action-plan.

Prukop, J., and R.J. Regan. 2005. In my opinion: The value of the North American Model of wildlife conservation: An International Association of Fish and Wildlife Agencies position. Wildlife Society Bulletin. 33:374–377.

Rihoy, E., and B. Maguranyanga. 2007. Devolution and democratisation of natural resource management in Southern Africa: A comparative analysis of CBNRM policy processes in Botswana and Zimbabwe. Centre for Applied Social Sciences and Programme for Land and Agrarian Studies, Bellville, South Africa and Harare, Zimbabwe.

Russell, B., and Grouper and Wrasse Specialist Group. 2004. *Cheilinus undulatus*. The IUCN Red List of Threatened Species 2004: e.T4592A11023949.

Schiffman, D. 2015. Trade in shark fins takes a plunge. Scientific American, 26 February.

Schiffman, R. 2016. Why Kenya is burning 100 tons of elephant ivory. Scientific American, 27 April.

Shyamsundar, P., E. Araral, and S. Weeraratne. 2005. Devolution of resource rights, poverty, and natural resource management: A review. World Bank,

Environmental Economics Series, Washington, DC, USA.

Smith, R.J., R.D. Muir, M.J. Walpole, A. Balmford, and N. Leader-Williams. 2003. Governance and the loss of biodiversity. Nature 426:263–279.

Underwood, F.M., R.W. Burn, and T. Milliken. 2013. Dissecting the illegal ivory trade: An analysis of ivory seizures data. PLOS One 8: e76539.

United Nations Office on Drugs and Crime. 2016. World wildlife crime report: Trafficking in protected species. United Nations, Vienna, Austria.

Weaver, L.C., E. Hamunyela, R. Diggle, G. Matongo, and T. Pietersen. 2011. The catalytic role and contributions of sustainable wildlife use to the Namibia CBNRM Programme. Pages 59–70 *in* M. Apensperg-Traun, D. Roe, and C. O'Criodain, editors. CITES and CBNRM: Proceedings of an international symposium, The Relevance of CBNRM to the Conservation and Sustainable Use of CITES-Listed Species in Exporting Countries. International Union for Conservation of Nature, Gland, Switzerland.

Wiig, O., S. Amstrup, T. Atwood, K. Laidre, N. Lunn, M. Obbard, E. Regehr, and G. Thiemann. 2015. *Ursus maritimus*. The IUCN Red List of Threatened Species 2015: e. T17026A1306343.

Woodworth, P. 2017. Can we keep the hen harrier dancing in Irish skies? Irish Times, 15 April.

3

SANDRO BERTOLINO
LUCAS A. WAUTERS
ADRIANO MARTINOLI

Invasive Species

The Challenges of Nonnative Species Establishment and Spread to Native Wildlife Populations

Introduction

Animal and plant species have always accompanied humans during their movements and migrations around the globe. In the past two centuries, however, this phenomenon has dramatically exploded without showing signs of slowing down (Seebens et al. 2017).

Biological invasions have received increasing attention within the last decades, and important progress has been made in our understanding of the effects of alien species (Ricciardi et al. 2013). Among these, invasive alien species (IAS) pose a significant threat to biodiversity, and since 2004, Gurevitch and Padilla (2004) have asked whether IAS are a major cause of extinction. Moreover, compelling evidence exists, based on global trade and movement patterns, that the magnitude of this threat is increasing worldwide (Hulme 2009). Invasive alien species alter ecosystem processes (Raizada et al. 2008); decrease native species abundance and richness via competition, predation, hybridization, and indirect effects (Blackburn et al. 2004, Gaertner et al. 2009); change community structure (Hejda et al. 2009); and alter genetic diversity (Ellstrand and Schierenbeck 2000). Effects of alien species can be substantial and costly (Pimentel et al. 2005, Stohlgren and Schnase 2006, Kettunen et al. 2009, Ricciardi et al. 2011), making

species invasions an environmental issue of global significance. Some environments are more vulnerable than others: for example, insular ecosystems have proven exceptionally susceptible to invasion by alien species and vulnerable to their negative effects. Of 680 known animal extinctions in the past 400 years, about half were of island species, and of the bird species that have become extinct in that period, about 90% were island dwellers; most of these extinctions were caused fully or in part by invasive animal species (Clavero and Garcia-Berthou 2005). The centers of IAS-threatened vertebrates are concentrated in the Americas, India, Indonesia, Australia, and New Zealand (Bellard et al. 2016).

Of the IAS introduction pathways, 39% of species were introduced intentionally and 26% unintentionally, 22% both intentionally and unintentionally, and 13% had no information available (Turbellin et al. 2017). There is still considerable debate, however, and uncertainty as to how alien species affect their environment (Richardson and Ricciardi 2013), and the lack of consensus as to the severity and significance of alien species effects has been attributed to differences in human perception of invasions (Simberloff et al. 2013). For example, being on a crossroad of many disciplines (ecology, social sciences, resource management, economics), invasion science is sometimes accused of being incapable of interacting with society and has even been tagged as xenophobic by extreme animal rights groups. Also, based on human culture, some highly invasive species may be a food resource (rats on Polynesian islands) or serve as exotic curiosities (Simberloff et al. 2013).

Biological invasions have subtle socioeconomic consequences, which are difficult to assess using traditional monetary approaches and market-based models, even though some surveys have revealed that more than 1,000 alien species are known to have caused ecological or economic effects in Europe (Vilà et al. 2010). Although these findings reflect the current state of knowledge, they are likely to change as more information is gathered. In addition, most risk assessments for alien species focus on environmental effects exclusively, even though many alien species have substantial effects on economy and human social life. For example, many of the harmful alien insects are crop pests, which pose harm not necessarily to biodiversity or the environment, but to agricultural production and thus to the economy (Kumschick et al. 2015). For managing biological invasions, it is important to identify the mechanisms through which alien species are affecting their surroundings, especially if certain ecosystems or ecosystem services are to be protected. An understanding of effect mechanisms can also shed light on how consistent an effect is likely to be over different regions. For example, if the main mechanism is hybridization, effect is dependent on the presence or absence of a closely related species (Kumschick et al. 2015).

In some areas of the world, contrasting measures are being defined: for example the European Union (EU) has recently adopted a new regulation on invasive alien species (EU Regulation 1143/2014), with a list of invasive alien species of concern that should be actively managed (Genovesi et al. 2014). By contrast, New Zealand and Australia have significantly reduced the number of new invasions with a biosecurity approach based on risk assessment (Simberloff et al. 2013).

Ecosystem Processes Alterations

Invasive alien species are a major influence on change in ecosystem composition and processes. Plants can devastate whole ecosystems, replacing highly diversified communities with a sort of monoculture. For example, in northern Australia, the introduced giant sensitive tree (*Mimosa pigra*) forms dense stands that outcompete all other forms of vegetation, replacing sedgelands, riparian vegetation, and monsoon forest communities (Braithwaite et al. 1989). As a result, many native birds and lizards are nearly completely excluded from these areas, resulting in local extinction of populations.

Vitousek (1990) identified three primary ways in which IAS can affect ecosystems: (1) invasive species can use resources, typically nutrients, differently from native species, affecting availability for native species; (2) they can change food webs, interfering with biomass and energy flow; and (3) they can alter the disturbance regime (fires, insect, or pathogen outbreaks) of an invaded area.

Here we provide some typical examples for each of these processes. The introduction of the nitrogen-fixing plant firetree (*Myrica faya*), native to the Canary Islands, into the volcanic lava flows of Hawai'i, USA, is an example of the first effect. The plant is fixing nitrogen, increasing this nutrient in the originally nitrogen-poor volcanic soils (Vitousek and Walker 1989). This negatively affects many native plant species and paves the way for colonization by other nonindigenous nitrophilous species. Moreover, the presence of firetree has altered the habitat to such an extent that the island has been subsequently colonized by a seed-dispersing bird, the Japanese white-eye (*Zosterops japonicus*), which has negative effects on native passerines (Mountainspring and Scott 1985).

Changes in ecosystem food webs often occur when an alien predator is introduced. A typical and dramatic example is the brown tree snake (*Boiga irregularis*) introduced to the island of Guam, which has caused the extinction of more than half of the native bird and lizard species and two out of three of Guam's native bats (Savidge 1987, Rodda and Savidge 2007). The American mink (*Neovison vison*) is one of the invasive mammal species with the highest effect on native fauna in Europe. The species preys on waterfowl, seabirds, small mammals, amphibians, and fish (Clode and MacDonald 2002, Ahola et al. 2006, Melero et al. 2012), affecting negatively at least 47 native species (Genovesi et al. 2012), some of which have great conservation or recreational-touristic value (bird-watching of seabirds on islands). The American mink not only acts as predator, but also competes with ecologically similar native mustelids such as the endangered European mink (*Mustela lutreola*), polecat (*Mustela putorius*), and stoat (*Mustela erminea*), affecting their populations or distribution (Sidorovich 2000).

Overgrazing from feral goats, sheep, other ungulates, and rabbits (*Oryctolagus cuniculus*) tends to suppress woody-species regeneration, reducing or altering herbaceous cover, thereby contributing to soil erosion (Klinger et al. 1994, Campbell and Donlan 2005). In most cases, communities of oceanic islands have evolved in the absence of large herbivores, and their introductions could lead to severe ecosystem degradation and subsequent loss of biodiversity (Coblentz 1978, Campbell and Donlan 2005).

Invasive species may affect ecosystems even more deeply, changing their physical structure, for example, when ecosystem engineers are introduced (American beavers, *Castor canadensis*; Jones et al. 1994). The American beaver has been introduced to Europe, Asia, and South America. The species is known as an ecosystem engineer for its ability to alter the physical and chemical nature of water bodies and their adjacent terrestrial systems (Naiman et al. 1986, 1988). Beavers reduce forest cover close to water bodies, exposing riverbanks to erosion. The dams constructed for lodging act as barriers to fish migration and create impounded water with higher temperature (Alexander 1998), which changes the physical-chemical parameters in the aquatic environ-

ment, with potential subsequent changes in flora and (invertebrate) fauna.

Some areas and habitats are more exposed to the negative effects of IAS than others. Though limited in global surface, islands hold a disproportionate percentage by unit of area of global biodiversity, and endemism richness of plants and vertebrates on islands exceeds those of mainland regions by a factor of 9.5 and 8.1, respectively (Kier et al. 2009). But islands are highly sensitive to any kind of disturbance and thus, also, to IAS. In fact, a high proportion of extinctions on islands has been caused by invasive vertebrates (Simberloff 1995, Courchamp et al. 2003). Because of their small size and high degree of isolation, many islands were never naturally colonized by certain taxa, like terrestrial mammals, until humans deliberately introduced them or favored their colonization (rats, *Rattus* spp.; mice, *Mus* spp.). Consequently, introduction of mammals to islands is often disruptive to part of the native fauna and/or flora.

Invasive alien species cause strong ecological effects in aquatic ecosystems, with consistent patterns among different functional groups. A meta-analysis conducted by Gallardo et al. (2015) showed that the introduction of alien invasive species determines a decrease in the abundance and diversity of aquatic communities, particularly severe in fish, zooplankton, and macrophytes, whereas the effects on the abundance of benthic invertebrates and phytoplankton are highly variable. Considering the trophic links that characterize aquatic ecosystems, the authors anticipate far-reaching consequences of IAS on the structure and functionality of these habitats.

One of the main criticisms of the field of invasion biology has been the lack of quantification of effects on species and ecosystems (McLaughlin et al. 2014). Recently, however, the International Union for Conservation of Nature (IUCN) adopted the Environmental Impact Classification of Alien Taxa (EICAT), an assessment process that classifies alien species according to the magnitude of their detrimental effects to the environment. The system is based in five categories and uses semiquantitative criteria to highlight progressively higher and more complex effects, from reductions in the fitness of native individuals to irreversible changes in the structure of communities (Hawkins et al. 2015). The EICAT will help in evaluating, comparing, and predicting the magnitudes of IAS effects, to determine and prioritize appropriate actions where necessary.

Effect on Native Species and Change of Community Structure

The introduction of IAS increases the number of species established locally, but this is not equal to an increase of biodiversity. Over time, introduced species could become part of the new ecosystems, and in some cases the new species may have also positive effects. The European wild rabbit performs significant ecosystem services, both in the native and in some part of the introduced range (Lees and Bell 2008). In the native Iberian Peninsula, rabbits are a keystone species, maintaining mosaics of differing vegetation through selective grazing and being an important prey for many vertebrates, including the endangered Iberian lynx (*Lynx pardinus*) and Spanish imperial eagle (*Aquila adalberti*), which prey almost exclusively on them. In Great Britain, where they are introduced, grazing from rabbits is fundamental in maintaining plant communities and floristic diversity regulated originally by large native herbivores. In many other places rabbits reduce the richness of plant communities and affect native species. Goodenough (2010) reviewed the range of effects—negative, neutral, or positive—that introduced species may have on native ones. Considering the complexity of the interaction between native and introduced species, biologists prefer to use the term "xenodiversity" to maintain a separation between introduced and native species, highlighting that diversity is not just the global amount of species, but how they interact with each other and with the environment.

The introduction of nonnative species breaks down biogeographical barriers and changes ecosystem composition and functioning, altering the rela-

tionship between species and energy flow and, consequently, affecting human well-being (Ehrenfeld 2010, Strayer 2012). During the process of invasion, introduced species should pass through different stages of transport and introduction into new areas, establishment, and spread (Blackburn et al. 2011). All these stages act as nonrandom filters, generally selecting generalist species (with a wide ecological niche) with the ability to cope with new habitats. By contrast, species groups more prone to extinction are those with traits associated with specialization. One of the results of the human-mediated expansion of cosmopolitan species, accompanied by the extinction of native species, is the increased taxonomic similarity among species assemblages. This process, called biotic homogenization (McKinney and Lockwood 1999), has induced an irreversible worldwide loss of distinctiveness in biota across different spatial scales (Olden 2006, Baiser et al. 2012). The loss of uniqueness in ecosystems is particularly impressive in freshwater fish communities. Leprieur et al. (2008) identified six major worldwide invasion hotspots where nonnative fish species represent more than one-quarter of the species.

Adding new species to native communities may differentiate them, but when the same species are largely introduced, homogenization will prevail (McKinney 2004). The effects on beta diversity (the degree of communities' differentiation in relation to a complex gradient of environment) after species introduction have been the object of many studies. Exotic and translocated species could generate different patterns, with locally translocated species promoting homogenization and exotic species introduced from faraway resulting in decreased similarity (McKinney 2005, Leprieur et al. 2007).

One of the most evident and direct changes in ecosystem composition is the loss of species. In the case of vertebrates, IAS are now considered one of the main influences on extinction, after habitat loss and degradation, through different mechanisms, such as predation, competition, or disease transmission (Table 3.1; Vié et al. 2009, Bellard et al. 2016).

Table 3.1 Breakdown of the major threats to vertebrates assessed as threatened (critically endangered, endangered, vulnerable) according to the IUCN Red List Criteria (data reanalyzed from Vié et al. 2009)

	Habitat loss / degradation[1]	Pollution	Diseases	Invasive species	Fires	Direct utilization[2]	Human disturbance	Urban development / other artifacts[3]	Intrinsic factors	Water extraction / drought
Amphibians	1°	2°	5°	4°	3°					
Reptiles	1°	4°		5°		3°			2°	
Birds	1°			2°	5°	3°		4°		
Mammals	1°			3°	4°	2°	5°			
Freshwater fishes[4]		1°		2°		5°		4°		3°

[1] Habitat loss and degradation include agriculture and logging.

[2] Includes hunting and professional fishing.

[3] Urban development includes residential and commercial development and energy production and mining.

[4] Assessment made only in Madagascar, Europe, the Mediterranean, and Eastern and Southern Africa.

Invasive mammals can cause the extinction of native species on islands (Courchamp et al. 2003), and the worst IAS are predators (feral cats, mustelids/mongooses, and omnivorous rats adapted to prey on small vertebrates or their eggs and young; McCreless et al. 2016). Cats are generalist predators that easily become feral and have been widely introduced to islands, where they prey on a variety of endemic birds and mammals, which often lack defenses against mammalian predators and can suffer severe population declines and extinction (Medina et al. 2011). Ship (*Rattus rattus*), Norway (*R. norvegicus*), and Pacific (*R. exulans*) rats are associated with extinctions or declines of flightless invertebrates, ground-dwelling reptiles, land birds, and burrowing seabirds (Towns et al. 2006, Jones et al. 2008). Preying on eggs and nestlings, rats can reduce to zero the reproductive output of seabird colonies (Jones et al. 2008). Predation, however, can be destructive also on the mainland. Several studies have demonstrated that the North American mink can affect native populations of ground-nesting birds, rodents, and mustelids (Woodroffe et al. 1990, Craik 1997, Macdonald and Harrington 2003).

Invasive alien species compete with native species in different ways. The competitive process could be the results of interference, involving direct interactions between species, or exploitation competition connected to the use of a common resource (Schoener 1983). Exploitation competition has been observed between introduced and native squirrels in Europe (Gurnell et al. 2004), and between marsupial and eutherian carnivores in Australia (Glen and Dickman 2008). Competitive interactions may be more complex, however, being mediated by predators, parasites, or pathogens (Zhang et al. 2006). The so-called apparent competition occurs as an indirect effect when species that do not directly compete for resources affect each other by being prey for the same predator (Courchamp et al. 2000). This has been observed, especially on islands, when both a prey and a predator are introduced, causing hyperpredation processes (Courchamp et al. 2000, Zhang et al. 2006), but also in mainland environments where an invasive mammal could modify the native predator-prey dynamics (Cerri et al. 2017). Other forms of indirect competition between native and invasive species include those mediated by diseases (Fournier-Chambrillon et al. 2004, Strauss et al. 2012). The Eastern gray squirrel (*Sciurus carolinensis*), for instance, is replacing the native Eurasian red squirrel (*S. vulgaris*) through exploitation competition

Figure 3.1   The Eastern gray squirrel (shown) is influencing the native Eurasian red squirrel in Europe. Sandro Bertolino

in Italy (Gurnell et al. 2004; Fig. 3.1), whereas in Great Britain the competition is mainly disease-mediated by healthy gray squirrels carrying and transmitting a squirrel poxvirus that kills most red squirrels that become infected (Tompkins et al. 2002).

The emergence of pathogenic infectious diseases represents a substantial global threat to biodiversity, often amplified by IAS introductions (Daszak et al. 2000, Roy et al. 2017). The fungal disease chytridiomycosis, caused by the chytrid fungus (*Batrachochytrium dendrobatidis*), has been associated with global amphibian declines and species extinctions. The fungus proliferates in epidermal cells, leading to hyperkeratosis, electrolyte loss, and ultimately death in susceptible amphibians (Berger et al. 1998, James et al. 2009). Evidence suggests that the amphibian trade, primarily of North American bullfrogs (*Rana catesbeiana*), is implicated in the emergence of chytridiomycosis by spreading infected animals worldwide (Garner et al. 2006, Fisher and Garner 2007).

Invasive Alien Species Management Success Stories

During the past decades, eradication efforts have delivered a suite of benefits to biodiversity that are gradually revealing themselves. Most successful eradications of IAS have occurred during island restoration management projects. In fact, over the past

30 years, the eradication of invasive mammals from islands has become one of the most powerful tools for preventing extinction of insular endemics and restoring insular ecosystems (Carrion et al. 2011). Three main factors heavily influence success and outcomes of these projects: degree of local support, ability to eradicate nonnative species cost-effectively, and ability to mitigate for nontarget effects. A good example of the importance of such efforts is the eradication of feral pigs (*Sus scrofa*) from Santiago Island in the Galápagos archipelago, Ecuador, which is the largest insular pig removal to date (Cruz et al. 2005). Feral pigs are omnivores and can have devastating effects on island plant and animal communities. Using a combination of ground hunting and poisoning, over 18,000 pigs were removed during a 30-year eradication campaign. The long-term sustained hunting effort was possible only with a high degree of support from the majority of local (island and archipelago-wide) stakeholders. Cost-effectiveness was only partly met; poisoning was efficient, and hunting success improved by increasing access to animals by cutting more trails, but limited conservation funds obliged managers to further increase eradication efficiency. Nontarget effects were mitigated by favoring intense rather than sustained control (Cruz et al. 2005).

An important aspect that must be considered is the risk of new introductions of alien species (deliberate or accidental ones). On the Galápagos Islands, this was done by eradicating the target invasive goats from the entire archipelago. Project Isabela, the world's largest island restoration effort to date, removed more than 140,000 goats from more than 500,000 ha for a cost of US$10.5 million (Carrion et al. 2011). This project demonstrated, as do many other eradication attempts, that for many invasive mammal species, island size (or IAS population size) is often not the limiting factor with respect to eradication, but bureaucratic processes, financing, political will, and stakeholder approval tend to be the harder challenges. Also on the Galápagos Islands, the technical, direct goat control to keep archipelago-wide goat density to low levels was complemented with long-term social activities focused on education and governance (Carrion et al. 2011).

These eradications have provided clear benefits to native species. One typical example is that of the endemic Galápagos rails (*Laterallus spilonotus*), whose populations had been reduced severely because of predation by pigs and habitat degradation by goats. Comparisons of rail density estimates in 1986–1987 and 2004–2005 between islands without invasive mammals (Fernandina Island, trend slight increase), islands with pig and goat eradication (Santiago Island, trend tenfold increase), and islands without control during the study period (Isabela Island, decrease in rail density) provided evidence that pig and goat eradication is essential for Rallidae conservation (Donlan et al. 2007).

Eradications of vertebrates might encounter more resistance from the public or animal welfare groups in more developed regions such as Europe. Also, lack of long-term financial commitment may be a problem. For example, although Europe is one of the richest regions of the globe, and despite its formal commitment to halting the regional loss of biodiversity by 2020, the level of action to prevent, eradicate, or control invasive alien species (even on islands) has been so far very scant (Genovesi and Carnevali 2011). The database (scientific literature review, unpublished data provided by experts, reports to Bern Convention) on invasive species on the islands of Europe provides an updated list of attempted eradication programs and the presence of alien species, the native species directly affected by these, and tools to support more efficient decision making (Genovesi and Carnevali 2011). In 2011, the database contained information on 224 eradication programs on 170 islands, 86% of which have been successfully completed, mostly targeting rats (68%).

These problems highlight the need to have priority rules for conservation actions. An example is the eradication of the black rat to protect Cory's shearwater (*Calonectris diomedea*) and Yelkouan shearwater (*Puffinus yelkouan*) on Mediterranean islands. Researchers and managers evaluated for each island the effectiveness of rat eradication by means of two indices, both based on the relative importance of the island's nesting population of the two species at the national and regional scale (Capizzi et al. 2010). The first considered the number of nesting pairs in rat-free islands, the second the number of islands occupied by shearwaters. Next the monetary cost of rat eradication on each island was analyzed, and cost-benefit models were used to compare costs and effectiveness of all possible combinations of control on a set of islands. These analyses revealed that the maximum increase in effectiveness (largest proportion of shearwaters protected with sustainable monetary costs) fell around a relatively small budget (US$230,000), which had the highest number of pairs protected per US$1,500 of investment (Capizzi et al. 2010).

The eradication of the coypu (*Myocastor coypus*) from Britain is what can be called a textbook example of efficient practice (Fig. 3.2). The elements that were essential for this successful campaign were reviewed by Baker (2006). Coypus, native to South America, have been introduced in many countries for fur farming. Escapes or deliberate releases have occurred in North America, the Middle East, Africa, Japan, the Asiatic part of the former Soviet Union, and in Europe. Of the two populations that became established in Britain (introduction 1929, end of farming 1945), one disappeared without any known control in 1956, while the second expanded to cover the whole of East Anglia (Gosling and Baker 1989), a distribution of approximately 190 by 150 km in the early 1960s, with an estimated peak population size of approximately 200,000 animals. During an early coordinated trapping campaign and an exceptionally cold winter, numbers dropped sharply, but eradication was not achieved. The strategies adopted during this early attempt had several flaws: the main trapper force spent much of its time clearing relatively low-density areas rather than attempting to maximize capture rates, and although the effect of immigration into cleared areas was considered, it was not given sufficient attention (Gosling and Baker 1989). As population size increased in the 1970s, the typical

Figure 3.2   A live-captured coypu in Italy. Sandro Bertolino

damages to riverbanks, flood defenses, native vegetation, and agricultural crops became consistent; a second campaign was started in 1981, and in line with the objectives, the coypu was eradicated by 1989 (Baker 2006).

At least seven key features to set up and bring to a successful conclusion an effective eradication campaign were identified (Baker 2006).

1. Clear damage to human activities and to native fauna and flora made a strong case for cost-efficient eradication.
2. Detailed technical assessments of the effort, costs, and likely chances of success were achieved by a long-term study of population ecology, targeted to a particular control application. Research into coypu biology and population dynamics, and the computer simulations that resulted, allowed the size of the trapping force and time frame neces-

sary to achieve eradication to be estimated. A successful trial eradication exercise, at a realistic scale, gave confidence to those recommending a way forward and those funding the exercise that it could achieve its objectives.

3. Cage trapping and subsequent killing by a single shot to the head was felt to be a humane and acceptable control technique. Interestingly, but creating preoccupation for wildlife managers, Baker (2006) argued that, although there was little interference with the trapping campaign in the 1980s, it is likely that protests from animal rights campaigners would be more significant if a similar campaign were to be repeated now.
4. Sound management structure and finances were in place, backed by the central government and key local stakeholders.
5. The progress of the campaign was intensively monitored by retrospective census: the response

of the population to changes in trapping strategy or improved trapping practice was quantified, and trapping force was altered where necessary.

6. There was an incentive for the trappers to achieve their objective. Those carrying out an eradication exercise will potentially be unemployed if they are successful, which may lead them to avoid the campaign achieving its objective. Also, effort and removal ratio will increase steeply during the final parts of a successful eradication campaign. Hence the continuous motivation of the trappers is very important. This was achieved by an incentive bonus offered at the start of the campaign, to be paid in the event of successful eradication within a ten-year period.

7. A clear definition of a successful end of campaign was in place: no more animals captured for a period of three years (Baker 2006).

Robertson et al. (2017) reviewed 15 attempts for larger-scale removals (mean area 2,627 km^2) of invasive mammals (edible dormouse, *Glis glis*; muskrat, *Ondatra zibethicus*; coypu; Himalayan porcupine, *Hystrix brachyura*; Pallas's squirrel, *Callosciurus erythraeus*; and eastern gray squirrels and American mink) in northern Europe using static traps. They reported that 12 were successful (true eradication or complete removal to a buffer zone; Table 3.2). A project's costs were best predicted by and increased with area of removal, while the number of animals removed had no significant effect on costs (Robertson et al. 2017). It must be said, however, that the numbers removed in these examples were relatively low. In comparison with eradications on islands, these larger-scale programs were characterized by extra challenges, such as defining boundaries and related uncertainties about monetary/manpower costs, the definition of clear objectives and confirmation of success, and a different approach to manage potential recolonization (Robertson et al. 2017). The examples show that a rapid response to new incursions increased the chance of success and is recommended as best practice; in contrast, large-scale (and longer-term) control will strongly increase the environmental, financial, and welfare costs.

Table 3.2 Data on large-scale mammal eradications in Europe (revised from Robertson et al. 2016 under Creative Commons license; see original paper for references and how data were extracted)

Species	Years	Region	Area (km2)	Trapper-years	Animals removed	Success
Edible dormouse	Early 1900s	Bedfordshire, England				No
Muskrat	1932–1935	Shropshire, England	1813	61	3052	Yes
Muskrat	1932–1937	Scotland	2815	35.5	1248	Yes
Muskrat	1933–1935	Surrey, England	96	8b	169	Yes
Muskrat	1932–1935	Sussex, England	81	18b	52	Yes
Muskrat	1933–1935	Clare/Tipperary, Ireland	414	21	487	Yes
American mink	1964–1969	Great Britain	184,000	77	ca. 5000	No
American mink	2001–2005	Hebrides/Uists, Scotland	850	23.5	228	Yes
American mink		Highland, Scotland	29,000			
American mink	2007–2013	Harris and Lewis, Scotland	2611	78	1514	Yes
Himalayan porcupine	?–1979c	Devon, England	280	9	6	Yes
Coypu	1981–1989	East Anglia, England	19,210a	192	34,822	Yes
Gray squirrel	1998–2001	Thetford, England	17–46d	1.6d	2209	No
Gray squirrel	1998–2013	Anglesey, Wales	710	30	6397	Yes
Pallas's squirrel	2005–2011	Belgium	2.7a	2.3	248	Yes

Finally, an excellent review paper (Jones et al. 2016) describes and discusses the conservation gains of invasive mammal eradication on islands. This paper underlines that, although annually more than US$21 billion is spent worldwide on biodiversity conservation, the global effect of many conservation and management actions is rarely systematically assessed. Jones et al. (2016) limit their review to islands because they house a much higher amount of biodiversity compared with mainlands, and because many species native to islands are highly threatened with extinction (two-thirds of recent extinctions). Using a combination of extensive literature, database review, and expert interviews, Jones et al. (2016) estimated to which extent the eradication of invasive mammals on islands produced global benefits to biodiversity conservation. Of 251 eradications of invasive mammals on 181 islands, positive demographic or distributional responses occurred in 596 populations of 236 native terrestrial faunal species. Of these 236 demonstrated beneficiary species, 62 (26%) are threatened with extinction (IUCN Red List categories: critically endangered, endangered, and vulnerable) and 20 (9%) are in the near-threatened category. In contrast, eradications had a negative effect in eight cases (populations) of seven native species. Four threatened species (island fox, *Urocyon littora-*

lis; Seychelles magpie-robin, *Copsychus sechellarum*; Cook's petrel, *Pterodroma cookii*; and black-vented shearwater, *Puffinus opisthomelas*) had their IUCN Red List extinction-risk categories reduced as a direct result of invasive mammal eradication, and it was predicted that more than a hundred highly threatened birds, mammals, and reptiles have benefited from these interventions (Jones et al. 2016). These authors argued that, because monitoring and publishing outcomes of eradication and control programs is limited, the effects of global eradications are likely greater than they reported in their review.

Summary

The global challenges of IAS are considerable with the increasing mobility of humans and the species that are transported intentionally or accidentally. Impacts of IAS are often more broad and insidious than initially appreciated. Management strategies to minimize or reverse the effect of IAS require decision making on tactics necessary to achieve a strategy of action or triage. All of the cases detailed in this chapter show that removing established IAS is possible, with great benefit for ecosystems and species conservation. Reducing IAS effects, however, requires stringent prevention measures (Fig. 3.3),

Figure 3.3 Avoiding the introduction of red-eared slider turtles by trade ban is only partly effective. Sandro Bertolino

effective early warning and rapid response, and advanced management approaches. Management of IAS is really a field in which policy, legislation, education, population dynamics research, and field activity must work together continuously and innovatively.

LITERATURE CITED

Ahola, M., M. Nordström, P.B. Banks, N. Laanetu, and E. Korpimäki. 2006. Alien mink predation induces prolonged declines in archipelago amphibians. Proceedings of the Royal Society, London B 273:1261–1265.

Alexander, M.D. 1998. Effects of beaver (*Castor canadensis*) impoundments on stream temperature and fish community species composition and growth in selected tributaries of Miramichi River, New Brunswick. Canadian Technical Report of Fisheries and Aquatic Sciences 2227:1–44.

Baker, S. 2006. The eradication of coypus (*Myocastor coypus*) from Britain: The elements required for a successful campaign. Pages 142–147 *in* F. Koike, M.N. Clout, M. Kawamichi, M. De Poorter, and K. Iwatsuki, editors. Assessment and control of biological invasion risks. Shoukadoh Book Sellers, Kyoto, Japan, and IUCN, Gland, Switzerland.

Berger, L., R. Speare, P. Daszak, D.E. Green, A.A. Cunningham, C.L. Goggin, R. Slocombe, et al. 1998. Chytridiomycosis causes amphibian mortality associated with population declines in the rain forests of Australia and Central America. Proceedings National Academy of Sciences United States of America 95:9031–9036.

Blackburn, T.M., P. Cassey, R.P. Duncan, K.J. Evans, and K.J. Gaston. 2004. Avian extinction and mammalian introductions on oceanic islands. Science 305:1955–1958.

Blackburn, T.M., P. Pyšek, S. Bacher, J.T. Carlton, R.P. Duncan, V. Jarošík, J.R.U. Wilson, and D.M. Richardson. 2011. A proposed unified framework for biological invasions. Trends in Ecology and Evolution 26:333–339.

Braithwaite, R.W., W.M. Lonsdale, and J.A. Estbergs. 1989. Alien vegetation and native biota in tropical Australia: The spread and impact of *Mimosa pigra*. Biological Conservation 48:189–210.

Campbell, K.J., and C.J. Donlan. 2005. A review of feral goat eradication on islands. Conservation Biology 19:1362–1374.

Capizzi, D., N. Baccetti, and P. Sposimo. 2010. Prioritizing rat eradication on islands by cost and effectiveness to protect nesting seabirds. Biological Conservation 143:1716–1727.

Carrion, V., J.S. Donlan, K.J. Campbell, C. Lavoie, and F. Cruz. 2011. Archipelago-wide island restoration in the Galápagos Islands: Reducing costs of invasive mammal eradication programs and reinvasion risk. https://doi.org/10.1371/journal.pone.0018835.

Cerri, J., M. Ferretti, and S. Bertolino. 2017. Rabbits killing hares: An invasive mammal modifies native predator-prey dynamics. Animal Conservation 20:511–519.

Clavero, M., and E. Garcia-Berthou. 2005. Invasive species are a leading cause of animal extinctions. Trends in Ecology and Evolution 20:110.

Clode, D., and D.W. MacDonald. 2002. Invasive predators and the conservation of island birds: The case of American mink *Mustela vison* and terns *Sterna* spp. in the Western Isles, Scotland. Bird Study 49:118–123.

Coblentz, B.E. 1978. The effects of feral goats (*Capra hircus*) on island ecosystems. Biological Conservation 13:279–286.

Courchamp, F., M. Langlais, and G. Sugihara. 2000. Rabbits killing birds: Modelling the hyperpredation process. Journal of Animal Ecology 69:154–164.

Courchamp, F., J.L. Chapuis, and M. Pascal. 2003. Mammal invaders on islands: Impact, control and control impact. Biological Reviews 78:347–383.

Craik, J.C.A. 1997. Long-term effects of North American mink *Mustela vison* on seabirds in western Scotland. Bird Study 44:303–309.

Cruz, F., J.S. Donlan, K.J. Campbell, and V. Carrion. 2005. Conservation action in the Galapagos: Feral pig (*Sus scrofa*) eradication from Santiago Island. Biological Conservation 121:473–478.

Daszak, P., A.A. Cunningham, and A.D. Hyatt. 2000. Emerging infectious diseases of wildlife: Threats to biodiversity and human health. Science 287:443–449.

Donlan, C.J., K. Campbell, W. Cabrera, C. Lavoi, V. Carrion, and F. Cruz. 2007. Recovery of the Galàpagos rail (*Laterallus spilonotus*) following the removal of invasive mammals. Biological Conservation 138:520–524.

Ehrenfeld, J.G. 2010. Ecosystem consequences of biological invasions. Annual Review in Ecology, Evolution and Systematics 41:59–80.

Ellstrand, N.C., and K.A. Schierenbeck. 2000. Hybridization as a stimulus for the evolution of invasiveness in plants? Proceedings of the National Academy of Sciences of the United States of America 97:7043–7050.

Fisher, M.C., and T.W.J. Garner. 2007. The relationship between the emergence of *Batrachochytrium dendrobatidis*, the international trade in amphibians and introduced amphibian species. Fungal Biological Review 21:2–9.

Fournier-Chambrillon, C., B. Aasted, A. Perrot, D. Pontier, F. Sauvage, M. Artois, J.M. Cassiède, et al. 2004. Antibodies to Aleutian mink disease parvovirus in free-ranging European mink (*Mustela lutreola*) and other small carnivores from southwestern France. Journal of Wildlife Diseases 40:394–402.

Gaertner, M., A. Den Bree, C. Hui, and D.M. Richardson. 2009. Impacts of alien plant invasions on species richness in Mediterranean-type ecosystems: A meta-analysis. Progress in Physical Geography 33:319–338.

Gallardo, B., M. Clavero, M.I. Sánchez, and M. Vilà. 2016. Global ecological impacts of invasive species in aquatic ecosystems. Global Change Biology 22:151–163.

Garner, T.W., M.W. Perkins, P. Govindarajulu, D. Seglie, S. Walker, A.A. Cunningham, and M.C. Fisher. 2006. The emerging amphibian pathogen *Batrachochytrium dendrobatidis* globally infects introduced populations of the North American bullfrog, *Rana catesbeiana*. Biology Letters 2:455–459.

Genovesi, P., and L. Carnevali. 2011. Invasive alien species on European islands: Eradications and priorities for future work. Pages 56–62 *in* C.R. Veitch, M.N. Clout, and D.R. Towns, editors. Island invasives: Eradication and management. IUCN, Gland, Switzerland.

Genovesi, P., L. Carnevali, A. Alonzi, and R. Scalera. 2012. Alien mammals in Europe: Updated numbers and trends, and assessment of the effects on biodiversity. Integrative Zoology 7:247–253.

Genovesi, P., C. Carboneras, M. Vilà, and P. Walton. 2014. EU adopts innovative legislation on invasive species: A step towards a global response to biological invasions? Biological Invasions 17:1307–1311.

Glen, A.S., and C.R. Dickman. 2008. Niche overlap between marsupial and eutherian carnivores: Does competition threaten the endangered spotted-tailed quoll? Journal of Applied Ecology 45:700–707.

Goodenough, A.E. 2010. Are the ecological impacts of alien species misrepresented? A review of the "native good, alien bad" philosophy. Community Ecology 11:13–21.

Gosling, L.M., and S.J. Baker. 1989. The eradication of muskrats and coypus from Britain. Biological Journal of the Linnean Society 38:39–51.

Gurevitch, J., and D.K. Padilla. 2004. Are invasive species a major cause of extinctions? Trends in Ecology and Evolution 19:470–474.

Gurnell, J., L.A. Wauters, P.W.W. Lurz, and G. Tosi. 2004. Alien species and interspecific competition: Effects of introduced eastern grey squirrels on red squirrel population dynamics. Journal of Animal Ecology 73:26–35.

Hawkins, C.L., S. Bacher, F. Essl, P.E. Hulme, J.M. Jeschke, I. Kühn, S. Kumschick, et al. 2015. Framework and guidelines for implementing the proposed IUCN Environmental Impact Classification for Alien Taxa (EICAT). Diversity Distribution 21:1360–1363.

Hejda, M., P. Pyšek, and V. Jarosik. 2009. Impact of invasive plants on the species richness, diversity and composition of invaded communities. Journal of Ecology 97:393–403.

Hulme, P.E. 2009. Trade, transport and trouble: Managing invasive species pathways in an era of globalization. Journal of Applied Ecology 46:10–18.

James, T.Y., A.P. Litvintseva, R. Vilgalys, J.A.T. Morgan, J.W. Taylor, M.C. Fisher, L. Berger, C. Weldon, L. du Preez, and J.E. Longcore. 2009. Rapid global expansion of the fungal disease chytridiomycosis into declining and healthy amphibian populations. PLOS Pathogens 5: e1000458. doi:10.1371/journal.ppat.1000458.

Jones, C.G., J.H. Lawton, and M. Shachak. 1994. Organisms as ecosystem engineers. Oikos 69:373–386.

Jones, H.P., B.R. Tershy, E.S. Zavaleta, D.A. Croll, B.S. Keitt, M.E. Finkelstein, and G.R. Howald. 2008. Severity of the effects of invasive rats on seabirds: A global review. Conservation Biology 22:16–26.

Jones, H.P., N.D. Holmes, S.H. Butchart, B.R. Tershy, P.J. Kappes, I. Corkery, A. Aguirre-Muñoz, et al. 2016. Invasive mammal eradication on islands results in substantial conservation gains. Proceedings of the National Academy of Sciences of the United States of America 113:4033–4038.

Kettunen, M., P. Genovesi, S. Gollasch, S. Pagad, U. Starfinger, P. ten Brink, and C. Shine. 2009. Technical support to EU strategy on invasive species (IAS): Assessment of the impacts of IAS in Europe and the EU (final module report for the European Commission). Institute for European Environmental Policy (IEEP), Brussels, Belgium.

Kier, G., H. Kreft, T.M. Lee, W. Jetz, P.L. Ibisch, C. Nowicki, J. Mutke, and W. Barthlott. 2009. A global assessment of endemism and species richness across island and mainland regions. Proceedings of the National Academy of Sciences of the United States of America 106:9322–9327.

Klinger, R.C., P.T. Schuyler, and J.D. Sterner. 1994. Vegetation response to the removal of feral sheep from Santa Cruz Island. Pages 341–350 *in* W.L. Halvorson and G.J. Maender, editors. The fourth annual California islands symposium: Update on the status of resources. Santa Barbara Museum of Natural History, Santa Barbara, California, USA.

Kumschick, S., S. Bacher, T. Evans, Z. Markov, J. Pergl, P. Pysek, S. Vaes-Petignat, G. van der Veer, M. Vila, and W. Nentwig. 2015. Comparing impacts of alien plants and animals in Europe using a standard scoring system. Journal of Applied Ecology 52:552–561.

Lees, A.C., and D.J. Bell. 2008. A conservation paradox for the 21st century: The European wild rabbit *Oryctolagus cuniculus*, an invasive alien and an endangered native species. Mammal Review 38:304–320.

Leprieur, F., O. Beauchard, B. Hugueny, G. Grenouillet, and S. Brosse. 2007. Null model of biotic of homogenization: A test with the European freshwater fish fauna. Diversity and Distributions 14:291–300.

Macdonald, D.W., and L.A. Harrington. 2003. The American mink: The triumph and tragedy of adaptation out of context. New Zealand Journal of Zoology 30:421–441.

McCreless, E.E., D.D. Huff, D.A. Croll, B.R. Tershy, D.R. Spatz, N.D. Holmes, S.H. Butchart, and C. Wilcox. 2016. Past and estimated future impact of invasive alien mammals on insular threatened vertebrate populations. Nature Communications 7:12488.

McKinney, M.L. 2004. Do exotics homogenize or differentiate communities? Roles of sampling and exotic species richness. Biological Invasions 6:495–504.

McKinney, M.L. 2005. Species introduced from nearby sources have a more homogenizing effect than species from distant sources: Evidence from plants and fishes in the USA. Diversity and Distributions 11:367–374.

McKinney, M.L., and J.L. Lockwood. 1999. Biotic homogenization: A few winners replacing many losers in the next mass extinction. Trends in Ecology and Evolution 14:450–453.

McLaughlin, C., B. Gallardo, and D.C. Aldridge. 2014. How complete is our knowledge of the ecosystem services impacts of Europe's top 10 invasive species? Acta Oecologica 54:119–130.

Medina, F.M., E. Bonnaud, E. Vidal, B.R. Tershy, E.S. Zavaleta, C. Josh Donlan, B.S. Keitt, M. Le Corre, S.V. Horwath, and M. Nogales. 2011. A global review of the impacts of invasive cats on island endangered vertebrates. Global Change Biology 17:3503–3510.

Melero, Y., M. Plaza, G. Santulli, D. Saavedra, J. Gosàlbez, J. Ruiz-Olmo, and S. Palazón. 2012. Evaluating the effect of American mink, an alien invasive species, on the abundance of a native community: Is coexistence possible? Biodiversity and Conservation 21:1795–1809.

Mountainspring, S., and J.M. Scott. 1985. Interspecific competition among Hawaiian forest birds. Ecological Monographs 55:219–239.

Naiman, R.J., J.M. Mellilo, and J.E. Hobbie. 1986. Ecosystem alteration of boreal forest streams by beaver (Castor canadensis). Ecology 67:1254–1269.

Naiman, R.J., C.A. Johnston, and J.C. Kelley. 1988. Alteration of North American streams by beaver. Bioscience 38:753–762.

Olden, J.D. 2006. Biotic homogenization: A new research agenda for conservation biogeography. Journal of Biogeography 33:2027–2039.

Pimentel, D., R. Zuniga, and D. Morrison. 2005. Update on the environmental and economic costs associated with alien-invasive species in the United States. Ecological Economics 52:273–288.

Raizada, P., A.S. Raghubanshi, and J.S. Singh. 2008. Impact of invasive alien plant species on soil processes: A review. Proceedings of the National Academy of Sciences India Section B, Biological Sciences 78:288–298.

Ricciardi, A., M.E. Palmer, and N.D. Yan. 2011. Should biological invasions be managed as natural disasters? Bioscience 61:312–317.

Ricciardi, A., M.F. Hoopes, M.P. Marchetti, and J.L. Lockwood. 2013. Progress toward understanding the ecological impacts of non-native species. Ecological Monographs 83:263–282.

Richardson, D.M., and A. Ricciardi. 2013. Misleading criticisms of invasion science: A field guide. Diversity and Distributions 19:1461–1467.

Robertson, P.A., T. Adriaens, X. Lambin, A. Mill, S. Roy, C.M. Shuttleworth, and M. Sutton-Croft. 2017. The large-scale removal of mammalian invasive alien species in northern Europe. Pest Management Science 73:273–279.

Rodda, G.H., and J.M. Savidge. 2007. Biology and impacts of Pacific Island invasive species 2: Boiga irregularis, the brown tree snake (Reptilia: Colubridae). Pacific Science 61:307–324.

Roy, H.E., H. Hesketh, B.V. Purse, J. Eilenberg, A. Santini, R. Scalera, G.D. Stentiford, et al. 2017. Alien pathogens on the horizon: Opportunities for predicting their threat to wildlife. Conservation Letters 10:477–484.

Savidge, J.A. 1987. Extinction of an island forest avifauna by an introduced snake. Ecology 68:660–668.

Schoener, T.W. 1983. Field experiments on interspecific competition. American Naturalist 122:240–285.

Seebens, H., T.M. Blackburn, E.E. Dyer, P. Genovesi, P.E. Hulme, J.M. Jeschke, S. Pagad, et al. 2017. No saturation in the accumulation of alien species worldwide. Nature Communications 8:14435.

Sidorovich, V.E. 2000. The ongoing decline of riparian mustelids (European mink, Mustela lutreola, polecat, Mustela putorius and stoat, Mustela erminea) in Eastern Europe: A review of the results to date and a hypothesis. Pages 295–317 in H.I. Griffiths, editor. The mustelids in a modern world: Management and conservation aspects of small carnivore-human interactions. Bachuys, Leiden, Netherlands.

Simberloff, D. 1995. Why do introduced species appear to devastate islands more than mainland areas? Pacific Science 49:87–97.

Simberloff, D., J.-L. Martin, P. Genovesi, V. Maris, D.A. Wardle, J. Aronson, F. Courchamp, et al. 2013. Impacts of biological invasions: What's what and the way forward. Trends in Ecology and Evolution 28:58–66.

Stohlgren, T.J., and J.L. Schnase. 2006. Risk analysis for biological hazards: What we need to know about invasive species. Risk Analysis 26:163–173.

Strauss, A., A. White, and M. Boots. 2012. Invading with biological weapons: The importance of disease-mediated invasions. Functional Ecology 26:1249–1261.

Strayer, D.L. 2012. Eight questions about invasions and ecosystem functioning. Ecology Letters 15:1199–1210.

Tompkins, D.M., A.W. Sainsbury, P. Nettleton, D. Buxton, and J. Gurnell. 2002. Parapoxvirus causes a deleterious disease in red squirrels associated with United Kingdom population declines. Proceedings of the Royal Society of London B 269:529–533.

Towns, D.R., I.A.E. Atkinson, C.H. Daugherty. 2006. Have the harmful effects of introduced rats on islands been exaggerated? Biological Invasions 8:863–891.

Turbelin, A.J., B.D. Malamud, and R.A. Francis. 2017. Mapping the global state of invasive alien species: Patterns of invasion and policy responses. Global Ecology and Biogeography 26:78–92.

Vié, J.C., C. Hilton-Taylor, and S.N. Stuart, editors. 2009. Wildlife in a changing world: An analysis of the 2008 IUCN Red List of Threatened Species. IUCN, Gland, Switzerland.

Vilà, M., C. Basnou, P. Pyšek, M. Josefsson, P. Genovesi, S. Gollasch, W. Nentwig, et al. 2010. How well do we understand the impacts of alien species on ecosystem services? A pan-European, cross-taxa assessment. Frontiers in Ecology and the Environment 8:135–144.

Vitousek, P.M. 1990. Biological invasions and ecosystem processes: Towards an integration of population biology and ecosystem studies. Oikos 57:7–13.

Vitousek, P.M., and L.R. Walker. 1989. Biological invasion by *Myrica faya* in Hawaii: Plant demography, nitrogen fixation, ecosystem effects. Ecological Monographs 59:247–265.

Woodroffe, G.L., J.H. Lawton, and W.L. Davidson. 1990. The impact of feral mink *Mustela vison* on water voles *Arvicola terrestris* in the North Yorkshire Moors National Park. Biological Conservation 51:49–62.

Zhang, J., M. Fan, and Y. Kuang. 2006. Rabbits killing birds revisited. Mathematical Biosciences 203:100–123.

4

Delwin E. Benson
Wouter van Hoven
Yves Lecocq
Bob van den Brink

Appreciation, Encouragement, and Rating of Wildlife and Nature Conservation on Private Lands

Introduction

We can generate more interest and actions to help landowners improve wildlife and nature conservation through greater awareness of private contributions and using common criteria conservation practices. Landowners can become more active with management and adopt sound conservation measures on lands for personal interests and benefits to society.

Privately controlled lands and initiatives provide opportunities for wildlife around the world (Hudson et al. 1989, Renecker and Hudson 1991, Dickson et al. 2009). In this chapter, we briefly outline conservation values and positive outcomes for wildlife management on private lands in Europe, Africa, and the United States, then present a rating system to evaluate, classify, and encourage wildlife conservation by the private sector.

Ecological processes or societies are not developing more natural areas; consequently, we need to enhance processes whereby people and nature coexist. Public ownership of parks and protected areas has become expensive, competitive with other land uses, and politically less desirable; these lands are inadequately managed, or the areas have become commons, with tragic results. More persons use lands more extensively and intensively than in the past; consequently, we need to better manage what exists, especially as human populations grow.

The World Bank reported more than a doubling of world human populations from 1960 to 2015, from 3.04 to 7.3 billion (http://data.worldbank.org/indicator/SP.POP.TOTL, accessed 20 November 2017), with most people (55% world, and 82% US) living in urban areas (http://data.worldbank.org/indicator/SP.URB.TOTL.IN.ZS, accessed 20 November 2017). Rural populations grew from 2 billion to 3.4 billion (http://data.worldbank.org/indicator/SP.RUR.TOTL?view=chart, accessed 20 November 2017). The World Bank emphasized the role of indigenous peoples as the natural but often forgotten partners in biodiversity conservation, yet it did not represent the vast landscapes under the control of private ranchers and farmers in first-world nations (http://documents.worldbank.org/curated/en/995271468177530126/The-role-of-indigenous-peoples-in-biodiversity-conservation-the-natural-but-often-forgotten-partners). Although indigenous peoples own, occupy, or use a quarter of the world's surface area and safeguard 80% of the biodiversity (http://www.worldbank.org/en/topic/indigenouspeoples), relatively modernized rural farmers and ranchers can be added to this safeguard system when properly enfranchised and motivated; therefore, we focus on Europe, Africa, and the United States.

Landscapes lose natural biodiversity as more humans occupy and alter wild places. Nature conservation for public or private benefit is practiced in various quantities and with various quality around the world, generally with some form of governmental authorization, but private contributions are underappreciated (Benson et al. 1999). Private lands are used for production ranging from subsistence to commercial agricultural products, and wildlife are viewed as liabilities if not as assets. Rural private landscapes and wildlife face increasing economic and social uncertainties, and they need more public and private support.

There are opportunities in nature conservation awaiting positive private actions. Private ranchers and farmers include commercial businesses, subsistence survivors, nouveaux rural enthusiasts, and modern landowners making decisions about landscape, social, and business uncertainties that affect conservation by their actions or inactions (Benson et al. 1999). Urbanizing societies, outdoor users, and conservation leaders must help manage private and communal lands by providing landowners more recognition, opportunities, and incentives, and less uncertain policies and practices. Policy can enfranchise the private sector toward meaningful stewardship for personal benefits and values for society (Benson et al. 1999) or discourage private contributions.

We acknowledge the relatively disenfranchised but growing private sector in nature conservation, featuring private land management approaches from Europe, Africa, and the United States. In that order, one can relate to effects of dense human populations, older to newer industrialized human history and conservation systems, European colonization, and the benefits of organized hunting for conservation. There are three common themes in these areas of the world.

1. Landowners manage wildlife as part of agriculture and nature conservation.
2. Landowners seek personal, economic, ecological, and social values from their work and outcomes.

They appreciate wildlife variously around the world and are burdened by nurturing wild animals privately for societal benefits that may not balance the social and economic costs of the wildlife production that they bear.
3. Aside from personal enjoyment, hunting is a powerful incentive encouraging landowners to manage wildlife and their habitats (Wall and Child 2009). Hunting benefits society directly through contributions to nature conservation and outdoor uses. Hunting provides a positive demand and thus incentives to supply and manage wildlife on private lands. Species that are not hunted have fewer incentives for landowners. Unfortunately, controversies about hunting and economic incentives cause some to overlook the benefits hunting generates for wildlife production and conservation, including for nonhunted species and biodiversity. Managed hunting helps balance animal numbers within landscape capabilities, has a demonstrated user demand, can constructively reduce human and wildlife conflicts, provides economic and social incentives for producers, and results in animal and land protection.

Wildlife is considered as a liability if problems outweigh benefits; consequently, nature conservation could be discouraged by private owners when incentives are absent. More enthusiasm, commitment, and policies for private wildlife and nature conservation incentives are needed by leaders in public and private sectors.

Public Politics, Laws, and Policies Enable or Restrict Conservation on Private Lands

Public policy influences opportunities and conservation actions on private land, whether markets can allocate natural resources or whether some form of governmental intervention is used (Wall and Child 2009, Cubbage et al. 2017). Uncontrolled

Table 4.1 Nature conservation and rating criteria to evaluate, document, and celebrate wildlife and landscape management. Highest numbers within categories and among all categories reflect most natural conditions.

Wildness factor	Domestic (range = 1–5)	Semidomestic (range = 6–10)	Semi-wild (range = 11–15)	Wild (range = 16–20)	Score/ land unit or species*
Land base**	Confined	Home ranges are restricted and require frequent intervention by humans	Home ranges for several breeding populations needing occasional genetic supplementation	Home ranges for several free-ranging and interactive breeding populations	
Herd size**			Not applicable and considered wild if animals are free ranging and part of regulated state or federal management systems. Calculations vary per species if private land only.	Not applicable and considered wild if animals are free ranging and part of regulated state or federal management systems. Calculations vary per species if private land only.	
Fencing	Tightly confined	Exterior and interior fences	Exterior fences only	Only in conflict areas	
Feeding	Constant	Seasonal	Emergency only	Not recommended; emergency only	
Population control	Human stabilization	Little natural fluctuation	Some natural fluctuation	Unregulated	
Culling, hunting, and live removal	Highly selective for sex, age, health, and commercial purposes	Selective and intensive, few die on the land	Extensive hunting and removal by age, sex, and inferior characteristics; balanced with natural deaths	Hunting randomly and selection by age, sex, and phenotypical conditions; natural deaths on the land	
Disease control	Annual, several diseases	Routine, few diseases	Rarely	Never	
Pasture control	As needed between confined areas unless totally supplemented	Routine	Rare	None	
Public access	None, recreational and commercial	Commercial and recreational; controlled public access	Commercial and recreational with some public access	Fewest public restrictions	
Distribution	Not applicable	Redistributed or in native ranges	Redistributed and mostly native ranges	Occupy native ranges	
Genetic purity	>5% of markers show introgression	<2%–5% show introgression	<2% show introgression	No known introgression	
Legal status	Livestock designation	Legal restrictions by law	Considered wild and part of public and private planning with public legal restrictions	Considered wild and free ranging over lands; landowner may have some controls by public	
Ecological interactions	Artificial, with incidental interactions with native systems	10%–50% native vertebrates and >20% human altered habitat systems	Most native vertebrates; >80% natural habitat systems	Complete natural systems including predators	

*Calculate number for each category within the five-point spread. Highest score within groups and sum of all groups is most natural.

**Varies and must be calculated per species and land base.

exploitation of wildlife before the 1900s in Western cultures and bushmeat uses worldwide (Chapter 12) prompted governments to enact laws regulating animal losses. Laws protected landscapes with parks and refuges and limited access to the supply of wildlife by restricting markets and legislating against demand.

Cubbage et al. (2017) reviewed how public and private policies emerge for natural resources. Government intervention can occur if natural resource issues are recognized, if leaders are amenable to making decisions, and if governments are adequately funded. Government intervention will not likely occur if markets are sufficient to sustainably manage resources, if governments are destitute, or if public agreement cannot be reached. Under poor or non-supported government management, the private sector might tend to intervene and assert individual or group controls (removing animal damage threats, poaching for food, using economic incentives to mitigate for liabilities).

Policy in the public sector can address government ownership, production, and planning; public regulation of private landowners' financial incentives or subsidies to encourage desirable social outcomes; public protection of private lands and goods; education; and research. Governments also set policy for market exchanges and local management on private lands. They can abdicate control, as with parts of Southern Africa; take control generally, as in the United States; or blend the two, as in Europe.

Nongovernmental organizations (NGOs), including hunter, landowner, and conservation groups, have filled gaps between governments and private interests (Fraser et al. 2009). Historically Europe and now Southern Africa represent strong private influences (landowner and hunter groups) for hunted species. Other NGOs focus more on less-abundant species. Growing influences of NGOs are prominent for all nature conservation generally.

Private natural resource policy tools may include training and self-governance through environmental or labor certifications (with wild meat production and sales), green labeling, organic or natural production methods, hunting (Benson and White 1995, Child and Wall 2009), or other voluntary environmental social programs. Certification for sale of regulated rhino (*Rhinoceros* spp.) horns and elephant (*Loxodonta* spp., *Elephas* spp.) tusks has been recommended in Africa to curb illegal sales of those products, to protect wildlife, and to have a business incentive to encourage those species on private lands. The classification suggestions at the end of this chapter aid rational certification of private contributions to conservation (Table 4.1).

Game Management on Private Land in Europe

Europe can be defined as all land west of the Ural Mountains, including 47 countries and almost 10 million km^2, about equal to the size of the United States. The European Union (EU) is another important political and economic entity to consider, as it currently covers 28 member states and 4.5 million km^2, with a common nature conservation policy and legal framework (called Nature Directives).

Europe has considerable biological diversity, and even rather densely populated and developed countries like France, Italy, or Spain harbor highly emblematic or conservation flagship species, such as brown bear (*Ursus arctos*), wolf (*Canis lupus*), and larger eagles (*Aquila* spp.). This diversity is important from a sociocultural point of view and is expressed in the way wildlife is managed. Hunting has contributed to conservation in Europe since the Middle Ages at least, mainly as recreational activity or sport. Hunting practices have developed and undergone an evolution across these centuries, reflecting national, regional, and even local traditions, customs, situations, constraints, and opportunities (Bubenik 1989).

An additional factor complicating the European overview is that Europe has almost 40 different official languages. For hunting, unlike most other domains in which English has become the leading idiom, there is no lingua franca, with hundreds, if not thousands, of relevant books, magazines, articles, and other sociological and anthropological publications dealing with hunting and its traditions written in German, French, Spanish, and other languages.

We describe a broad European wildlife management system in four generalized forms. The Federation of Associations for Hunting and Conservation of the European Union is a useful resource for those systems (http://www.face.eu/). J. M. Pinet summarized hunters in Europe (http://www.kora.ch/malme /05_library/5_1_publications/P_and_Q/Pinet_1995 _The_hunters_in_Europe.pdf):

1. Nordic or Scandinavian countries, such as Sweden, Norway, Finland, and Denmark, have healthy approaches to hunting and wisely use renewable natural resources through activities such as fishing, picking berries, and harvesting mushrooms. Wild meats matter much more than trophies or large numbers of animals taken. Hunting is widespread, popular, and democratic, receiving public support from across the political spectrum. Legislation is prescriptive but rather flexible, based on sound biological principles and reliable data (Bergstrom et al. 1992).

2. In the United Kingdom, hunting is considered a country sport for driven game birds (in which case the term "shooting" is generally used) and preferably for larger harvests (>500 pheasants [*Phasianus colchicus*] in one day for eight guns), to be achieved through intensive game management, including rearing and releasing; hence, it is rather costly and exclusive. Country sports are perceived by the public and mainstream media as elitist, receiving little or no support from political parties left of the center. The legal framework is relatively strict but rather generous, leaving certain discretion to landowners, estate managers, and wildlife professionals, and providing for individuals' responsibilities and rights.

3. The Germanic approach to hunting is most typical for countries such as Austria, Germany, Poland, Hungary, and several other Central European countries, where managing wildlife is selective (in more than one sense of the word) and proactive, with a focus on ungulate species. Hunting is a well-planned tool for regulating animal populations by skilled and trained operators within a rigid administrative and legal frame (Wolfe 1970).

4. The Latin or Mediterranean countries have roughly 50% of all seven million European hunters. In Southern Europe, hunting is a popular leisure activity, very much part of the social fabric of the countryside and of the democratic right to enjoy nature and its products. Laws, regulations and rules are strict and restrictive, above all to protect wildlife from overharvesting (Chantelat 1985).

These four systems are not necessarily linked to strict geographical boundaries or governmental frames. Rather, they have resulted from socioeconomic, cultural, and political developments throughout Europe in the eighteenth, nineteenth, and early twentieth centuries.

In Europe, there is a closer link between hunting and landownership, and therefore to wildlife management, than in many parts of the world (Fig. 4.1). Hunting rights are among the fundamental rights of landowners, as explicitly recognized by several judgments of the European Court of Human Rights in Strasbourg since 1999 (http://www.echr.coe.int). Even in those countries where wildlife is legally considered as res nullius (belonging to nobody) or as res communis (the property of the state and its people), these classifications have no significance for the right to hunt, which in most cases belongs to the landowner. Landowners may use this right personally, may decide not to hunt at all, or may allow other people to hunt the land, subject to rules and conditions set in place by competent authorities. Conditions may include the obligation to have a permit or license (in most cases subject to passing a serious test or exam); to respect closed seasons and protected species; to follow, in some cases, specific quantitative and qualitative guidelines (number of specimens

Figure 4.1   Private lands in Europe provide opportunities for outdoor recreation, habitat conservation, and hunting for species such as this chamois in the Bavarian Alps. There is also a need to control chamois populations to reduce damage to mountain forests, which play a key role in preventing avalanches and erosion. Pixabay.com

of a given species allowed to be taken); or to combine lands to achieve a minimum surface area for managing wildlife. This, however, does not affect the fundamental rights of landowners.

Landownership parameters in Central and Eastern Europe, areas that belonged to the sphere of influence of the former Soviet Union, have returned largely to pre–World War II status, namely that private individuals and companies can be proprietors of farmland, woodland, and other areas in the countryside. A significant number of the original owners of sometimes very large estates had their land returned to them or were compensated, and many rural entrepreneurs coming from Western Europe have bought large to very large farms to exploit with modern agricultural techniques.

Hunting practices in Europe are increasingly affected by international biodiversity conservation legislation, including directives for wild birds in 1979 (http://eur-lex.europa.eu/legal-content/EN/TXT/?uri=CELEX:32009L0147) and habitats in 1992 (http://eur-lex.europa.eu/legal-content/EN/TXT/?uri=CELEX:31992L0043, accessed 20 November 2017), which are legally binding for all EU member states. Although the directives do not interfere directly with national hunting systems and practices, they pro-

vide common rules on species that may be hunted, on bird hunting seasons, and on prohibited hunting methods. Thus, harmonization of hunting activities is progressively taking place within Europe, and this has not resulted—and is highly unlikely ever to do so—in the loss of the typical diversity or the multiplicity of hunting and wildlife management protocols at national and regional levels.

Concern by some local communities, certain political parties, and the conservation movement that private use of wildlife would lead to socioeconomic conflicts with other interests and loss of biodiversity did not materialize. Markets regulated these developments in a balanced way, and EU legislation prevented overexploitation and irreversible habitat degradation. Hunting and game management have adapted to these changes in landownership and constitute assets contributing directly to the preservation of suitable wildlife habitats. Local hunters' clubs continue to operate next to private commercial hunting estates and large blocks of state- or public-owned forests and other land. Hunting guests from abroad who pay to hunt continue to be welcomed, although this income is now more equally distributed, including to private landowners.

There are hardly any public or free hunting locations anywhere in Europe, with Greece and some other regions (in the Mediterranean) being the only exceptions. Generally, economic or service incentives are in place for landowners in Europe to allow and encourage hunting and therefore also to conserve and manage their property for wildlife management purposes (http://www.wildlife-estates.eu/).

Landowners in Anglo-Saxon regions have almost absolute freedom and power over what happens with wildlife. This means that no legal obstacle exists for owners of small woodlands in England to harvest (during the open season and with the means foreseen by the law—a suitable rifle) every red deer (*Cervus elaphus*) on that land, just as the owner of 500 ha in Scotland may decide not to kill any animals. This system works most of the time because it developed over long periods of time in a satisfactory manner. Also, it would not function properly elsewhere (in Southern Europe, for example, where hunting culture has not developed to value such self-discipline).

The positive contribution made by the traditional rural sports to the shape, texture, and general attractiveness of the English countryside, which would not have come about if food and timber production alone had been the concern of the land-user, is described by Vandervell and Coles (1980). They clearly demonstrate how much the countryside has benefited from landowners and users being sympathetic to conserving and managing the habitats of the wild creatures they enjoy hunting. The European Federation of Associations for Hunting and Conservation published a compilation of documented case studies from 14 European countries, illustrating how hunters maintain and improve habitats and conserve and manage populations of wild species for the five main habitats—forests, wetlands, coastlines, farmland, and mountains—precisely because they enjoy hunting on these lands (Lecocq 2004).

Unfortunately, there are other instances in Europe where incentives for landowners, hunters, and other users are lacking, and therefore seminatural habitats (particularly on farmland) and the species associated with them are deteriorating. This is notably the case in the Netherlands, the most densely populated and urbanized country in Europe, where, for purely emotional and ideological reasons, recreational hunting opportunities and even the concept of wise use through hunting have been almost entirely abolished (but not angling, because the Netherlands has about fifty times more anglers than hunters). In the neighboring Belgian region of Flanders, the equally unjustified strict protection of snipe (*Gallinago gallinago*) has resulted in the loss of hundreds of small wetlands that had been created for that species by hunters on farmland. These micromarshes were returned to poor-quality grassland because the reward of bagging a few snipes each autumn is no longer allowed.

The bottom line is simple: the axiom regularly cited for wanting, encouraging, and harvesting wild species in Africa—if it pays, it stays—applies fully to the situation in Europe, and as we elaborate below on the United States, it applies there also (Benson 1998a,b, Benson et al. 1999, Benson 2001a).

Wildlife and Nature Conservation on Private Land in Africa

Africa is a large continent comprising 54 independent countries, which were colonies of Europe from the 1900s. The youngest country is South Sudan, which has not yet experienced peace during its years of independence since 2011. This is the country where the northern white rhino (*Ceratotherium simum cottoni*) once thrived and now is locally extinct; only three nonreproductive individuals are left under close guard in Kenya. Leadership conflicts with colonial governments and local peoples led to national and civil wars with a devastating effect on local communities and wildlife.

Many countries under colonial rule established national parks to protect wildlife, and most of these famously exist today (van Hoven 2015a). Some parks have virtually no wildlife left because of poor law enforcement and land conflicts. Colonists in Africa also exploited wildlife with minimal controls, as was the practice in the United States in the 1800s.

Today African nations are transitioning from European colonies without management authority to sovereign nations exerting some local involvement with wildlife (Adams and McShane 1992) when incentives are positive, notably in Southern Africa. Independence was accompanied by extremely violent wars in many cases, such as Angola, the Congo, Mozambique, Zimbabwe, and others. The breakdown of law and order during wars led to more uncontrolled hunting and harvesting of bushmeat and other natural resources.

Craigie et al. (2010) suggest how, in the past 40 years, wildlife has declined in Africa. On average, the West African countries lost 80% of their wildlife in national parks and 100% of their wildlife outside of parks. East Africa, including Kenya and Tanzania, lost 52% of the wildlife in their national parks, and 75% of all wildlife outside of parks. South Africa, over the same forty years since 1970, had a net increase in wildlife in the national parks and a thirty-fold increase of wildlife on private and communal lands (Craigie et al. 2010). The key to this success, particularly on private lands, is the legal private ownership of wildlife, a system that is not supported by the public trust doctrine in the United States (Cantor 2016).

Wildlife management on private lands in Africa is fostered from love of wildlife, enjoyment of the outdoors, and the potential of economic gain from consumptive or nonconsumptive uses of the wildlife (Benson 1991a, van Hoven 2015a). Three labels for categories of management are generally employed. Larger properties with sustainable hunting as the major product are referred to as "wildlife ranches." Where more intensive management and breeding of rare, endangered, or exotic species takes place, the term "farming" is used. Places where tourism and photographic safaris are the focus are called "private nature reserves." In all three forms of managing wildlife on private land, making a profit from wildlife enables the landowners to practice conservation of

wildlife species and nature management. Landowners pay taxes to the state and become labeled as part of the wildlife industry (van Hoven 2015b).

The fundamental difference between conservation in Southern Africa and the rest of Africa lies with the private ownership of wildlife outside of public protected areas, creating the largest privately owned wildlife industry in the world (Cloete et al. 2015). Private wildlife management started in Zimbabwe, but the government allowed and encouraged the seizure of these lands during the past twenty years, with the result that very few such lands remain today; this seizure was accompanied by an uncontrolled decline in wildlife numbers outside national parks. Botswana, Namibia, and Zambia allow development of private wildlife initiatives, and these are currently growing.

In 1960, the estimated number of wildlife in South Africa on private and public land was 500,000. In 1985, 8,200 persons were estimated to participate in some form of wildlife use on private lands, and the first countrywide study of private management was conducted (Benson 1991a). Wildlife numbers had grown to 23,000,000 by 2015 (van Hoven 2015b). The economic value of wildlife was the single biggest driving force in the rapid growth of wildlife management on private lands in Southern Africa (van Hoven and Viljoen 1995). Wildlife roaming freely under various natural forces of ecology and artificial management became successful because it generated income, provided a profit to owners, and became an acceptable land management alternative within the private sector. In fewer than five decades, this industry grew from a zero base to a US$300 million local and US$120 million foreign hunting industry today. In 2012 the income on taxidermy was 30 million; translocation, 13 million; wing shooting, 20 million; birding, 10 million; bowhunting, 10 million; wildlife auctions, 20 million; firearm, infrastructure, and vehicles, 25 million; and lodges, tourism, and provincial permits, another 12 million (Dry 2013). The South African gross domestic product was 400 billion for 2012, of which the agricultural sector contributed 2.5% and the private wildlife sector 0.25%, for a total of US$1 billion (van Hoven 2013).

About 75% of all land under wildlife conservation in South Africa is private property, owned not only by individuals or corporations but also by communities. These systems work well in South Africa and some neighboring countries like Namibia, Botswana, and Zambia and could be exported to other African countries. Until the populace of a country wants to protect wildlife as a natural and aesthetic common inheritance, having an economic value will remain the pragmatic road to conservation.

Wildlife Stewardship and Recreation on Private Lands in the United States

The populace and governmental leaders in the United States have developed management systems for wildlife that are dominated by governmental wildlife management authority on behalf of all people, unlike European colonial founders who focused more on allocation based on social status and land ownership. The United States has fifty unique state wildlife agencies plus city, county, and federal jurisdictions with added powers over local, national, and international environmental management and trade. Logically, private wildlife management has garnered more policy and management attention in recent years by interested landowners and by professionals in governmental agencies (Benson 1991a, Benson 1992a,b, Benson et al. 1999). Recently, increasing numbers of nongovernmental organizations represent specific species interest groups and wildlife users and care for environmental management, biodiversity, and land protection (Benson 2001a,b,c,d, Benson 2014, Moore 2014).

Since national independence from Europe, wildlife and human interactions in the United States continue to evolve and adapt to changing times. In nearly 250 years, the relationship has changed from subsistence uses by early peoples and pioneers to unmanaged market overexploitation in the 1800s and early 1900s fostered by more efficient harvest methods and transportation (Trefethen 1975). Later, lands were managed and animal populations were rebuilt by increased funding from users for governmental institutions. Governmental and nongovernmental conservation organizations grew during the 1930s and presently have increasing roles and influence (Manfredo et al. 2009). A progressive period of extensive environmental concern and ecosystem management resurged in the 1960s. Hunters and other conservationists led early changes, but it was government at all levels that enacted the laws and policies and created agencies to manage wildlife on behalf of the people in society at large, who were the official owners.

Landowners in the United States do not own wildlife on their lands, but have considerable informal power on their lands because they control land uses and human access to wildlife. Neither landowners nor wildlife users can buy or sell wildlife except for some nonnative or domesticated and legally authorized populations. Private landowners legally control access and can desire to support wildlife, but they have limited and regulated ability to manage wildlife resources or the demands for its uses on their properties, leading to voids that landowners address in their own ways among the states (Benson 2001b). Private management has evolved in response to different historical uses, amount of private land in the state, governmental encouragement or discouragement of private practices, user demand, and personal beliefs (Benson et al. 1999).

Nearly 66% of lands in the United States are private, with higher percentages found in the eastern and southern states. Landowners exert de facto management of wildlife by how land and wildlife are treated, beneficially or negatively. Interest by landowners ranges from little attention and inaction to tolerance, resistance to wildlife conflicts, and uncertainty about roles and responsibilities, to use of animals by family and friends for hunting and enjoyment, fee-based access to wildlife, and other forms of active conservation (Benson et al. 1999).

Landowners exert substantial power to control access to private lands, with regional and legal differences. Landowners must construct fences to delineate boundaries in some states. Posting signs against ingress is sometimes required. Mere ownership is all that is necessary to restrict access in other states. Laws against trespass are enforced by governments, and state wildlife agencies have additional statutes against hunting without permission. Traditionally access to private lands was less restricted in the East, where the earliest settlers became accustomed to gaining ingress and egress. Access is likely by permission only in the West, where public land percentages are higher. In the South, where lands are mostly private, landowners have greater influences over wildlife and access, and fees are more likely to be paid to access private land and its wildlife. Fee-based access is growing around the United States (Benson 1989a,b, Benson 1992a), indicative of user demands, their willingness to pay, and the greater influences that landowners are having on wildlife use, management, and conservation.

Wildlife management is a business decision regardless of who manages, who pays, or who benefits. US governmental income has varied historically, and prospects for funding are in question (Hutchins et al. 2009). Governments manage wildlife and user-pay systems commercially by selling licenses, permits, and user fees, primarily for anglers and hunters. Public wildlife management benefits from excise taxes on some outdoor product sales, notably for hunting and fishing. Some state incomes use general tax revenues, gambling proceeds, and license plate and other product sales to manage wildlife. Most expenditures by users do not directly support management. The status of hunting, fishing, and wildlife-associated recreation in the United States for persons 16 years and older is reported every five years (2017 US Fish and Wildlife Service National Overview, https://wsfrprograms.fws.gov/Subpages/NationalSurvey/nat_survey2016.pdf, accessed 20 November 2017).

Preliminary findings for participation in 2016, compared with 2011 (in parenthesis below; US Fish and Wildlife Service 2012), indicate that 101.6 million hunters, anglers, and wildlife watchers contributed US$156.3 billion—1% of the gross national product—to the US economy. Hunters spent US$25.6 billion ($36.3 billion); anglers spent $46.1 billion ($45.0 billion); wildlife watchers spent $75.9 billion ($59.1 billion). In 2016, the most money was spent on the demand and service sides of economics for equipment (wildlife watching 73%, hunting 48%, fishing 46%) and trip-related expenses (fishing 47%, hunting 36%, wildlife watching 16%). The smallest proportion of expenditures was related to the supply and production costs of wildlife: hunting licenses and fees 3%; fishing licenses and fees 1%.

Nearly three times the number of hunters use private land than use public land. In 2011, 77% of days (219 million) were spent on private land. Amounts spent on leasing or buying private lands included hunting US$7.1 billion, wildlife watching US$5.7 billion, and fishing US$3.4 billion. In 2016, 16% of hunters spent US$2.9 billion, and 3% of anglers spent US$2.3 billion on land leasing and ownership.

Wildlife production and use are a cost to private producers. Including private lands and private management actions into nature conservation policy and practices is difficult, but necessary if we are to affect conservation on lands where people and wild animals must live together. Having a classification and rating system to review, appreciate, certify, and document wildlife management contributions from private lands and landowners should help to establish clearer relevance, suggest improvements, and bring private wildlife and user management into the mainstream of nature conservation.

Rating Classification of Private Lands and Wildlife to Understand and Improve Nature Conservation

Wildlife management strategies on private lands vary around the world, as reported in Europe, Africa, and the United States. A rating system should help wildlife professionals and landowners generate a common vocabulary, understanding, evaluation criteria, certification system, and means to document and celebrate wildlife management contributions on private lands. A common vocabulary, rating system, and certifications could improve policy agreements, cooperatives, landowner and user associations, and ethical approaches to resource management and help assess intervention outcomes. Missing in wildlife policy is agreement on what constitutes acceptable conservation measures practiced on private and even public lands. Ratings should benefit wildlife, habitats, resource producers, neighboring persons and jurisdictions, users, society, and nature conservation generally. A rating system should help to reduce ecological and management fragmentation and promote more holistic and integrated wildlife management partnerships (Benson 1991b). Below is a rubric to describe and promote conservation, inspired by Bailey (2013) for bison (*Bos bison*), but which should have a much broader application.

The Classification System

There is a continuum between public and personal control of wildlife and hunting systems as incentives to have animals on the land (Wall and Child 2009). Wildlife management on private lands needs to be placed into the perspectives of overall land availability, ecological suitability, ownership, and values to persons on the land and to society. The policies of public sectors affect feasible nature conservation outcomes in private sectors, enabling or denying a range of possible interventions. Not all lands are natural wildernesses, and some land uses are not compatible with wildlife. Ecological management is ultimately the outcome of incentives and governance (Child and Wall 2009) that enfranchise or deny most actions; thus, our rating system qualifies and perhaps ultimately can certify conservation actions.

On one end of the continuum are animals in zoos and small community parks, wildlife reared in captivity with artificial supplements of food and veterinary care, or other intensively managed species that

have no freedoms to interact with wildlands. We acknowledge that these properties and management systems have value to landowners and society, with contributions ranging from education to endangered species protection, and from live sales or meat sales to genetic pools, disease manipulation, and human enjoyment. These wildlife and properties are classified as domestic on one end of our rating system; the other end of the continuum is classified as wild and free-ranging, with minimal inputs needed by humans (Table 4.1). Between the extremes are contributions to nature conservation and to the livelihoods of humans if manager-landowners are given sufficient authority and responsibility to be effective.

Simplicity in land and animal classification systems is important if they are to be used. Trained ecologists use their mental computers to rapidly assess land and wildlife situations on small and large scales, knowing that more detailed temporal and spatial data might be better. We propose rapid assessment parameters that could become universal criteria for scoring land units and individual species or groups therein.

Essentially, the goal is to obtain a conservation–wise use factor and rating for private (or public) lands and wildlife. The range has five quality points of variability in each category of wildness from domestic and semidomestic to semi-wild and wild. Intensively managed, domestic ratings result from the lowest numbers within categories and among all categories combined. Extensively managed, wilder situations are given the highest numbers within categories and among all categories. Habitat suitability, animal capability, and human interventions are summed for an index of the following parameters:

Land base needs vary by species, with some animals confined while others have operational home ranges that enable free-ranging and interactive breeding populations. Herd size can be evaluated in relation to habitat availability or by the supplements artificially provided in the domestic range of factors.

Fencing is used by some to keep animals within the limitations of habitats, property ownership boundaries, and legal authorizations. Maintaining a proper balance of animals in relation to available habitats or supplemental feed is important. In truly wild systems, free-ranging animals are not affected by fences and have unrestricted movements.

Feeding is not part of wild systems, except perhaps in emergencies. Feeding is practiced more in systems removed from the wild, either by intensive habitat improvements or with supplemental and artificial feeds.

Population control is necessary in domestic and confined systems to balance food, shelter, water, and social needs with animal numbers. Wild systems can operate without human controls because predators, diseases, weather, and other limiting factors tend to keep animal numbers within some sustainable level.

Culling, hunting, and live removal can take place in all systems if policy allows. The need for removal is highest with domestic systems and lowest for wild systems. The aesthetic quality and ethical implications of hunts vary within the range from domestic to wild.

Disease control is generally expected with animals that are confined in small areas where disease transmission is exacerbated. In wild populations, diseases should be part of normal biological processes that can be tolerated except in drastic circumstances.

Pasture control is more related to domestic production, wherein animals are moved from one confined area to another to accommodate the varied needs of species therein. Wild systems have free-ranging animal movements that do not require human intervention.

Public access is not a need for animals or the land, but in many systems, it is a social and policy consideration. From the perspective of wildlife, access to domestic situations can have physiological and social consequences resulting in fight or flight reactions. These consequences exist in wild systems also, but could be mitigated to promote access. In all cases, the quality and rating for access depends on the effects that are made to wildlife, landscapes, and other users.

Distribution of animals in space and time is a factor of being natural. Having polar bear (*Ursus maritimus*) in the desert is not natural. Having bison in a city pasture is not natural. Encouraging white rhinoceros where black rhinoceros (*Diceros bicornis*) are found in Africa is not the normal distribution, but it is more natural than if they are introduced into Australia. Animals might also adapt naturally as climates change, a change that could be classified as a normal distribution. Moving animals to nonindigenous habitats is not a normal distribution that ecologists or policy makers might allow, and some conservationists will likely be opposed. For the purposes of this rating system, the true natural ecological distribution should be rated highest. Producers, policy makers, and society can determine whether other distributions are acceptable.

Legal status is a policy decision, but if a thoughtful policy is established, then the rating is easy to determine. Bison, for example, in the United States are generally considered to be domestic and privately owned, yet in Yellowstone National Park, they would receive a relatively wild rating because they interact with nature quite differently than bison that reside on farms and are being raised for live sales or meat production. Bison serve as a practical example and a social metaphor for human abilities to enable rewilding iconic animals (Bailey 2013) or relegating them to novelties in artificial parks and in the artificial thinking of humans. Answers lie on a philosophical and ecological continuum that only human policies can enable.

Ecological interaction is a category that some might consider simple to separate between domestic and wild, but others might have more difficulty. Essentially, consider whether the land base or species being considered is on the site with other parts of the system still in place. Conversely, management for a single species (abundant or endangered) could come at the expense of other species (insects, predators, diseases).

The rating system could classify and determine conservation contributions from private lands with less emotional judgment. Conservationists and policy makers could use the ratings of private, or public, land in context with the surrounding land uses and conservation ideals. Without some artificiality, Europe could not have the biomass of wild ungulates on landscapes for human enjoyment, considering the large human populations. Africa and other locations can rank some levels of wildness as more artificial if animals are nurtured in fenced pastures; yet, reintroducing wildlife to lands formally used for agriculture is a positive outcome compared with not nurturing wildlife. Policy makers in the United States can rate lands and wildlife on private lands as both free-ranging and more restricted, depending on situations. In all places, thoughtful analyses and high ratings (Table 4.1) can be used to appreciate and reward good conservation. Lower ratings can also induce management guidance to those who want higher ratings. A uniform procedure, in use worldwide, enables landowners, policy makers, conservationists, and society to appreciate and compare private land contributions toward all wildlife production and use.

Summary

Rural lands occupied and influenced by private landowners are much larger than formal protected areas. If conservation is to be achieved on most lands, then private land conservation is vital. Landowners make prudent decisions to combat the uncertainty of their livelihoods and enterprises. They need incentives that are ecological, personal, social, legal/political, administrative, and economic for promoting wildlife and nature conservation and uses on private lands (Benson 2014). Costs for production need to be balanced with benefits: if it pays, it stays!

Society can help landowners by recognizing and valuing their stewardship toward wildlife management, nature conservation, provision of open spaces, and access thereon. A rating system is recommended to evaluate and provide common criteria that improve wildlife conservation in relation to the landscape. The rating system can be used by deci-sion makers to document the quantity and quality of contributions from landowner partners.

Landowners should be enfranchised and supported with public or private incentives to practice conservation for the good of owners, users, and society. We have explored situations and contributions from Europe, Africa, and the United States with an emphasis on how hunting has been a major incentive for landowners to keep wildlife on their properties. Beyond the personal interests of landowners, the demand for using wildlife on private lands has been bolstered by economic incentives, sometimes at the resistance of public wildlife managers.

Prudent decisions to nurture wildlife and recreation are logically balanced against economic costs for producing wildlife, such as uncertainty for actions, depredations and other conflicts with humans, or the nuisance of repairing facilities or dealing with recreationists who want access to wildlife. Landowners can decide to become angry about wildlife, complacent, disinterested, or aggressive. Conversely, landowners can find ways to tolerate wildlife, benefit from their presence, and, ideally, to proactively manage and successfully encourage wildlife uses and nature conservation generally.

LITERATURE CITED

Adams, J.S., and T.O. McShane. 1992. The myth of wild Africa: Conservation without illusion. W. W. Norton and Company, New York, USA.

Bailey, J.A. 2013. American plains bison: Rewilding an icon. Sweetgrass Books, Helena, Montana, USA.

Benson, D.E. 1989a. What fee hunting means to sportsmen in the U.S.A: A preliminary analysis. International Wildlife Ranching Symposium 1:296–302.

Benson, D.E. 1989b. Changes from free to fee hunting. Rangelands 11:176–180.

Benson, D.E. 1991a. Values and management of wildlife and recreation on private land in South Africa. Wildlife Society Bulletin 19:497–510.

Benson, D.E. 1991b. Integrated wildlife management partnerships among agriculture, natural resources professions and business. Pages 344–348 *in* L.A. Renecker and R.J. Hudson, editors. Wildlife production: Conservation and sustainable development. University of Alaska, Fairbanks, Alaska, USA.

Benson, D.E. 1992a. Commercialization of wildlife: A value-added incentive for conservation. Pages 539–553 in R.D. Brown, editor. The biology of deer. Springer-Verlag, New York, USA.

Benson, D.E. 1992b. Socio/economic values of wildlife on private land in Republic of South Africa and the benefits to conservation. Transactions of the Congress of the International Union of Game Biologists 18:507–509.

Benson, D.E. 1998a. Wildlife conservation strategies in southern Africa that empower the people. Human Dimensions of Wildlife 3:47–58.

Benson, D.E. 1998b. Enfranchise landowners for land and wildlife stewardship: Examples from the western United States. Human Dimensions of Wildlife 3:59–68.

Benson, D.E. 2001a. Wildlife and recreation management on private lands in the United States. Wildlife Society Bulletin 29:359–371.

Benson, D.E. 2001b. Survey of state programs for habitat, hunting and nongame management on private lands in the United States. Wildlife Society Bulletin 29:354–358.

Benson, D.E. 2001c. Wildlife stewardship and recreation on private lands: What now? Transactions of the North American Wildlife and Natural Resources Conference 66:110–125.

Benson, D.E. 2001d. Developing a philosophy about wildlife stewardship and recreation on private lands. Pages 3–14 in L.A. Renecker and T.A. Renecker, editors. Game conservation and sustainability: Biodiversity, management, ecotourism, traditional medicine and health. Proceedings of the International Wildlife Ranching Symposium. Toronto, Canada.

Benson, D.E. 2014. New focus on the back forty: Achieving effective wildlife management on private lands. Wildlife Professional 8:20–25.

Benson, D.E., and S. White. 1995. The status of advanced hunter education programs in North America. Wildlife Society Bulletin 23:600–608.

Benson, D.E., R. Shelton, and D. Steinbach. 1999. Wildlife stewardship and recreation on private lands. Texas A&M University Press, College Station, Texas, USA.

Bergström, R., H. Huldt, and U.L.F. Nilsson. 1992. Swedish game: Biology and management. Svenska jägareförbundet, Spånga, Sweden.

Bubenik, A.B. 1989. Sport hunting in continental Europe. Pages 115–146 in R. Hudson, K.R. Drew, and L.M. Baskin, editors. Wildlife production systems economic utilization of wild ungulates. Cambridge University Press, Cambridge, UK.

Cantor, A. 2016. The public trust doctrine and critical legal geographies of water in California. Geoforum 72:49–57.

Chantelat, J.C. 1985. Vivre et chasser au pays. Solar, Paris, France.

Child, B., and B. Wall. 2009. The application of certification to hunting: A case for simplicity. Pages 31–359 in B. Dickson, J. Hutton, and W.M. Adams, editors. Recreational hunting conservation and rural livelihoods: Science and practice. Wiley-Blackwell, Hoboken, New Jersey, USA.

Cloete, P.C., P. van der Merwe, and M. Saayman, editors. 2015. Game ranch profitability in South Africa. Amalgamated Banks of South Africa, Farmers Weekly, CTP Printers, Cape Town, South Africa.

Craigie, I.D., J.E.M. Baillie, A. Balmford, C. Carbone, B. Collen, R.E. Green, and J.M. Hutton. 2010. Large mammal population declines in Africa's protected areas. Biological Conservation 143:2221–2228.

Cubbage, F., J. O'Laughlin, and M.N. Peterson. 2017. Natural resource policy. Waveland Press, Long Grove, Illinois, USA.

Dickson, B., J. Hutton, and W.M. Adams, editors. 2009. Recreational hunting, conservation and rural livelihoods: Science and practice. Wiley-Blackwell, Hoboken, New Jersey, USA.

Dry, G.C. 2013. Strategic repositioning of WRSA. Wildlife Ranching 6:1.

Fraser J., D. Wilkie, R. Wallace, P. Coppolillo, R.B. McNab, R. Lilian, E. Painter, P. Zahler, and I. Buechsel. 2009. The emergence of conservation NGOs as catalysts for local democracy. Pages 44–56 in M.J. Manfredo, J.J. Vaske, P.J. Brown, D.J. Decker, and E.A. Duke, editors. Wildlife and society: The science of human dimensions. Island Press, Washington, DC, USA.

Hudson, R., K.R. Drew, and L.M. Baskin, editors. 1989. Wildlife production systems: Economic utilisation of wild ungulates. Cambridge University Press, Cambridge, UK.

Hutchins, M., H.E. Eves, and C.G. Mittermeier. 2009. Fueling the conservation engine: Where will the money come from to drive fish and wildlife management and conservation? Island Press, Washington, DC, USA.

Lecocq, Y., editor. 2004. Hunting, an added value for biodiversity. FACE, Brussels, Belgium.

Manfredo, M.J., J.J. Vaske, P.J. Brown, D.J. Decker, and E.A. Duke, editors. 2009. Wildlife and society: The science of human dimensions. Island Press, Washington, DC, USA.

Moore, L. 2014. Vital role of private lands. Wildlife Professional 872.

Renecker, L.A., and R.J. Hudson, editors. 1991. Wildlife production: Conservation and sustainable development. Agricultural and Forestry Experiment Station Miscellaneous Publication 91-6. University of Alaska, Fairbanks, Alaska, USA.

Trefethen, J.B. 1975. An American crusade for wildlife. Winchester Press, New York, USA.

US Fish and Wildlife Service. 2012. 2011 National survey of fishing, hunting, and wildlife-associated recreation. Washington, DC, USA.

Vandervell, A., and C. Coles. 1980. Game and the English landscape. Debrett's Peerage, London, UK.

Van Hoven, W. 2013. Wildlife ranching is a real "green economy." Agriland 27:30–31.

Van Hoven, W. 2015a. Private game reserves in Southern Africa. Pages 119–137 *in* R. van der Duim, M. Lamers, and J. van Wijk, editors. Institutional arrangements for conservation, development and tourism in Eastern and Southern Africa. Springer Science, New York, USA.

Van Hoven, W. 2015b. Economic value chain impact. Wildlife Ranching 3:25–30.

Van Hoven, W., and H. Viljoen. 1995. Fair game: Economics of game ranch investment. Debrett's international collection. Sterling, London, UK.

Wall, B., and B. Child. 2009. When does hunting contribute to conservation and rural development? Pages 255–265 *in* B. Dickson, J. Hutton, and W.M. Adams, editors. Recreational hunting conservation and rural livelihoods: Science and practice, Wiley-Blackwell, Hoboken, New Jersey, USA.

Wolfe, M.L. 1970. The history of German game administration. Forest History 94:6–16.

5 — Habitat Loss and Fragmentation

Hsiang Ling Chen
Gabrielle Beca
Mauro Galetti
Chiachun Tsai
Wei Hua Xu
Jing Jing Zhang
Patrick Zollner

Introduction

Habitat loss and fragmentation of natural ecosystems are the most serious threats to global biodiversity and the primary causes of the present extinction crisis (Newbold et al. 2015). Human activity is the main cause of these changes in land cover. Approximately 53% of the world's land area has already been converted to anthropogenic land uses (Fig. 5.1); in particular, forest has been converted to agricultural systems, which occupy about 38% of Earth's terrestrial surface (Foley et al. 2011, Hooke et al. 2012). Growing global demands for food, biofuel, and other commodities are influencing the rapid expansion of agriculture, infrastructure, and energy development (Sodhi et al. 2010). This conversion will continue in the future, as the human population and economy in developing countries are still growing (Sloan and Sayer 2015). By 2030, the area of agricultural land is expected to increase an additional 10% (Haines-Young 2009).

Preserving biodiversity with this ongoing habitat loss is one of the greatest challenges for conservation this century. In this chapter we review the changing worldwide patterns in forests, grasslands, and wetlands, and summarize recent findings on landscape conversion and novel approaches to investi-gating the effects of habitat loss and fragmentation on wildlife diversity, population, and movements. We then discuss management strategies and actions to mitigate habitat loss and fragmentation and identify needs and challenges to provide effective approaches for improving connectivity in landscapes experiencing habitat loss and fragmentation.

Primary Forests Are Deforested, Degraded, and Fragmented, Especially in Tropics

Globally, total forest area has declined by 3%, from 4,128 million ha in 1990 to 3,999 million ha in 2015. During this same time period, while primary forests were decreasing from 3,961 million ha to 3,721 million ha, planted forests were expanding because of growing demand for forest products (Hansen et al. 2013, Keenan et al. 2015). Total forest area declined in Central America, South America, South and Southeast Asia, and Africa, but expanded in Europe, North America, the Caribbean, East Asia, and Western-Central Asia (Keenan et al. 2015). In nations where forest recovery occurs, however, it is often based largely on exotic tree plantations, including monocultures of rubber, eucalyptus (*Eucalyptus* spp.), acacia (*Acacia* spp.), and oil palm (*Elaeis* spp.; Laurance et al. 2014b). Furthermore, increases in the

Figure 5.1 Humans have influenced most landscapes on Earth, including this stretch of the Gobi Desert near Mandelgovi, Mongolia. John Koprowski

amount of exotic forest cover are causing losses of primary forest (Bradshaw 2012).

Primary forests are globally irreplaceable and support more diverse and structurally complex biotas than degraded forests, even when degradation is mild (Gibson et al. 2011). Yet, more than 70% of the world's forests are within 1 km of a forest edge, near human activities that threaten the integrity of natural forest (Haddad et al. 2015). Deforestation, degradation, and fragmentation of primary forest are most evident in the tropics. The deforestation of tropical and subtropical forests has increased by 62% from the 1990s to the 2000s, and 67 million ha of primary forest have been lost (Kim et al. 2015). About half of the tropical and subtropical forest that still persists today has been altered, and one-quarter

of the tropical forest remaining is fragmented (Laurance et al. 2014b). The largest extension of remaining tropical forests is found in the humid tropical regions of the Amazon and Congo River Basins (Haddad et al. 2015). The remnants of native forest in Neotropical forests are mostly distributed in very small (fewer than 50 ha) and isolated patches (Ribeiro et al. 2009), embedded in anthropogenic matrices such as urban centers, agriculture, and cattle pasture (Beca et al. 2017).

Global forest area is projected to continue to decrease slightly during the next 30 years as a result of extensive deforestation in tropical forests and subtropical woodlands, although this decrease is partially offset by the increase of forest cover in the Northern Hemisphere (Pereira et al. 2010, d'Annunzio et al.

2015). The rate of overall loss is projected to slow from 0.13% per year at the beginning of the century to 0.06% per year by 2030. South America is projected to lose 9% of its forest during the next 15 years (d'Annunzio et al. 2015).

Agriculture Is the Major Cause of Deforestation

The conversion of natural forest into agricultural areas, including cropland and pasture, accounts for 80% of global deforestation (Kissinger et al. 2002). About 75% of forests have been converted to agriculture in Africa, around 70% in subtropical Asia, and more than 90% in Latin America (Food and Agriculture Organization of the United Nations 2010). In addition to agriculture, other important causes of deforestation include desertification, forest fire, mining, and energy development. For instance, combined industrial and fuelwood removals in the tropics increased by 35% between 1990 and 2015 (Sloan and Sayer 2015). Meanwhile, destruction of mangrove (*Rhizophora* spp.) forests for agriculture, urbanization, and resource exploration is also occurring worldwide (Wolanski et al. 2000). Globally, the area of mangrove forests has been reduced by about 35% (Valiela et al. 2001). The greatest loss of mangroves has occurred in Southeast Asia, where approximately one-third of the planet's mangrove forest area is located (Thomas et al. 2017).

Grasslands Are Under Severe Degradation

At a global scale, 50% of grassland ecosystems experienced degradation from 2000 to 2010, mainly due to human activities such as farming and livestock overgrazing, especially in arid and semiarid places (Liu and Diamond 2005, Verón et al. 2006). While climate changes are the main causes of grassland degradation in Oceania and South America, human activities played a dominant role in driving degradation in other areas of the world (Gang et al. 2014). As a result of increased timber plantations, grasslands have decreased significantly in South Africa, with only 6% of the original area remaining in 2008 (Niemandt and Greve 2016). In the United States and Canada, oil extraction has undergone rapid expansion on the grasslands (Thompson et al. 2015). Although large areas of indigenous grasslands remain in the South Island of New Zealand, grassland loss has been ongoing (Weeks et al. 2013). To reverse the degradation of grassland, large-scale restoration projects, such as the Grain for Green program in China (Liu et al. 2008), have been implemented in Asia, North America, South America, and Europe.

Wetlands Are Drying

Wetlands have been lost, degraded, or strongly modified worldwide, with agricultural development as the main proximate cause (Fig. 5.2), and growths in

Figure 5.2   Development of wetlands in Asia: (A) natural riparian area before construction; (B) riparian vegetation was removed during construction along the Nanshi River, Taichung, Taiwan. Jinghong Lee

human population and economy are underlying forces (van Asselen et al. 2013). The reported long-term loss of natural wetlands worldwide averages between 54% and 57%, reaching up to 90% in some regions of the world (Junk et al. 2013). Although riparian wetlands in South America and Africa are only slightly modified, new dam initiatives and projected climate change pose important challenges to the conservation of riparian wetlands in South America (Zarfl et al. 2014, Schneider et al. 2017). About 89% of wetlands are unprotected, especially in Asia, which contains the largest wetland area of the world (Reis et al. 2017). Deforestation, drainage, and reclamation of wetlands (including peat swamps) for conversion to agriculture, fish culture, and dam construction on rivers with floodplains have been increasing threats to Asian wetlands since 1960 (Gopal 2013, Sasidhran et al. 2016).

Effects of Habitat Loss and Fragmentation on Wildlife

Effects of habitat loss and fragmentation on wildlife are influenced by three broadly defined mechanisms: loss of habitat area; changes in the spatial configuration of the landscape, such as isolation; and interaction between habitat patches and the surrounding matrix (Didham et al. 2012, Wilson et al. 2016). As habitat fragmentation is often concurrent with habitat loss in natural settings, large-scale field-based fragmentation experiments have been implemented to further understand the relative influence of the loss of habitat, patch isolation, and edge effects on communities and ecosystems (Haddad et al. 2015). In the past, empirical studies have suggested that habitat loss has a large and consistently negative effect on biodiversity, whereas habitat fragmentation per se has much weaker effects (Fahrig 2003). For example, the size of a natural forest area was the most important variable affecting mammal richness (Laidlaw 2000, Nupp and Swihart 2000). Current studies, however, suggest that interactions of loss and fragmentation are the major determinants of

species responses, including population declines and trophic pyramid shifts (Bartlett et al. 2016).

Species Diversity

Habitat loss and fragmentation lead to loss of biodiversity (Cardinale et al. 2012). The likelihood of a species to persist in fragmented habitat is influenced primarily by the habitat and by specialization, with a consideration for the taxonomic class, fecundity, life span, and body mass (Keinath et al. 2017). Vertebrates like birds and mammals need large areas, which make them susceptible to fragmentation and local extinction (Jorge et al. 2013). Mammals living in more fragmented landscapes are at greater risk of extinction (Crooks et al. 2017), and large species experience more severe population declines at lower extents of habitat loss than do smaller animals (Bartlett et al. 2016). For example, massive deforestation in Indonesia has caused Sumatran elephant (*Elephas maximus sumatranus*) and tiger (*Panthera tigris sumatrae*) populations in Riau to plummet to around 216 (84% reduction) and 192 (70% reduction) individuals, respectively (Uryu et al. 2008).

The global rate of biodiversity loss may exceed the leveling trend of habitat loss for numerous reasons. Low elevations suffer the greatest threat of extinction because of the higher pressure of deforestation (Hall et al. 2009). Furthermore, habitat loss in the tropics cannot be directly compensated by habitat gain in other ecological zones (Pereira et al. 2010). For example, forest fragmentation and removal of large trees in the Philippines has resulted in the loss of 16.4%–24.6% of the 61 species of amphibians and reptiles in forest fragments (Alcala et al. 2004). The creation of forest edges is another cause of physical and biotic changes that affect biodiversity (Bierregaard et al. 2001). Habitat fragmentation often interacts synergistically with other pressures associated with forest edges, such as roads and human settlements (Pattanavibool and Dearden 2002). These human activities can influence the decline of native species through hunting (Peres and Palacios 2007), plant harvesting (Muler et al. 2014), avoidance of human settlement (Rovero et al. 2017), logging (Broadbent et al. 2008), and the invasion of exotic species (Pedrosa et al. 2015).

Delays in the response of species to habitat modification can occur after the loss or fragmentation of habitat. In this context, species appear to resist and persist, but the extinction debt dictates that many current species are doomed to eventual extinction (Tilman et al. 1994). Extinction debts are more likely to happen in landscapes that had an intermediary perturbation (Uezu and Metzger 2016), because in these conditions some species can persist for only a limited amount of time, close to the minimal conditions required to maintain their populations (Hanski and Ovaskainen 2002).

Movement

Habitat loss and fragmentation strongly influence animal movement patterns, which in turn have a considerable influence on the exchange of genetic material between distant locations, the rescue of small populations at risk of extinction, and, ultimately, on population dynamics (Knowlton and Graham 2010). Road construction not only causes destruction and loss of habitat but also facilitates deforestation and landscape fragmentation. A growing body of work suggests that roads and traffic can serve as barriers that impede animal movement; decrease accessibility to resources such as food, shelter, or mates; reduce reproductive success and gene flow; and ultimately threaten population persistence (Trombulak and Frissell 2000, Bennett et al. 2013, Chen and Koprowski 2016a). To what extent roads affect populations depends on accumulated barrier effects within movement range, mechanisms of road avoidance, and the impacts on reproductive activities. Large mammal species are less susceptible to the barrier effects of individual roads, but are more vulnerable to mortality caused by wildlife-vehicle

collisions and their negative effects on population (Rytwinski and Fahrig 2011, 2012). Small mammals are more susceptible to the barrier effects of roads, because even narrow roads can reduce the probability of crossing significantly. In addition to body size, life history characteristics such as being a generalist or a specialist and circumstances such as being a native or introduced species can also alter how species respond to roads (Chen and Koprowski 2016b).

Studies of landscape and movement ecology have addressed animal movement processes in patchy landscapes, focusing on the influence of distinct movement strategies on the efficiency of searching for resources and on population and community consequences of animal movement in changing landscapes (Zollner and Lima 1999, Nathan et al. 2008). Such studies may attempt to elucidate critical thresholds for the proportion of landscapes that need to retain intact natural habitat to support connectivity (With and King 1999). Such thresholds are often species and circumstance specific, rendering broad generalizations meaningless, and potentially counterproductive, for management when such thresholds fail to provide proper consideration in the requisite context (Swift and Hannon 2010). A combination of classic wildlife techniques and innovative tools provides an expanding means to improve understanding of how habitat amounts and configurations influence dispersal, survival, and population dynamics. For example, studies combining molecular genetics and spatial modeling of organism movements have provided novel insights into spatial patterns of dispersal (Cushman et al. 2006). Spatial capture-recapture models provide opportunities to estimate animal abundance, density (Efford 2004, Royle and Young 2008), and landscape connectivity (Fuller et al. 2016, Morin et al. 2017) by including movement information from mark-release-recapture studies. Advances in global positioning system technology provide fine-scale insights on animal locations for long durations, which can provide invaluable inputs for state-space models (Smouse et al. 2010) and mechanistic home range models (Moorcroft et al. 1999). The incorporation of remotely sensed imagery and animal movements also permits exciting hypotheses on seasonal movements, migration, and climate shifts in response to plant phenology (Merkle et al. 2016, Aikens et al. 2017). Finally, the combination of individual-based ecology and pattern-oriented modeling provides the means to investigate the behavioral rules that underlie animal movements (Grimm and Railsback 2012).

Management Strategies to Mitigate Habitat Loss and Fragmentation

To offset the loss of native habitats and the associated ecosystem services, governmental and nongovern-mental organizations have developed management programs to restore landscape connectivity and minimize further conversion of natural vegetation. At broad spatial and temporal scales, conserving landscape connectivity is necessary to maintain the continuity of ecological processes. Several countries recognize biodiversity and landscape management as top priorities and base their national conservation strategies on large-scale connectivity (DeClerck et al. 2010, European Union 2011). Integration of protected areas, corridors, and wildlife passages (Fig. 5.3), and enhanced matrix quality within human-dominated landscape, can be used to establish an effective conservation system (Nopper et al. 2017). For instance, implementing corridors along matrix streams connecting forest habitats and

Figure 5.3 Examples of wildlife crossing structures: (A) underpass to maintain connectivity along a riparian area in the Netherlands; (B) overpass over a highway in Japan; (C) logs provide shelters and enhance movements across open areas for small mammals; (D) small, open-topped underpass with funnel fencing to help amphibians cross roads. Hsiang Ling Chen

established buffer zones for natural forest patches or protected areas (banana plantations) might be an important contribution to amphibian conservation in fragmented landscapes in Madagascar (Ndriantsoa et al. 2017).

Legislation and Protected Areas

Legislation and protected areas are the most important approaches for biodiversity conservation and mitigation of habitat loss and fragmentation. Countries are increasingly protecting forests of ecological significance at the global scale. For instance, protected areas now cover more than 10% of China, and a series of laws and regulations concerning biodiversity conservation, such as Regulations on Nature Reserves Management, have been issued (Xu et al. 2017). Currently, 16.3% of global forests and 26.3% of tropical forests are protected (Morales-Hidalgo et al. 2015). The protected area system faces several limitations, however, such as isolation, location bias, and lack of societal support, and has been inadequate to safeguard biodiversity and ecological services generally (Miura et al. 2015). Many threatened birds, mammals, and amphibian species are not found in any protected areas, and large proportions of high-quality habitat are unprotected and experiencing high rates of deforestation (Crooks et al. 2011, Venter et al. 2014). Combining both biodiversity conservation and ecosystem services when planning protected areas can enhance public awareness and attract more government capital investment. Incorporating a social-ecological approach to protected areas and complementing ecological with social analyses will optimize the overall functions and benefits of protected areas (Palomo et al. 2014). Furthermore, government land-use policies can take environmental concerns into account. For example, proactively planning and zoning new roads can limit the footprint and impacts of agriculture. High-quality roads that link existing agricultural areas with markets might reduce overall deforestation. Such efforts will help to ensure that biologically critical lands are spared for nature conservation (Laurance et al. 2014a).

Habitat Restoration and Conservation Act

In addition to protected areas and legislation, a variety of conservation practices have been implemented to enhance environments in human-used landscapes. Compared to the legal system, these programs or plans are flexible in conservation of species and ecosystems; they can include polycultures, agroforestry practices, payments for ecosystem services programs, and other means of increasing the structural diversity of used land (Tscharntke et al. 2012, Asare et al. 2014). For example, planting forests without clearing natural forests and reforestation of agricultural land may help reduce the effects of deforestation and support biodiversity conservation (Brady et al. 2011). Planted forest can serve as high-quality matrix, increase functional connectivity among remnants, buffer edge effects, and provide habitat (Barlow et al. 2007). With plantation forest in the matrix, reduction in bird species richness by deforestation decreases by 75% (Ruffell et al. 2017). Plantations with monocultures or exotic species, however, tend to have much lower species richness than natural forests and are typically dominated by a few abundant generalists and non-forest species, including exotic species (Edwards et al. 2010). In contrast, plantations with a mixture of indigenous tree species and low-intensity grazing by cattle support species richness similar to natural areas (Mann et al. 2015). Moreover, farmers and ranchers can also plan their properties to encourage the flow of animals by preserving the areas of riparian forest and hilltops, thus creating a more permeable environment and enhancing connectivity. For example, a continuous network of hedges and associated trees and shrubs in pastures enhances landscape connectivity and contributes to the maintenance of reptile biodiversity in dry southwestern Madagascar (Nopper et al. 2017).

Corridors and Wildlife Passages

Establishing ecological corridors in anthropogenic landscapes (Fig. 5.4) can reduce the effects of fragmentation and improve the connectivity between forest fragments, allowing genetic exchange and faunal displacement (Magioli et al. 2016). In the Brazilian Atlantic Forest, a large corridor was developed, establishing about 30 million ha of forest to connect fragmented landscapes (Tabarelli et al. 2005). To create effective corridors, it is necessary to have information about the home range and movement of the species, so the rules of landscape use can be established. For example, particular matrix types, such as matrix streams and banana plantations, might serve

as corridors for most forest frog species, and secondary vegetation and rice fields are dispersal barriers (Ndriantsoa et al. 2017).

A variety of modeling techniques can provide invaluable insights to delineate likely travel routes for animals that may merit prioritization for conservation because of their contribution to landscape connectivity. For example, expert-opinion-based least-cost analysis has been used to model the quantity and configuration of available suitable elephant habitat in the Lower Kinabatangan (Estes et al. 2012). Similar analytical tools have been used in combination with spatial genetic data to identify major movement corridors and barriers for bears in western North America. Circuit theory (McRae et al. 2008) and graph theoretic approaches (Pinto and Keitt 2009) both build on the least-cost principles illustrated in the above examples. Individual-based

Figure 5.4   Crossing structure for ocelots (*Leopardus pardalis*), small endangered felids, in Texas, USA. Amanda Veals

models provide an alternative approach to examining connectivity that can provide invaluable perspectives on asymmetries in dispersal related to landscape characteristics (Gustafson and Gardner 1996) or behavioral characteristics of the dispersing species (Pauli et al. 2013). These insights can have important conservation implications, such as identifying how road-based mortality isolates many patches of suitable habitat for lynx in Europe that might otherwise be connected (Kramer-Schadt et al. 2004). Such individual-based modeling approaches are also promising for their ability to predict the potential for shifts in species distributions in response to factors such as climate change (Bocedi et al. 2014).

Summary

Looking forward, we are facing one of the greatest challenges of the twenty-first century. The area of Earth's land surface devoted to cropland already occupies 1,530 million ha (Foley et al. 2011), and the area committed to urban centers is predicted to triple to 1,800 million ha by 2030 (Seto et al. 2012). For instance, China's new urbanization plan (2014–2020) predicts that about 100 million people will migrate from rural to urban areas in China by 2020. We can take these challenges as opportunities for conservation. On the one hand, such urbanization, infrastructure construction, and agricultural expansion may accelerate habitat loss and fragmentation, including deforestation, reclamation of wetland, and loss of grassland. On the other hand, the rural-urban migration from biodiversity hotspot areas to coldspot areas through urbanization may provide opportunities for reducing the dependence of rural populations on ecosystems and natural resources, lessening pressures on biodiversity (Xu et al. 2017). The effort to mitigate global habitat loss and biodiversity requires international collaboration. Many migratory birds (red-crowned crane, *Grus japonensis*) and large mammals have extensive geographic distributions that span multiple countries. The long-term survival of these species and the capacity of the remaining habitats to sustain biodiversity and ecosystem services depend on the amount and quality of habitat remaining globally, and its degree of connectivity across countries (Haddad et al. 2015). It is necessary to strengthen the cooperation among countries to reduce habitat loss and fragmentation and achieve global targets such as the Aichi targets set forth for biodiversity in the United Nations Convention on Biodiversity (Chapter 13).

LITERATURE CITED

Aikens, E.O., M.J. Kauffman, J.A. Merkle, S.P. Dwinnell, G.L. Fralick, and K.L. Monteith. 2017. The greenscape shapes surfing of resource waves in a large migratory herbivore. Ecology Letters 20:741–750.

Alcala, E.L., A.C. Alacala, and C.N. Dolino. 2004. Amphibians and reptiles in tropical rainforest fragments on Negros Island, the Philippines. Environmental Conservation 31:254–261.

Asare, R., V. Afari-Sefa, Y. Osei-Owusu, and O. Pabi. 2014. Cocoa agroforestry for increasing forest connectivity in a fragmented landscape in Ghana. Agroforestry Systems 88:1143–1156.

Barlow, J., T.A. Gardner, I.S. Araujo, T. C. Avila-Pires, A.B. Bonaldo, J.E. Costa, M.C. Esposito, et al. 2007. Quantifying the biodiversity value of tropical primary, secondary, and plantation forests. Proceedings of the National Academy of Sciences of the United States of America 104:18555–18560.

Bartlett, L.J., T. Newbold, D.W. Purves, D.P. Tittensor, and M.B.J. Harfoot. 2016. Synergistic impacts of habitat loss and fragmentation on model ecosystems. Proceedings of the Royal Society B 283:20161027.

Beca, G., M.H. Vancine, C.S. Carvalho, F. Pedrosa, R.S.C. Alves, D. Buscariol, C.A. Peres, M.C. Ribeiro, and M. Galetti. 2017. High mammal species turnover in forest patches immersed in biofuel plantations. Biological Conservation 210:352–359.

Bennett, V.J., D.W. Sparks, and P.A. Zollner. 2013. Modeling the indirect effects of road networks on the foraging activities of bats. Landscape Ecology 28:979–991.

Bierregaard, R.O. Jr., W.F. Laurance, C. Gascon, J. Benitez-Malvido, P.M. Fearnside, C.R. Fonseca, G. Ganade, et al. 2001. Principles of forest fragmentation and conservation in the Amazon. Pages 371–385 in R.O. Bierregaard Jr., C. Gascon, T.E. Lovejoy, and R. Mesquita, editors. Lessons from Amazonia: The ecology and conservation of a fragmented forest. Yale University Press, New Haven, Connecticut, USA.

Bocedi, G., S.C.F. Palmer, G. Pe'er, R.K. Heikkinen, Y.G. Matsinos, K. Watts, and J.M. J. Travis. 2014. RangeShifter: A platform for modelling spatial eco-evolutionary dynamics and species' responses to environmental changes. Methods in Ecology and Evolution 5:388–396.

Bradshaw, C.J.A. 2012. Little left to lose: Deforestation and forest degradation in Australia since European colonization. Journal of Plant Ecology 5:109–120.

Brady, M.J., C.A. McAlpine, H.P. Possingham, C.J. Miller, and G.S. Baxter. 2011. Matrix is important for mammals in landscapes with small amounts of native forest habitat. Landscape Ecology 26:617–628.

Broadbent, E.N., G.P. Asner, M. Keller, D.E. Knapp, P.J.C. Oliveira, and J.N. Silva. 2008. Forest fragmentation and edge effects from deforestation and selective logging in the Brazilian Amazon. Biological Conservation 141:1745–1757.

Cardinale, B.J., J.E. Duffy, A. Gonzalez, D.U. Hooper, C. Perrings, P. Venail, A. Narwani, et al. 2012. Biodiversity loss and its impact on humanity. Nature 489:326–326.

Chen, H.L., and J.L. Koprowski. 2016a. Barrier effects of roads on an endangered forest obligate: Influences of traffic, road edges, and gaps. Biological Conservation 199:33–40.

Chen, H.L., and J.L. Koprowski. 2016b. Differential effects of roads and traffic on space use and movements of native forest–dependent and introduced edge-tolerant species. PLOS One 11:e0148121.

Crooks, K.R., C.L. Burdett, D.M. Theobald, C. Rondinini, and L. Boitani. 2011. Global patterns of fragmentation and connectivity of mammalian carnivore habitat. Philosophical Transactions of the Royal Society of London B, Biological Sciences 366:2642–2651.

Crooks, K.R., C.L. Burdett, D.M. Theobald, S.R.B. King, M. DiMarco, C. Rondinini, and L. Boitani. 2017. Quantification of habitat fragmentation reveals extinction risk in terrestrial mammals. Proceedings of the National Academy of Sciences of the United States of America 114:7635–7640.

Cushman, S.A., K.S. McKelvey, J. Hayden, and M.K. Schwartz. 2006. Gene flow in complex landscapes: Testing multiple hypotheses with causal modeling. American Naturalist 168:486–499.

d'Annunzio, R., M. Sandker, Y. Finegold, and Z. Min. 2015. Projecting global forest area towards 2030. Forest Ecology and Management 352:124–133.

DeClerck, F.A.J., R. Chazdon, K.D. Holl, J.C. Milder, B. Finegan, A. Martinez-Salinas, P. Imbach, L. Canet, and Z. Ramos. 2010. Biodiversity conservation in human-modified landscapes of Mesoamerica: Past, present and future. Biological Conservation 143:2301–2313.

Didham, R.K., V. Kapos, and R.M. Ewers. 2012. Rethinking the conceptual foundations of habitat fragmentation research. Oikos 121:161–170.

Edwards, D.P., J.A. Hodgson, K.C. Hamer, S.L. Mitchell, A.H. Ahmad, S.J. Cornell, and D.S. Wilcove. 2010. Wildlife-friendly oil palm plantations fail to protect biodiversity effectively. Conservation Letters 3:236–242.

Efford, M. 2004. Density estimation in live-trapping. Oikos 106:598–610.

Estes, J.G., N. Othman, S. Ismail, M. Ancrenaz, B. Goossens, L. N. Ambu, A.B. Estes, and P.A. Palmiotto. 2012. Quantity and configuration of available elephant habitat and related conservation concerns in the lower Kinabatangan floodplain of Sabah, Malaysia. PLOS One 7:e44601.

European Union. 2011. The EU biodiversity strategy to 2020. European Union. http://ec.europa.eu/environment/nature/biodiversity/comm2006/2020.htm.

Fahrig, L. 2003. Effects of habitat fragmentation on biodiversity. Annual Review of Ecology, Evolution, and Systematics 34:487–515.

Fahrig, L., and T. Rytwinski. 2009. Effects of roads on animal abundance: An empirical review and synthesis. Ecology and Society 14:21.

Foley, J.A., N. Ramankutty, K.A. Brauman, E.S. Cassidy, J.S. Gerber, M. Johnston, N.D. Mueller, et al. 2011. Solutions for a cultivated planet. Nature 478:337–342.

Food and Agriculture Organization of the United Nations. 2010. Global forest resources assessment 2010: Main report. Forestry Paper 163. FAO, Rome, Italy.

Fuller, A.K., C.S. Sutherland, J.A. Royle, and M.P. Hare. 2016. Estimating population density and connectivity of American mink using spatial capture-recapture. Ecological Applications 26:1125–1135.

Gang, C., W. Zhou, Y. Chen, Z. Wang, Z. Sun, J. Li, J. Qi, and I. Odeh. 2014. Quantitative assessment of the contributions of climate change and human activities on global grassland degradation. Environmental Earth Sciences 72:4273–4282.

Gibson, L., T.M. Lee, L.P. Koh, B.W. Brook, T.A. Gardner, J. Barlow, C.A. Peres, et al. 2011. Primary forests are irreplaceable for sustaining tropical biodiversity. Nature 478:378–381.

Gopal, B. 2013. Future of wetlands in tropical and subtropical Asia, especially in the face of climate change. Aquatic Sciences 75:39–61.

Gustafson, E.J., and R.H. Gardner. 1996. The effect of landscape heterogeneity on the probability of patch colonization. Ecology 77:94–107.

Haddad, N.M., L.A. Brudvig, J. Clobert, K.F. Davies, A. Gonzalez, R.D. Holt, T.E. Lovejoy, et al. 2015. Habitat fragmentation and its lasting impact on Earth's ecosystems. Science Advances 1:e1500052.

Haines-Young, R. 2009. Land use and biodiversity relationships. Land Use Policy 26:S178–S186.

Hall, J., N.D. Burgess, J. Lovett, B. Mbilinyi, and R.E. Gereau. 2009. Conservation implications of deforestation across an elevational gradient in the Eastern Arc Mountains, Tanzania. Biological Conservation 142:2510–2521.

Hansen, M.C., P.V. Potapov, R. Moore, M. Hancher, S.A. Turubanova, A. Tyukavina, D. Thau, et al. 2013. High-resolution global maps of 21st-century forest cover change. Science 342:850–853.

Hanski, I., and O. Ovaskainen. 2002. Extinction debt at extinction threshold. Conservation Biology 16:666–673.

Holderegger, R., and M. DiGiulio. 2010. The genetic effects of roads: A review of empirical evidence. Basic and Applied Ecology 11:522–531.

Hooke, R.L.B., J.F. Martín-Duque, and J. Pedraza. 2012. Land transformation by humans: A review. GSA Today 22:4–10.

Jorge, M.L.S.P., M. Galetti, M.C. Ribeiro, and K.M.P.M.B. Ferraz. 2013. Mammal defaunation as surrogate of trophic cascades in a biodiversity hotspot. Biological Conservation 163:49–57.

Junk, W.J., S. An, C.M. Finlayson, B. Gopal, J. Květ, S.A. Mitchell, W.J. Mitsch, and R.D. Robarts. 2013. Current state of knowledge regarding the world's wetlands and their future under global climate change: A synthesis. Aquatic Sciences 75:151–167.

Keenan, R.J., G.A. Reams, F. Achard, J.V. deFreitas, A. Grainger, and E. Lindquist. 2015. Dynamics of global forest area: Results from the FAO Global Forest Resources Assessment 2015. Forest Ecology and Management 352:9–20.

Keinath, D.A., D.F. Doak, K.E. Hodges, L.R. Prugh, W. Fagan, C.H. Sekercioglu, S.H.M. Buchart, and M. Kauffman. 2017. A global analysis of traits predicting species sensitivity to habitat fragmentation. Global Ecology and Biogeography 26:115–127.

Kim, D.H., J.O. Sexton, and J.R. Townshend. 2015. Accelerated deforestation in the humid tropics from the 1990s to the 2000s. Geophysical Research Letters 42:3495–3501.

Kissinger, G., M. Herold, and V. DeSy. 2002. Drivers of deforestation and forest degradation: A synthesis report for REDD+ Policymakers. Lexeme Consulting, Vancouver, Canada.

Knowlton, J.L., and C.H. Graham. 2010. Using behavioral landscape ecology to predict species' responses to land-use and climate change. Biological Conservation 143:1342–1354.

Kramer-Schadt, S., E. Revilla, T. Wiegand, and U. Breitenmoser. 2004. Fragmented landscapes, road mortality and patch connectivity: Modelling influences on the dispersal of Eurasian lynx. Journal of Applied Ecology 41:711–723.

Laidlaw, R.K. 2000. Effects of habitat disturbance and protected areas on mammals of peninsular Malaysia. Conservation Biology 14:1639–1648.

Laurance, W.F., M. Goosem, and S.G.W. Laurance. 2009. Impacts of roads and linear clearings on tropical forests. Trends in Ecology and Evolution 24:659–669.

Laurance, W.F., G.R. Clements, S. Sloan, C.S.O. Connell, N.D. Mueller, M. Goosem, O. Venter, et al. 2014a. A global strategy for road building. Nature 513:229–232.

Laurance, W.F., J. Sayer, and K.G. Cassman. 2014b. Agricultural expansion and its impacts on tropical nature. Trends in Ecology and Evolution 29:107–116.

Liu, J., and J. Diamond. 2005. China's environment in a globalizing world. Nature 435:1179–1186.

Liu, J., S. Li, Z. Ouyang, C. Tam, and X. Chen. 2008. Ecological and socioeconomic effects of China's policies for ecosystem services. Proceedings of the National Academy of Sciences of the United States of America 105:9477–9482.

Magioli, M., K.M.P.M. de B. Ferraz, E.Z.F. Setz, A.R. Percequillo, M.V. de S.S. Rondon, V.V. Kuhnen, M.C. da S. Canhoto, et al. 2016. Connectivity maintain mammal assemblages functional diversity within agricultural and fragmented landscapes. European Journal of Wildlife Research 62:431–446.

Mann, G.K.H., J.V. Lagesse, M.J.O. Riain, and D.M. Parker. 2015. Beefing up species richness? The effect of land-use on mammal diversity in an arid biodiversity hotspot. African Journal of Wildlife Research 45:321–331.

McRae, B.H., B.G. Dickson, T.H. Keitt, and V.B. Shah. 2008. Using circuit theory to model connectivity in ecology, evolution, and conservation. Ecology 89:2712–2724.

Merkle, J.A., K.L. Monteith, E.O. Aikens, M.M. Hayes, K.R. Hersey, A.D. Middleton, B.A. Oates, H. Sawyer, B.M. Scurlock, and M.J. Kauffman. 2016. Large herbivores surf waves of green-up during spring. Proceedings of the Royal Society of London B, Biological Sciences 283:p20160456.

Miura, S., M. Amacher, T. Hofer, J. San-Miguel-Ayanz, Ernawati, and R. Thackway. 2015. Protective functions and ecosystem services of global forests in the past quarter-century. Forest Ecology and Management 352:35–46.

Moorcroft, P.R., M.A. Lewis, and R.L. Crabtree. 1999. Home range analysis using a mechanistic home range model. Ecology 80:1656–1665.

Morales-Hidalgo, D., S.N. Oswalt, and E. Somanathan. 2015. Status and trends in global primary forest, protected areas, and areas designated for conservation of biodiversity from the Global Forest Resources Assessment 2015. Forest Ecology and Management 352:68–77.

Morin, D.J., A.K. Fuller, J.A. Royle, and C. Sutherland. 2017. Model-based estimators of density and connectivity to inform conservation of spatially structured populations. Ecosphere 8:e01623.

Muler, A.E., D.C. Rother, P.S. Brancalion, R.P. Naves, R.R. Rodrigues, and M.A. Pizo. 2014. Can overharvesting of a non-timber-forest-product change the regeneration dynamics of a tropical rainforest? The case study of *Euterpe edulis*. Forest Ecology and Management 324:117–125.

Nathan, R., W.M. Getz, E. Revilla, M. Holyoak, R. Kadmon, D. Saltz, and P.E. Smouse. 2008. A movement ecology paradigm for unifying organismal movement research. Proceedings of the National Academy of Sciences of the United States of America 105:19052–19059.

Ndriantsoa, S.H., J.C. Riemann, N. Raminosoa, M.O. Rödel, and J.S. Glos. 2017. Amphibian diversity in the matrix of a fragmented landscape around Ranomafana in Madagascar depends on matrix quality. Tropical Conservation Science 10:1–16.

Newbold, T., L.N. Hudson, S.L.L. Hill, S. Contu, I. Lysenko, R.A. Senior, L. Börger, et al. 2015. Global effects of land use on local terrestrial biodiversity. Nature 520:45–50.

Niemandt, C., and M. Greve. 2016. Fragmentation metric proxies provide insights into historical biodiversity loss in critically endangered grassland. Agriculture, Ecosystems and Environment 235:172–181.

Nopper, J., B. Lauströer, M.O. Rödel, and J.U. Ganzhorn. 2017. A structurally enriched agricultural landscape maintains high reptile diversity in sub-arid southwestern Madagascar. Journal of Applied Ecology 54:480–488.

Nupp, T.E., and R.K. Swihart. 2000. Landscape-level correlates of small-mammal assemblages in forest fragments of farmland. Journal of Mammalogy 81:512–526.

Palomo, I., C. Montes, B. Martin-Lopez, J.A. Gonzalez, M. Garcia-Llorente, P. Alcorlo, and M.R.G. Mora. 2014. Incorporating the social-ecological approach in protected areas in the Anthropocene. BioScience 64:181–191.

Pattanavibool, A., and P. Dearden. 2002. Fragmentation and wildlife in montane evergreen forests, northern Thailand. Biological Conservation 107:155–164.

Pauli, B.P., N.P. McCann, P.A. Zollner, R. Cummings, J.H. Gilbert, and E.J. Gustafson. 2013. SEARCH: Spatially explicit animal response to composition of habitat. M. Convertino, editor. PLOS One 8:e64656.

Pedrosa, F., R. Salerno, F.V.B. Padilha, and M. Galetti. 2015. Current distribution of invasive feral pigs in Brazil: Economic impacts and ecological uncertainty. Natureza e Conservacao 13:84–87.

Pereira, H.M., P.W. Leadley, V. Proença, R. Alkemade, J.P.W. Scharlemann, J.F. Fernandez-Manjarrés, M.B. Araújo, et al. 2010. Scenarios for global biodiversity in the 21st century. Science 330:1496–1501.

Peres, C.A., and E. Palacios. 2007. Basin-wide effects of game harvest on vertebrate population densities in Amazonian forests: Implications for animal-mediated seed dispersal. Biotropica 39:304–315.

Pinto, N., and T.H. Keitt. 2009. Beyond the least-cost path: Evaluating corridor redundancy using a graph-theoretic approach. Landscape Ecology 24:253–266.

Reis, V., V. Hermoso, S.K. Hamilton, D. Ward, E. Fluet-Chouinard, B. Lehner, and S. Linke. 2017. A global assessment of inland wetland conservation status. BioScience 67:523–533.

Ribeiro, M.C., J.P. Metzger, A.C. Martensen, F.J. Ponzoni, and M.M. Hirota. 2009. The Brazilian Atlantic Forest: How much is left, and how is the remaining forest distributed? Implications for conservation. Biological Conservation 142:1141–1153.

Rovero, F., N. Owen, T. Jones, E. Canteri, A. Iemma, and C. Tattoni. 2017. Camera trapping surveys of forest mammal communities in the Eastern Arc Mountains reveal generalized habitat and human disturbance responses. Biodiversity and Conservation 26:1103–1119.

Royle, J.A., and K.V. Young. 2008. A hierarchical model for spatial capture recapture data. Ecology 89:2281–2289.

Ruffell, J., M.N. Clout, and R.K. Didham. 2017. The matrix matters, but how should we manage it? Estimating the amount of high-quality matrix required to maintain biodiversity in fragmented landscapes. Ecography 40:171–178.

Rytwinski, T., and L. Fahrig. 2011. Reproductive rate and body size predict road impacts on mammal abundance. Ecological Applications 21:589–600.

Rytwinski, T., and L. Fahrig. 2012. Do species life history traits explain population responses to roads? A meta-analysis. Biological Conservation 147:87–98.

Sasidhran, S., N. Adila, M.S. Hamdan, L.D. Samantha, N. Aziz, N. Kamarudin, C.L. Puan, E. Turner, and B. Azhar. 2016. Habitat occupancy patterns and activity rate of native mammals in tropical fragmented peat swamp reserves in peninsular Malaysia. Forest Ecology and Management 363:140–148.

Schneider, C., M. Flörke, L. DeStefano, and J.D. Petersen-Perlman. 2017. Hydrological threats to riparian wetlands of international importance: A global quantitative and qualitative analysis. Hydrology and Earth System Sciences 21:2799–2815.

Seto, K.C., B. Güneralp, and L.R. Hutyra. 2012. Global forecasts of urban expansion to 2030 and direct impacts on biodiversity and carbon pools. Proceedings of the National Academy of Sciences of the United States of America 109:16083–16088.

Sloan, S., and J.A. Sayer. 2015. Forest Resources Assessment of 2015 shows positive global trends but forest loss and degradation persist in poor tropical countries. Forest Ecology and Management 352:134–145.

Smouse, P.E., S. Focardi, P.R. Moorcroft, J.G. Kie, J.D. Forester, and J.M. Morales. 2010. Stochastic modelling of animal movement. Philosophical Transactions of the Royal Society B: Biological Sciences 365:2201–2211.

Sodhi, N.S., L.P. Koh, R. Clements, T.C. Wanger, J.K. Hill, K.C. Hamer, Y. Clough, T. Tscharntke, M.R.C. Posa, and T.M. Lee. 2010. Conserving Southeast Asian forest biodiversity in human-modified landscapes. Biological Conservation 143:2375–2384.

Swift, T.L., and S.J. Hannon. 2010. Critical thresholds associated with habitat loss: A review of the concepts, evidence, and applications. Biological Reviews 85:35–53.

Tabarelli, M., L.P. Pinto, J.M.C. Silva, M. Hirota, and L. Bedê. 2005. Challenges and opportunities for biodiversity conservation in the Brazilian Atlantic Forest. Conservation Biology 19:695–700.

Thomas, N., R. Lucas, P. Bunting, A. Hardy, A. Rosenqvist, and M. Simard. 2017. Distribution and drivers of global mangrove forest change, 1996–2010. PLOS One 12:e0179302.

Thompson, S.J., D.H. Johnson, N.D. Niemuth, and C.A. Ribic. 2015. Avoidance of unconventional oil wells and roads exacerbates habitat loss for grassland birds in the North American Great Plains. Biological Conservation 192:82–90.

Tilman, D., R.M. May, C.L. Lehman, and M.A. Nowak. 1994. Habitat destruction and the extinction debt. Nature 371:65–66.

Trombulak, S.C., and C.A. Frissell. 2000. Review of ecological effects of roads on terrestrial and aquatic communities. Conservation Biology 14:18–30.

Tscharntke, T., J.M. Tylianakis, T.A. Rand, R.K. Didham, L. Fahrig, P. Batáry, J. Bengtsson, et al. 2012. Landscape moderation of biodiversity patterns and processes: Eight hypotheses. Biological Reviews 87:661–685.

Uezu, A., and J.P. Metzger. 2016. Time-lag in responses of birds to Atlantic Forest fragmentation: Restoration opportunity and urgency. PLOS One 11:e0147909.

Uryu, Y., C. Mott, N. Foead, K. Yulianto, A. Budiman, B. Setiabudi, F. Takakai, et al. 2008. Deforestation, forest degradation, biodiversity loss and CO2 emissions in Riau, Sumatra, Indonesia. World Wildlife Fund (WWF) Indonesia Technical Report. Jakarta, Indonesia.

Valiela, I., J.L. Bowen, and J.K. York. 2001. Mangrove forests: One of the world's threatened major tropical environments. BioScience 51:807–815.

van Asselen, S., P.H. Verburg, J.E. Vermaat, and J.H. Janse. 2013. Drivers of wetland conversion: A global meta-analysis. PLOS One 8:e81292.

Venter, O., R.A. Fuller, D.B. Segan, J. Carwardine, T. Brooks, S.H.M. Butchart, M. DiMarco, et al. 2014. Targeting global protected area expansion for imperiled biodiversity. PLOS Biology 12:e1001891.

Verón, S.R., J.M. Paruelo, and M. Oesterheld. 2006. Assessing desertification. Journal of Arid Environments 66:751–763.

Weeks, E.S., S. Walker, J.R. Dymond, J.D. Shepherd, and B.D. Clarkson. 2013. Patterns of past and recent conversion of indigenous grasslands in the South Island, New Zealand. New Zealand Journal of Ecology 37:127–138.

Wilson, M.C., X.Y. Chen, R.T. Corlett, R.K. Didham, P. Ding, R.D. Holt, M. Holyoak, et al. 2016. Habitat fragmentation and biodiversity conservation: Key findings and future challenges. Landscape Ecology 31:219–227.

With, K.A., and A.W. King. 1999. Extinction thresholds for species in fractal landscapes. Conservation Biology 13:314–326.

Wolanski, E., S. Spagnol, S. Thomas, K. Moore, D.M. Alongi, L. Trott, and A. Davidson. 2000. Modelling and visualizing the fate of shrimp pond effluent in a mangrove-fringed tidal creek. Estuarine, Coastal and Shelf Science 50:85–97.

Xu, W., Y. Xiao, J. Zhang, W. Yang, L. Zhang, V. Hull, Z. Wang, et al. 2017. Strengthening protected areas for biodiversity and ecosystem services in China. Proceedings of the National Academy of Sciences of the United States of America 114:1601–1606.

Zarfl, C., A.E. Lumsden, J. Berlekamp, L. Tydecks, and K. Tockner. 2014. A global boom in hydropower dam construction. Aquatic Sciences 77:161–170.

Zollner, P.A., and S.L. Lima. 1999. Search strategies for landscape-level interpatch movements. Ecology 80:1019–1030.

6

Marta A. Jarzyna
Victoria L. Atkin Dahm
Benjamin Zuckerberg
William F. Porter

Consequences of Climate Change for Wildlife

Introduction

Climate change is a prolonged and significant change in weather patterns that is measured in statistical terms, such as changes in averages or variations of temperature or precipitation over decades or centuries. In recent years, modern climate change is of increasing concern to wildlife biologists tasked with the conservation and management of vulnerable ecosystems and species.

The Intergovernmental Panel on Climate Change (IPCC) reports that "warming of the climate system is unequivocal, and since the 1950s, many of the observed changes are unprecedented over decades to millennia. The atmosphere and ocean have warmed, the amounts of snow and ice have diminished, and sea level has risen" (IPCC 2014: 40). The IPCC, a Nobel Peace Prize–winning international organization of scientists, established by the United Nations to review and assess the most recent scientific, technical, and socioeconomic information relevant to the understanding of climate change, reported that changing climate and human influence on the climate system is clear.

Climate change is not a new phenomenon. Earth's climate has changed repeatedly over the course of its history, each time significantly altering the abundance and distribution of species. Modern climate change, however, is distinguished from past events by at least two factors: the rate of climatic change is greater than has been experienced since the rise of modern civilization, and the change is unequivocally human induced. Increased emissions of greenhouse gases (GHGs), mostly carbon dioxide (CO_2) but also other gases such as methane (CH_4), and nitrous oxide (N_2O), are the main cause of the modern climate change. Concentrations of CO_2, CH_4, and N_2O have increased by 40%, 150%, and 20%, respectively, since 1750 and are currently at levels that are unprecedented in at least 800,000 years (IPCC 2014). Greenhouse gasses affect climate by modifying the absorption, scattering, and emission of radiation within the atmosphere (IPCC 2014). Two main causes of recent global increases in concentrations of GHGs are burning fossil fuels and shifting land-use practices.

Though climate change is often thought of as a simple increase in temperatures that will result in species shifting their distributions poleward or higher in elevation, the reality is a more complex suite of changes in global-scale processes. Declines in the extent of snow cover and sea (Fig. 6.1) and land ice, sea level rise caused by thermal expansion of water and melting of glaciers (Fig. 6.2) and ice

Figure 6.1 Sea ice conditions in the southern Chukchi Sea in April 2017. Typically, the sea ice edge is >150 km south of this area in April; however, in 2017, higher temperatures, severe winter storms, and thinner sea ice caused the ice to retreat significantly earlier than normal. Ryan Wilson, US Fish & Wildlife Service

Figure 6.2 Glaciers have receded markedly, a direct and observable effect of climate change. Exit Glacier in Alaska has declined dramatically in fewer than 15 years, as shown here by the 2005 location denoted by the sign in the lower left. Marta Jarzyna

caps, ocean acidification, and increases in the frequency and intensity of extreme events such as heat waves, floods, and droughts are other physical consequences of climate change (IPCC 2014). As such, climate change has the potential to significantly alter the composition of ecological communities and functioning of ecosystems, affecting wildlife populations worldwide. As the climate changes, each species will adapt in different ways, leading to an extensive restructuring of communities. Wildlife management will need to adapt as well.

In this chapter we examine the causes of climate change and consequences of changing climatic conditions for wildlife and ecosystems. Specifically, we ask two central questions: What are the biological implications of climate change to wildlife populations and ecosystems? What will these implications mean for wildlife managers?

Consequences of Climate Change for Wildlife

Climate change will have far-reaching consequences for wildlife, and most taxonomic groups have already been affected by changing climatic conditions (Parmesan and Yohe 2003, Pecl et al. 2017). The effects will vary greatly depending on species, with some benefiting and others being negatively affected. The implications of changing climatic conditions for species will be determined by their exposure, sensitivity, and adaptive capacity, which together determine vulnerability of species to climate change (Stein et al. 2014). To adjust quickly to rapidly changing climate conditions, vulnerable species and populations will have to display high levels of phenotypic plasticity (the ability of individuals to modify their behavior, morphology, or physiology to altered environmental conditions) to adapt in situ or be able to move to favorable habitats outside their current range (Dunn and Winkler 2010). Alternatively, vulnerable species might be able to adjust through evolu-

tionary adaptation, though genetic adaptation will be typically important over multiple generations.

Evolutionary Consequences

Increasing evidence suggests that life history traits with strong genetic components (breeding phenology, clutch size, body size) are sensitive to long-term trends in climate (Sheldon 2010) and that evolutionary adaptation could be an important way for natural populations to counter rapid climate change (Hoffmann and Sgro 2011).

Most examples of evolutionary consequences of climate change come from studies of migratory behavior, a trait that has strong heritability and is particularly sensitive to changing climatic conditions. For example, Pulido and Berthold (2010) demonstrated genetic reduction in migratory activity and an evolutionary change in the onset of migration in blackcaps (*Sylvia atricapilla*). Selection for shorter migration distance in blackcaps might lead to residency in completely migratory bird populations, and climate change might favor genotypes that can winter closer to the breeding grounds. Work on migratory fish corroborates these findings. Kovach et al. (2012) reported that there has been directional selection for earlier migration timing in pink salmon (*Oncorhynchus gorbuscha*), resulting in a substantial decrease in the late-migrating phenotype (from more than 30% to less than 10% of the total abundance). From 1983 to 2011, the frequency of a genetic marker for late-migration timing declined threefold, which provides evidence for rapid microevolution for earlier migration timing in this population.

Evolutionary change is also common in species shifting their geographic distributions. Whitney and Gabler (2008) documented 38 species in which the specific traits commonly associated with invasive potential (growth rate, dispersal ability, generation time) have themselves undergone evolutionary change following introduction, in some cases over very short (≤10 year) timescales. Karell et al. (2011)

reported a significant increase in brown-plumed tawny owls (*Strix aluco*) over 28 years, a species that usually is pale gray or brown. The strong viability selection against the brown morph occurs only under snow-rich winters, and as winter conditions became milder and less snowy in the last decades, selection against the brown morph diminished. Shifts in genetic variability have been observed even in populations of the long-lived Canadian lynx (*Lynx canadensis*) and have been associated with snow depth and winter precipitation (Row et al. 2014).

Phenology

Phenological changes include changes in timing of migration, breeding, and hibernation. Evidence suggests that seasonality of many temperate regions has already been altered (Hansen et al. 2010), ultimately leading to altered phenology of plants and animals. Adding to the complexity, species or even populations of the same species might vary in the magnitude of the phenological change (Both et al. 2009). Phenological mismatch is when the timing of one phenomenon changes at a different rate than that of another phenomenon. It is particularly problematic when consumers are not able to track their food resources, which may be shifting more rapidly than their own phenology, such as when young are raised.

Most (80%) species experienced significant climate change–induced shifts of phenological events (Root et al. 2003). The best evidence that wildlife populations are exhibiting phenological shifts in response to climate warming comes from long-term studies of butterflies, birds, and mammals, though phenological changes for amphibians and reptiles have also been reported.

Changes in Timing of Migration

Migration is a behavioral adaptation to seasonal environments through which individuals can successfully exploit temporarily abundant resources in a given region, but can escape difficult physical conditions prevalent in that region at other times. Given that migration is tightly linked to environmental and climatic seasonality, it is not surprising that climate change will affect migration timing.

Birds

Birds may alter their migration timing in two ways: by advancing their arrival date to the breeding sites (either by leaving their wintering grounds earlier or by shortening the migration distance) or by delaying the onset of their fall migration (Lehikoinen and Sparks 2010). The timing of migration of short- and long-distance migratory birds depends on different environmental and biotic factors. Whereas the migration of short-distant migrants relies strongly on weather conditions, long-distance migrants are under endogenous control and mostly cued by photoperiod (Gwinner 1996). Consequently, short-distance migrants are more likely to attune the timing of their migration to increasing temperatures, while long-distance migrants are less likely to do so. As a result, long-distance migrants might be more vulnerable to population losses than short-distance migrants or resident species.

The evidence for birds advancing their arrival to the breeding grounds is mounting (Lehikoinen and Sparks 2010, Hurlbert and Liang 2012). Meta-analyses by Lehikoinen and Sparks (2010) revealed that 82% of trends in first arrival dates and 76% of trends in mean arrival dates to the breeding grounds were toward earliness. On average, the first and mean arrival dates advanced by 2.8 and 1.8 days/decade, respectively. Hurlbert and Liang (2012) reported that 18 North American bird species shifted their arrival times by 0.8 days earlier for every 1°C of warming of spring temperature. These changes might seem small, but they have a profound effect on the reproductive success of migratory birds.

The location and duration of overwintering is also important to managers. Some European long-distance migratory birds, formerly wintering entirely in tropical and southern Africa, are now wintering in the Mediterranean region (Berthold 2001).

Short-distance migratory raptors had a delayed mean passage date of the fall migration across five European regions in response to changes in temperature (Jaffré et al. 2013). Specifically, increasing fall temperatures caused delays of the fall migration by an average 3.5 days, while temperature declines resulted in migration advancement by approximately 4.5 days. Some populations of migrants, such as Canada goose (*Branta canadensis*) in North America and white stork (*Ciconia ciconia*) in Spain, now winter in their stopover sites or have ceased their migration and stay year-round on their breeding grounds (Flack et al. 2016).

Mammals

Migratory mammals are likely to respond to climate change in ways analogous to birds. For example, the seasonal movement of white-tailed deer (*Odocoileus virginianus*) in northern latitudes is cued by the snow depth; in the absence of snow, deer cease their migration and remain resident on summer range throughout the year (Tierson et al. 1985). In northern Sweden, where snowfall is projected to increase because of warmer winters, moose (*Alces alces*) are expected to migrate to their winter ranges earlier and leave their winter grounds later (Ball et al. 1999). The timing of migration of migratory caribou (*Rangifer tarandus*) will be altered by delays in or incomplete formation of ice in the Arctic (Fig. 6.3; Sharma et al. 2009). To ensure safe passage over large water bodies, migratory caribou will have to advance the onset of spring migration and delay fall migration or shift their ranges farther north (Sharma et al. 2009).

Changes in Timing of Breeding and Reproductive Success

The timing of breeding of wildlife populations can also be altered and is another example of phenotypic plasticity as a response to climate change. Weishampel et al. (2004) reported that the median nesting data of loggerhead sea turtle (*Caretta caretta*) on Florida's Atlantic coast advanced by approximately 10 days between 1989 and 2003. The timing of nesting was significantly correlated with nearshore May sea surface temperatures that warmed an average of 0.8°C over this period. For amphibians, changes in breeding phenology are often closely tied to changes in their migration to breeding ponds to reproduce.

Figure 6.3 Caribou migrations are influenced by changing patterns of snow and ice that influence conditions along traditional migration pathways. Marta Jarzyna

For example, Todd et al. (2010) examined the phenology of migrations to breeding ponds in 10 amphibian species and reported that the autumn-breeding dwarf (*Eurycea quadridigitata*) and marbled (*Ambystoma opacum*) salamanders delayed their breeding over a 30-year period in coincidence with an estimated 1.2°C increase in local overnight temperatures, while the winter-breeding tiger salamander (*Ambystoma tigrinum*) and ornate chorus frog (*Pseudacris ornata*) advanced their breeding. For those species, the reproductive timing shifted by an estimated 15.3 to 76.4 days, resulting in rates of phenological change that range from 5.9 to 37.2 days/decade. It is worth noting, however, that the remaining six studied species did not shift their breeding phenologies (Todd et al. 2010). The timing of calling for four out of six studied frog species in New York State has advanced by 10–13 days since 1900 (Gibbs and Breisch 2001).

Data from long-term studies indicate that 59% of 68 species of birds have significantly advanced their laying date and 79% advanced their laying date in warmer years (Dunn and Winkler 2010). Laying dates for the species that advanced their phenology have shifted by 0.13 days/year, which amounts to a shift of 2.4 days earlier for every degree Celsius in temperature increase. Other reproductive traits in birds (number and size of clutches, incubation behavior, recruitment) are directly linked to the tim-ing of breeding and thus indirectly affected by climate change (Dunn and Winkler 2010).

Studies report similar reproductive shifts in wild mammals (Réale et al. 2003, Moyes et al. 2011). For example, Daubenton's bats (*Myotis daubentonii*) advanced their breeding by approximately 11 days between 1970 and 2012, likely as a result of increases in spring temperature (Lučan et al. 2013). Canadian populations of red squirrel (*Tamiasciurus hudsonicus*) advanced their breeding by 18 days over a decade (Réale et al. 2003), a trend partly explained by increased food availability associated with increased temperatures, although evolutionary adaptation toward earlier breeders also played a role. Moyes et al. (2011) reported advances in six phenological traits in a population of red deer (*Cervus elaphus*).

Changes in Timing of Hibernation

Hibernation allows animals to conserve energy during winter, and hibernation patterns of some amphibians and mammals have already been affected by climate change (Fig. 6.4). By simulating hibernation conditions under projected temperature increase, Üveges et al. (2016) showed that shorter winters and milder hibernation temperature increased survival of common toads (*Bufo bufo*) during hibernation. Hibernation behavior of black-spotted pond frogs (*Pelophylax nigromaculatus*) was altered through

Figure 6.4   Hibernating species, especially those at high latitude or elevation, such as (A) arctic ground squirrels and (B) yellow-bellied marmots, are strongly influenced by temperature changes. A, Marta Jarzyna; B, Ken Armitage

outdoor experiments, with individuals entering hibernation later and emerging from hibernation earlier when temperatures were experimentally increased (Gao et al. 2015).

Hibernation patterns of Arctic ground squirrels (*Spermophilus parryii*) in northern Alaska depended on the environmental conditions that each population was exposed to (Sheriff et al. 2012). Lane et al. (2012) reported a significant delay (0.47 days/year over a 20-year period) in the hibernation emergence date of adult females in a wild population of Columbian ground squirrels (*Urocitellus columbianus*) in Canada. The delayed emergence was related to the climatic conditions, particularly delayed snowmelt.

Long-term studies (Inouye et al. 2000, Ozgul et al. 2010) of yellow-bellied marmots (*Marmota flaviventris*) in Colorado confirm advanced emergence from hibernation. Over 28 years, yellow-bellied marmots advanced their phenology by 38 days (Inouye et al. 2000), likely as a response to increasing spring air temperatures (Inouye et al. 2000). Rising spring temperatures were also a likely cause of the earlier emergence from hibernation of the edible dormouse (*Glis glis*) (Adamík and Král 2008), whose hibernation emergence was advanced by 8 days/decade.

Trophic Mismatches

Phenological changes are generally considered evidence that species are successfully adjusting to changing environmental conditions (Hetem et al. 2014). Some studies suggest that species that have advanced or delayed their phenology along with changing climatic conditions will be more likely to persist in a warming world (Moller et al. 2008, Cleland et al. 2012). There is also evidence that changes in phenology are sometimes associated with declines in fitness. While causes of such declines are varied, one of the main culprits might be increasing asynchrony, or mistiming of key ecological events (Hetem et al. 2014).

The responses of individual species to climate change will depend on species' traits (species' physi-

ological tolerance, dispersal abilities). Because species exhibit high variability in life history traits, they are likely to respond to changing climatic conditions in different ways (Parmesan and Yohe 2003, Jarzyna et al. 2015). What do such differential responses mean for species and communities? Species currently living in synchrony may find relationships altered through intra- and interspecific interactions, such as predator-prey or parasitoid-host interactions, and may be disrupted through the differential effects of climate change.

Evidence for altered synchrony in wildlife populations is mounting. Thackeray et al. (2010) reported differences in phenological changes among trophic levels across marine, freshwater, and terrestrial environments; advances in phenology were slowest for secondary consumers, thus increasing the risk of temporal mismatch in key trophic interactions. Visser and Both (2005) reported that in seven of eleven cases of predator-prey and insect host–plant interactions, responses of interacting species to climate change were different enough to put them out of synchrony with one another. In another study, populations of pied flycatcher (*Ficedula hypoleuca*), a long-distance migrant, declined significantly in seasonal environments characterized by short food peak but remained stable in less seasonal wetlands where the peak of food availability was longer (Both et al. 2009). In addition, resident and short-distance migratory birds remained stable in all the habitats, pointing to the increasing trophic mismatch between pied flycatcher and their food, rather than habitat quality, as the culprit for population declines of the long-distance migrant. In addition, stronger declines were reported for Western European populations, where spring temperatures increased more than in Northern Europe (Both et al. 2009).

Mammals are also likely to face trophic mismatch. For example, migratory herbivores are likely to develop trophic mismatches, because the timing of their spring migration to the breeding grounds is cued by photoperiod, while increasing spring temperatures cue the onset of plants. Post and Forchhammer (2008) reported a trophic mismatch between migratory caribou and their foraging plants. Caribou lagged the start of plant growth and were unable to keep pace with advancement in plant-growing season on their breeding ranges. Consequently, offspring production decreased fourfold and mortality of those young that were born increased.

Anatomy

The majority of responses to climate change relate to phenology or range shifts, but other traits are also affected by changing climatic conditions. For example, a decline in body mass as a response to increasing temperatures has been reported (Hetem et al. 2014). Small animals cannot prevent heat loss in cold environments as well as animals of the same shape but higher body mass. In an increasingly warmer world, smaller animals are expected to lose the disadvantage associated with higher heat loss. Ozgul et al. (2009) reported declines in body mass of up to 0.8%/year for Soay sheep on St Kilda, Scotland. They suggested that milder winters resulted in less reliance on fat reserves, which in turn enabled more of the small individuals to survive the winter. Reading (2007) reported a reduction in female body size for the common toad in the United Kingdom between 1983 and 2005, which was significantly correlated with the occurrence of mild winters. Over 55 years, six species of salamanders (*Plethodon* spp.) underwent reductions in body size, likely as a result of increased metabolism in regions experiencing the greatest drying and warming (Caruso et al. 2014).

A decline in body mass, however, does not appear to be a universal response to climate change. In fact, less than 10% of recently reported changes in mammalian body masses provide support for an advantage to smaller mammals (Teplitsky and Millien 2014). For example, body mass of otters (*Lutra lutra*) in Norway has increased between 1975 and 2000, presumably as a result of increased food availability. Ozgul et al. (2010) demonstrated increased body mass of the yellow-bellied marmot resulting from earlier emergence from hibernation and earlier weaning of young, ultimately leading to declines in adult mortality and increases in population size.

Distributional Responses

The location of species ranges is influenced by abiotic and biotic factors, with evidence pointing to climatic conditions as partial determinants of species range boundaries (Baselga 2012). As climatic conditions change, species will move their distributions in the direction of changing climate. Most range shifts have been northward in the Northern Hemisphere, southward in the Southern Hemisphere, and up in elevation, although there have been distributional changes in other directions associated with regional or local climate changes and the life history of certain species (Parmesan and Yohe 2003). Other species studied have shown no observed change in distribution (Parmesan and Yohe 2003). Species' ecological tolerances along with physical and biotic barriers will dictate how, and if, species will adjust their distributions. Ultimately, extinctions might occur if species are unable to track changes in climate, and if they cannot adjust in situ via phenotypic plasticity or genetic change.

How far can we expect species to move? Predictably, the answer to this question will depend on the species, because the responses will likely vary between different taxonomic groups. Range shifts averaging 6.1 km/decade toward the poles were reported for the range of birds, butterflies, and alpine herbs (Parmesan and Yohe 2003). Hickling et al. (2006) reported average poleward shifts of 31–60 km and elevational shifts of 25 m over a 40-year period for 16 different vertebrate and invertebrate taxonomic groups, including dragonflies and damselflies, grasshoppers, butterflies, spiders, freshwater fish, amphibians and reptiles, birds, and mammals. Most of the 329 species studied by Hickling et al. (2006) showed a significant shift toward higher latitudes. Songbirds in New York State shifted their ranges northward an average of 3.6 km between 1980 and 2000 (Zuckerberg et al. 2009). Similarly, the breeding range of Adelie penguins (*Pygoscelis adeliae*) in the Ross Sea (Antarctica) expanded 3 km southward in 4 years (Taylor and Wilson 1990).

Some of the most convincing pieces of evidence for climate change–related elevational range shifts come from studies of small mountain mammals. Throughout the Great Basin ecoregion, populations of American pika (*Ochotona princeps*) have been monitored since 1898 and are now showing large changes in their elevational ranges (Beever et al. 2011). The rate of upslope range retraction for the American pika (Fig. 6.5) accelerated elevenfold between 1998 and 2008, and the species' elevational range was recently reported shifting at an average rate of 145 m/decade. Rowe et al. (2015) resurveyed small mammals in historical collecting localities along elevational gradients in northern, central, and southern regions of montane California. Twenty-five of the 34 mammal species shifted their ranges upslope or downslope in at least one region of montane California, with high-elevation species typically contracting their lower range boundaries upslope and low-elevation species having heterogeneous responses. Temperature change also predicted responses of high-elevation species, indicating that climate change will ultimately lead to population declines and potential extinctions as the suitable climatic habitat for these species dwindles. As habitats and species move upslope, higher-altitude populations or species may be squeezed out of existence as their habitat disappears altogether with no further place to move.

Charismatic megafauna are also undergoing distributional shifts in ranges. For example, polar bears (*Ursus maritimus*) in Hudson Bay shifted their distributions poleward and eastward between 1986 and 2004, as a result of the 3-week advancement of

Figure 6.5   The American pika, a generalist herbivore that inhabits talus patches typically at high elevation, has been a model organism for numerous studies on behavior, vocal communication, metapopulation dynamics, source-sink dynamics, island biogeography, species-climate relationships, and local extinction dynamics. Though trends are variable across populations (some stable, others slightly to markedly declining), pikas provide a striking counterexample to the pattern that wildlife declines have typically been effected (or "precipitated") by habitat loss and degradation, overharvest, and invasive species; in their case, climatic aspect(s) are often the strongest single influence on the pattern of distributional change. This vocalizing individual was photographed in the East Humboldt Range, northeastern Nevada, USA. Shana Weber

spring sea ice breakup since the late 1970s (Towns et al. 2010). Tape et al. (2016) reported that increased riparian shrub habitat and warming led to the northward expansion of Alaskan moose (*Alces alces gigas*).

Changes in species distributions will have dramatic consequences for ecosystem stability and for human health, culture, food security, trade, and economics (Pecl et al. 2017). For example, changes in distributions of fish, wild reindeer, and caribou are affecting the food security, traditional knowledge systems, and endemic cosmologies of indigenous societies in the Arctic (Pecl et al. 2017). Mosquitoes shifting their ranges in response to global warming create a threat to human health through predicted increases in the occurrence of known and potentially new diseases (Pecl et al. 2017). Species redistribution will also increasingly lead to substantial conflict; the recent expansion of mackerel into Icelandic waters, as a case in point, resulted in mackerel wars between Iceland and competing countries that have traditionally been allocated mackerel quotas (Astthorsson et al. 2012).

Population and Community Dynamics

Climate change–mediated changes in the abundance and distributions of species will disrupt biotic interactions such as predation, parasitism, competition, and mutualism, with serious implications for community composition, biodiversity, and ecosystem functioning. Indirect effects of climate change, mediated through effects on intra- and interspecific interactions, may have more severe consequences than the direct effects of warming (Cahill et al. 2013, Ockendon et al. 2014).

Predator-Prey Dynamics

Mismatches in the predator-prey trophic system have been increasingly reported. For example, increasing temporal mismatch was reported for a trophic system of plants, caterpillars, four species of passerine birds that prey on caterpillars, and a raptor that preys on passerines (Both et al. 2009). Increasing temperatures contributed to annual advances of 0.17 days in budburst, 0.75 days in caterpillar emergence, and 0.36–0.5 days in passerine hatching dates. The raptor in this system showed no trend in phenology. Overall, the higher trophic levels (passerines and the raptor) showed weaker phenological responses than their prey, resulting in increasing asynchrony between the food demand and food availability.

A study by Hamilton et al. (2017) reported that the degree of spatial overlap, and hence the strength of the predator-prey relationship between polar bears (Fig. 6.6) and ringed seals (*Pusa hispida*), has

Figure 6.6   Polar bears in the southern Beaufort Sea subpopulation are coming to shore in higher numbers and for significantly longer than they were as recently as two decades ago because of sea ice loss. The mean length of time bears spent onshore between 1986 and 1999 was 20 days, whereas, between 2000 and 2014, it was 56 days (Atwood et al. 2016). Ryan Wilson, US Fish & Wildlife Service

declined following sea ice reduction. Polar bears were reported to instead spend more time feeding on eggs and chicks of ground-nesting birds such as snow goose (*Chen caerulescens*) (Rockwell et al. 2011), thus contributing to the emergence of novel predator-prey interactions. Another study reported that the implications of climate change for migratory caribou include increased predation pressure (Vors and Boyce 2009). Currently, predation is not a contributing factor to caribou population regulation, because the distributional range of potential predators does not generally overlap with the distribution of the caribou. The geographic overlap between wolf (*Canis lupus*) and migratory caribou, however, will increase if the boreal forest biome shifts poleward, resulting in significant caribou mortality on its winter range. Predation pressure might also increase for mammal species with seasonal coat color changes from brown to white, such as collared lemming (*Dicrostonyx groenlandicus*), long-tailed weasel (*Mustela frenata*), snowshoe hare (*Lepus americanus*), mountain hare (*Lepus timidus*), Arctic hare (*Lepus arcticus*), white-tailed jackrabbit (*Lepus townsendii*), or Siberian hamster (*Phodopus sungorus*). Mills et al. (2013) reported that the initiation dates of color change of the fall brown-to-white molt in snowshoe hare is influenced by photoperiod, rather than snow fall, which causes increased prey visibility in winters with delayed snow fall.

Predator-prey dynamics might also be altered because of changing population cycles. For example, population dynamics of lemmings (*Lemmus lemmus*) are partly influenced by winter weather conditions. Increasingly warm winters cause irregular peaks in the lemming cycles, affecting entire communities that are linked to these rodents across local and regional scales (Kausrud et al. 2008). In the United Kingdom, dampened population cycles of the field vole (*Microtus agrestis*) drastically reduced abundance of their predators, the tawny owl, potentially influencing local owl populations toward extirpation (Millon et al. 2014). In the US Upper Midwest, milder winters are predicted to promote widespread dampening of population cycling for ruffed grouse (*Bonasa umbellus*) (Pomara and Zuckerberg 2017).

Host-Parasite Dynamics

Similarly to predator-prey interactions, host-parasite interactions are likely to be affected by climate change through changes in phenology of parasite or host, changes in spatial distribution of parasite or host, changes in the prevalence and intensity of parasitism, and changes in virulence or antiparasite defenses of hosts (Merino and Moller 2010).

Increasing temperatures affect the timing of emergence of parasites and vectors (Mouritsen and Poulin 2002, Moller 2010). The parasitoid hippoboscid fly has advanced its phenology to emerge during egg laying and early incubation period of its host, the barn swallow (*Hirundo rustica*), rather than during late stages of the swallow breeding cycle (Moller 2010). These shifts in the hippoboscid fly phenology might result in lower reproductive success and recruitment of barn swallow. On the other hand, a long-distance migratory bird, the brood parasitic cuckoo (*Cuculus canorus*), will likely be affected by changes in migratory patterns of its host, short-distance migratory birds (Saino et al. 2011). Short-distance migrants have advanced the timing of their spring migration on average more than long-distance migratory birds, and are now starting their breeding season prior to cuckoo's arrival. This mismatch might result in lower opportunities for successful parasitism and, consequently, declines in cuckoo populations.

The prevalence of parasites will also change with increasing temperatures. The castor bean tick (*Ixodes ricinus*) has been increasing in Europe (Gilbert 2010) and is one of the primary vectors of zoonotic diseases, transmitting *Borrelia burgdorferi*, tick-borne encephalitis virus, and louping ill virus (Gilbert 2010). In the United States, *B. burgdorferi*, noted for causing Lyme disease, is carried by ticks of the

Ixodes genus (*I. dammini*). Gilbert (2010) reported that the abundance of *I. ricinus* in Scotland was negatively associated with elevation; elevation is often used as proxy for climatic conditions. As climate change continues and temperatures increase, *I. ricinus* is likely to become more abundant at higher elevations, which might result in higher prevalence of louping ill virus, because additional competent hosts (red grouse, *Lagopus lagopus scoticus*; and mountain hares) increase in abundance at higher elevations. Similar elevational shifts and increased densities of *I. ricinus* were reported over a 15-year study period in Sweden (Lindgren et al. 2000) and attributed to increasingly mild climatic conditions.

In the United States, moose population growth is influenced by winter ticks (*Dermacentor albipictus*). Winter ticks have undergone northward range expansion in response to landscape and climate change (Kutz et al. 2014) and mild winters favor higher tick survival (Musante et al. 2010). Consequently, mortality in wild ungulates, especially moose, has increased.

Inter- and Intraspecific Competition

By affecting phenology and spatial distributions, climate change will likely disrupt competitive interactions. Competition between short-distance migratory and resident birds might be disrupted if short-distance migrants advance the onset of their spring migration and return earlier to the breeding grounds (Forchhammer et al. 2002, Hubalek 2003), and if survival of birds wintering in Europe (the resident and short-distance migrants) is enhanced by increasingly mild winters (Lemoine and Böhning-Gaese 2003). This could leave long-distance migrants at a competitive disadvantage. Furthermore, shifts in species distributions might create a spatial overlap among species potentially competing for the same resource.

Climate change will alter reproductive phenology, including interspecific laying dates, potentially lead-

ing to altered competitive interactions among species (Ahola et al. 2007, Adamík and Král 2008). Species belonging to different taxonomic groups might also see their interactions affected.

Competitive inter- and intraspecific interactions among mammals will also be altered when climate changes. Populations of Arctic fox (*Vulpes lagopus*) declined along the southern range boundary of the Arctic tundra because of the invasion of red foxes (*Vulpes vulpes*), whose range shifted northward because of increasing temperatures. Different populations of migratory caribou might come in contact as a result of increased temperatures and earlier sea ice breakup (Sharma et al. 2009), leading to increased competition on the calving grounds.

Management and Conservation Actions in the Face of Climate Change

Climate change adaptation is a unique and difficult challenge for wildlife management (Mawdsley et al. 2009) because, to be effective, it must account for present and future needs and be robust to future climate change (Bonebrake et al. 2017). For many wildlife managers, the effects of climate change are viewed as a distant concern, and incorporating climate change effects into planning and management represents a distraction from seemingly more-pressing problems facing wildlife populations (habitat loss, invasive species management). This perspective is flawed and worrisome. The ecological, social, and economic consequences of climate change are unquestionable (Parmesan 2006, Pecl et al. 2017), and the role of wildlife managers in addressing them will only become increasingly important.

In 2014, a group of leaders in climate adaptation from federal and state agencies and nongovernmental organizations led by the National Wildlife Federation constructed a guideline document for designing and carrying out conservation in the face of climate change (Stein et al. 2014). They advocate for a climate-smart cycle, a generalized framework composed of discrete steps for climate-smart conservation. Below we discuss each of the cycle steps suggested by Stein et al. (2014). The steps in the climate-smart cycle are designed to work together, with each one building on previous ones and providing input to subsequent stages.

Step 1. Define Planning Purpose and Scope

Clearly defining the purpose and scope of conservation and/or management is crucial for designing effective conservation strategies. This step ensures that conservation and management plans are aligned with stakeholders' needs and can be accomplished with available resources. In this step, the wildlife managers should articulate their planning purpose, clarify existing conservation goals and objectives, identify conservation targets (species, communities, ecosystem services.), identify geographic and temporal scope, engage key stakeholders, and determine resource needs and availability (time, money, staff, expertise).

Step 2. Assess Climate Effects and Vulnerabilities

Effective adaptation strategies rely on understanding vulnerabilities associated with climate change. Vulnerability to climate change is defined as the extent to which a species, community, habitat, or ecosystem is susceptible to harm from climate change effects. The components of vulnerability (sensitivity, exposure, adaptive capacity) provide a useful framework for linking actions to effects. Sensitivity captures how the target species or system is likely to be affected, in terms of demography, physiology, or behavior, by climate variability. Exposure quantifies the magnitude of historic or future climate change affecting a given population. Finally, adaptive capacity refers to species' or system's ability to adapt to or cope with change.

Step 3. Review or Revise Conservation Goals and Objectives

All conservation and management goals should be climate informed, forward looking, and continually reevaluated to ensure their continued relevance or to identify the need for refinement. Four components of goals and objectives may help in conducting refinement of conservation goals: what (what is the conservation target: species, ecosystem), why (what is the intended outcome: an increase in population size of a vulnerable species), where (what is the relevant geographic scope), and when (what is the relevant time frame).

Step 4. Identify Possible Adaptation Options

In this step, managers identify a range of possible adaptation strategies and actions to reduce climate-related vulnerabilities. Clearly defined vulnerabilities (Step 2) serve as a basis for identifying possible adaptation options. The suggested adaptation strategies and actions should be based primarily on their purported ecological effectiveness and their potential for accomplishing the conservation goals and management objectives (Step 3).

Step 5. Evaluate and Select Adaptation Actions

All options identified in Step 4 can now be evaluated based on their effectiveness from ecological, social, technical, and financial perspectives. The selected actions should address imminent conservation challenges and advance long-term conservation goals. Four general criteria can be used to evaluate the alternative adaptation scenarios: how well the proposed options help accomplish conservation goals, how well the proposed options help achieve broader societal (social, cultural) goals, how feasible the proposed options are, and whether the proposed options are climate smart.

Step 6. Implement Priority Adaptation Actions

This step requires individual leadership, institutional commitment, and resources; its success depends on engaging diverse stakeholders early in the process. Unfortunately, uncertainties associated with climate change often prevent putting adaptation plans into action. Conservationists and wildlife managers, however, are familiar with uncertainty associated with environmental variability and have a diverse toolbox for incorporating it in natural resources management (Yoccoz et al. 2001, Williams et al. 2002). Managers should view climate change as a factor likely to exacerbate other sources of uncertainty affecting target species or systems, rather than as an entirely new source of uncertainty (Nichols et al. 2011). To implement adaptation actions, conservationists and wildlife managers must embrace the uncertainty associated with climate change.

Step 7. Track Action Effectiveness and Ecological Responses

Long-term monitoring of wildlife populations is a key component in an appropriate measurement of biodiversity change (Jarzyna and Jetz 2017) and as such is crucial to successful management in the face of climate change (Lepetz et al. 2009, Beever and Woodward 2011, Conroy et al. 2011). Indeed, the flexibility of existing monitoring programs will be tested by climate change. With many species shifting their range distributions and migratory patterns (Pecl et al. 2017), wildlife managers will be forced to reconsider their survey strata and boundaries. Shifting species' phenologies add another level of complication. Monitoring in the face of climate change will have to account for species that might be arriving to their breeding grounds earlier or leaving them later or not at all, singing or displaying at a different time, or using alternative food resources. Monitoring approaches should also be carefully designed to ensure that they properly guide needed adjustments in strategies and actions (Stein et al. 2014).

There may be more than one appropriate method or technique to completing the climate-smart cycle. In fact, multiple approaches or methods might be suitable for each step. The approaches should be selected based on their ability to accomplish the conservation goals and produce the expected outputs (Stein et al. 2014).

Summary

Despite the political and social debate surrounding the causes of climate change, the scientific evidence and conclusions are clear. The observed warming is human caused, and many species and ecosystems are adapting (or failing to adapt) to changing environmental conditions.

The consequences of climate change for wildlife include geographic redistribution of species, changes in species' phenology (timing of migration, breeding, hibernation), and even evolutionary adaptation. The different responses of species will result in changes in community composition and dynamics and, ultimately, restructuring of ecosystems.

Wildlife managers now face the challenge of adapting to the ongoing and future changes in climate and its consequences for wildlife. There is no single, optimal solution to managing wildlife (and their habitats) under climate change. Instead, methodologies must be implemented to identify likely consequences of climate change followed by developing and implementing management programs specific to an area or species. Even conservation objectives may require adjustment to account for the consequences of climate change.

LITERATURE CITED

Adamík, P., and M. Král. 2008. Climate- and resource-driven long-term changes in dormice populations negatively affect hole-nesting songbirds. Journal of Zoology 275:209–215.

Astthorsson, O.S., H. Valdimarsson, A. Gudmundsdottir, and G.J. Óskarsson. 2012. Climate-related variations in the occurrence and distribution of mackerel (*Scomber scombrus*) in Icelandic waters. ICES Journal of Marine Science 69:1289–1297.

Atwood, T.C., E. Peacock, M.A. McKinney, K. Lillie, R. Wilson, D.C. Douglas, S. Miller, and P. Terletzky. 2016. Rapid environmental change drives increased land use by an Arctic marine predator. PLOS One 11:e0155932.

Ball, J.P., G. Ericsson, and K. Wallin. 1999. Climate changes, moose and their human predators. Ecological Bulletins 47:178–187.

Baselga, A. 2012. The relationship between species replacement, dissimilarity derived from nestedness, and nestedness. Global Ecology and Biogeography 21:1223–1232.

Beever, E.A., and A. Woodward. 2011. Ecoregional-scale monitoring within conservation areas, in a rapidly changing climate. Biological Conservation 144:1255–1257.

Beever, E.A., C. Ray, J.L. Wilkening, P.F. Brussard, and P.W. Mote. 2011. Contemporary climate change alters the pace and drivers of extinction. Global Change Biology 17:2054–2070.

Berthold, P. 2001. Bird migration: A general survey. Second edition. Oxford University Press, Oxford, UK.

Bonebrake, T.C., C.J. Brown, J.D. Bell, J.L. Blanchard, A. Chauvenet, C. Champion, I.C. Chen, et al. 2017. Managing consequences of climate-driven species redistribution requires integration of ecology, conservation and social science. Biological Reviews 93:284–305.

Both, C., C.A.M. Van Turnhout, R.G. Bijlsma, H. Siepel, A.J. Van Strien, and R.P.B. Foppen. 2009. Avian population consequences of climate change are most severe for long-distance migrants in seasonal habitats. Proceedings of the Royal Society of London B: Biological Sciences 277:1259–1266.

Caruso, N.M., M.W. Sears, D.C. Adams, and K.R. Lips. 2014. Widespread rapid reductions in body size of adult salamanders in response to climate change. Global Change Biology 20:1751–1759.

Conroy, M.J., M.C. Runge, J.D. Nichols, K.W. Stodola, and R.J. Cooper. 2011. Conservation in the face of climate change: The roles of alternative models, monitoring, and adaptation in confronting and reducing uncertainty. Biological Conservation 144:1204–1213.

Dunn, P.O., and D.W. Winkler. 2010. Effects of climate change on timing of breeding and reproductive success in birds. Pages 113–128 *in* A.P. Moller, W. Fiedler, and P. Berthold, editors. Effects of climate change on birds. Oxford University Press, Oxford, UK.

Flack, A., W. Fiedler, J. Blas, I. Pokrovsky, M. Kaatz, M. Mitropolsky, K. Aghababyan, et al. 2016. Costs of migratory decisions: A comparison across eight white stork populations. Science Advances 2:e1500931.

Forchhammer, M.C., E. Post, and N.C.H.R. Stenseth. 2002. North Atlantic Oscillation timing of long- and short-distance migration. Journal of Animal Ecology 71:1002–1014.

Gao, X., C. Jin, D. Llusia, and Y. Li. 2015. Temperature-induced shifts in hibernation behavior in experimental amphibian populations. Scientific Reports 5:11580.

Gibbs, J.P., and A.R. Breisch. 2001. Climate warming and calling phenology of frogs near Ithaca, New York, 1900–1999. Conservation Biology 15:1175–1178.

Gilbert, L. 2010. Altitudinal patterns of tick and host abundance: A potential role for climate change in regulating tick-borne diseases? Oecologia 162:217–225.

Gwinner, E. 1996. Circannual clocks in avian reproduction and migration. Ibis 138:47–63.

Hamilton, C.D., K.M. Kovacs, R.A. Ims, J. Aars, and C. Lydersen. 2017. An Arctic predator-prey system in flux: Climate change impacts on coastal space use by polar bears and ringed seals. Journal of Animal Ecology 86:1054–1064.

Hansen, J., R. Ruedy, M. Sato, and K. Lo. 2010. Global surface temperature change. Reviews of Geophysics 48:4.

Hetem, R.S., A. Fuller, S.K. Maloney, and D. Mitchell. 2014. Responses of large mammals to climate change. Temperature: Multidisciplinary Biomedical Journal 1:115–127.

Hickling, R., D.B. Roy, J.K. Hill, R. Fox, and C.D. Thomas. 2006. The distributions of a wide range of taxonomic groups are expanding polewards. Global Change Biology 12:450–455.

Hoffmann, A.A., and C.M. Sgro. 2011. Climate change and evolutionary adaptation. Nature 470:479–485.

Hubalek, Z. 2003. Spring migration of birds in relation to North Atlantic Oscillation. Folia Zoologica 52:287–298.

Hurlbert, A.H., and Z. Liang. 2012. Spatiotemporal variation in avian migration phenology: Citizen science reveals effects of climate change. PLOS One 7:e31662.

Inouye, D.W., B. Barr, K.B. Armitage, and B.D. Inouye. 2000. Climate change is affecting altitudinal migrants and hibernating species. Proceedings of the National Academy of Sciences of the United States of America 97:1630–1633.

IPCC (Intergovernmental Panel on Climate Change). 2014. Climate change 2014: Synthesis report.

Contributions of Working Groups I, II, and III to the Fifth Assessment Report of the Intergovernmental Panel on Climate Change. Core Writing Team, R.K. Pachauri, and L.A. Meyer, editors. IPCC, Geneva, Switzerland. https://epic.awi.de/id/eprint/37530/1/IPCC_AR5_SYR_Final.pdf.

Jaffré, M., G. Beaugrand, É. Goberville, F. Jiguet, N. Kjellén, G. Troost, P.J. Dubois, A. Leprêtre, and C. Luczak. 2013. Long-term phenological shifts in raptor migration and climate. PLOS One 8:e79112.

Jarzyna, M.A., and W. Jetz. 2017. A near half-century of temporal change in different facets of avian diversity. Global Change Biology 23:2999–3011.

Jarzyna, M.A., W.F. Porter, B.A. Maurer, B. Zuckerberg, and A.O. Finley. 2015. Landscape fragmentation affects responses of avian communities to climate change. Global Change Biology 21:2942–2953.

Karell, P., K. Ahola, T. Karstinen, J. Valkama, and J.E. Brommer. 2011. Climate change drives microevolution in a wild bird. Nature Communications 2:208.

Kausrud, K.L., A. Mysterud, H. Steen, J.O. Vik, E. Ostbye, B. Cazelles, E. Framstad, et al. 2008. Linking climate change to lemming cycles. Nature 456:93–97.

Kovach, R.P., A.J. Gharrett, and D.A. Tallmon. 2012. Genetic change for earlier migration timing in a pink salmon population. Proceedings of the Royal Society B: Biological Sciences 279:3870–3878.

Kutz, S.J., E.P. Hoberg, P.K. Molnár, A. Dobson, and G.G. Verocai. 2014. A walk on the tundra: Host-parasite interactions in an extreme environment. International Journal for Parasitology: Parasites and Wildlife 3:198–208.

Lane, J.E., L.E.B. Kruuk, A. Charmantier, J.O. Murie, and F.S. Dobson. 2012. Delayed phenology and reduced fitness associated with climate change in a wild hibernator. Nature 489:554–557.

Lehikoinen, E., and T.H. Sparks. 2010. Changes in migration. Pages 89–112 in A. P. Moller, W. Fiedler, and P. Berthold, editors. Effects of climate change on birds. Oxford University Press, Oxford, UK.

Lemoine, N., and K. Böhning-Gaese. 2003. Potential impact of global climate change on species richness of long-distance migrants. Conservation Biology 17:577–586.

Lepetz, V., M. Massot, D.S. Schmeller, and J. Clobert. 2009. Biodiversity monitoring: Some proposals to adequately study species' responses to climate change. Biodiversity and Conservation 18:3185.

Lindgren, E., L. Tälleklint, and T. Polfeldt. 2000. Impact of climatic change on the northern latitude limit and population density of the disease-transmitting European tick Ixodes ricinus. Environmental Health Perspectives 108:119–123.

Lučan, R.K., M. Weiser, and V. Hanák. 2013. Contrasting effects of climate change on the timing of reproduction and reproductive success of a temperate insectivorous bat. Journal of Zoology 290:151–159.

Mawdsley, J.R., R. O'Malley, and D.S. Ojima. 2009. A review of climate-change adaptation strategies for wildlife management and biodiversity conservation. Conservation Biology 23:1080–1089.

Merino, S., and A.P. Moller. 2010. Host-parasite interactions and climate change. Pages 213–226 in A.P. Moller, W. Fiedler, and P. Berthold, editors. Effects of climate change on birds. Oxford University Press, Oxford, UK.

Millon, A., S.J. Petty, B. Little, O. Gimenez, T. Cornulier, and X. Lambin. 2014. Dampening prey cycle overrides the impact of climate change on predator population dynamics: A long-term demographic study on tawny owls. Global Change Biology 20:1770–1781.

Mills, L.S., M. Zimova, J. Oyler, S. Running, J.T. Abatzoglou, and P.M. Lukacs. 2013. Camouflage mismatch in seasonal coat color due to decreased snow duration. Proceedings of the National Academy of Sciences of the United States of America 110:7360–7365.

Moller, A.P. 2010. Host-parasite interactions and vectors in the barn swallow in relation to climate change. Global Change Biology 16:1158–1170.

Mouritsen, K.N., and R. Poulin. 2002. Parasitism, climate oscillations and the structure of natural communities. Oikos 97:462–468.

Moyes, K., D.H. Nussey, M.N. Clements, F.E. Guinness, A. Morris, S. Morris, J.M. Pemberton, L.E.B. Kruuk, and T.H. Clutton-Brock. 2011. Advancing breeding phenology in response to environmental change in a wild red deer population. Global Change Biology 17:2455–2469.

Musante, A.R., P.J. Pekins, and D.L. Scarpitti. 2010. Characteristics and dynamics of a regional moose Alces alces population in the northeastern United States. Wildlife Biology 16:185–204.

Nichols, J.D., M.D. Koneff, P.J. Heglund, M.G. Knutson, M.E. Seamans, J.E. Lyons, J.M. Morton, M.T. Jones, G.S. Boomer, and B.K. Williams. 2011. Climate change, uncertainty, and natural resource management. Journal of Wildlife Management 75:6–18.

Ockendon, N., D.J. Baker, J.A. Carr, E.C. White, R.E.A. Almond, T. Amano, E. Bertram, et al. 2014. Mechanisms underpinning climatic impacts on natural populations: Altered species interactions are more important than direct effects. Global Change Biology 20:2221–2229.

Ozgul, A., S. Tuljapurkar, T.G. Benton, J.M. Pemberton, T.H. Clutton-Brock, and T. Coulson. 2009. The dynamics of phenotypic change and the shrinking sheep of St Kilda. Science 325:464–467.

Ozgul, A., D.Z. Childs, M.K. Oli, K.B. Armitage, D.T. Blumstein, L.E. Olson, S. Tuljapurkar, and T. Coulson. 2010. Coupled dynamics of body mass and population growth in response to environmental change. Nature 466:482–485.

Parmesan, C. 2006. Ecological and evolutionary responses to recent climate change. Annual Review of Ecology, Evolution, and Systematics 37:637–669.

Parmesan, C., and G. Yohe. 2003. A globally coherent fingerprint of climate change impacts across natural systems. Nature 421:37–42.

Pecl, G.T., M.B. Araújo, J.D. Bell, J. Blanchard, T.C. Bonebrake, I.-C. Chen, T.D. Clark, et al. 2017. Biodiversity redistribution under climate change: Impacts on ecosystems and human well-being. Science 355:eaai9214.

Pomara, L.Y., and B. Zuckerberg. 2017. Climate variability drives population cycling and synchrony. Diversity and Distributions 23:421–434.

Post, E., and M.C. Forchhammer. 2008. Climate change reduces reproductive success of an Arctic herbivore through trophic mismatch. Philosophical Transactions of the Royal Society B: Biological Sciences 363:2367–2373.

Pulido, F., and P. Berthold. 2010. Current selection for lower migratory activity will drive the evolution of residency in a migratory bird population. Proceedings of the National Academy of Sciences of the United States of America 107:7341–7346.

Reading, C.J. 2007. Linking global warming to amphibian declines through its effects on female body condition and survivorship. Oecologia 151:125–131.

Réale, D., A.G. McAdam, S. Boutin, and D. Berteaux. 2003. Genetic and plastic responses of a northern mammal to climate change. Proceedings of the Royal Society of London, Series B: Biological Sciences 270:591–596.

Rockwell, R.F., L.J. Gormezano, and D.N. Koons. 2011. Trophic matches and mismatches: Can polar bears reduce the abundance of nesting snow geese in western Hudson Bay? Oikos 120:696–709.

Root, T.L., J.T. Price, K.R. Hall, S.H. Schneider, C. Rosenzweig, and J.A. Pounds. 2003. Fingerprints of global warming on wild animals and plants. Nature 421:57–60.

Row, J.R., P.J. Wilson, C. Gomez, E.L. Koen, J. Bowman, D. Thornton, and D.L. Murray. 2014. The subtle role of climate change on population genetic structure in Canada lynx. Global Change Biology 20:2076–2086.

Rowe, K.C., K.M.C. Rowe, M.W. Tingley, M.S. Koo, J.L. Patton, C.J. Conroy, J.D. Perrine, S.R. Beissinger, and C. Moritz. 2015. Spatially heterogeneous impact of climate change on small mammals of montane California. Proceedings of the Royal Society B: Biological Sciences 282:20141857.

Saino, N., R. Ambrosini, D. Rubolini, J. von Hardenberg, A. Provenzale, K. Hüppop, O. Hüppop, et al. 2011. Climate warming, ecological mismatch at arrival and population decline in migratory birds. Proceedings of the Royal Society of London B: Biological Sciences 278:835–842.

Sharma, S., S. Couturier, and S.D. Côté. 2009. Impacts of climate change on the seasonal distribution of migratory caribou. Global Change Biology 15:2549–2562.

Sheriff, M.J., C.T. Williams, G.J. Kenagy, C.L. Buck, and B.M. Barnes. 2012. Thermoregulatory changes anticipate hibernation onset by 45 days: Data from free-living arctic ground squirrels. Journal of Comparative Physiology B 182:841–847.

Stein, B.A., P. Glick, N. Edelson, and A. Staudt. 2014. Climate-smart conservation: Putting adaptation principles into practice. National Wildlife Federation, Washington, DC, USA.

Tape, K.D., D.D. Gustine, R.W. Ruess, L.G. Adams, and J.A. Clark. 2016. Range expansion of moose in arctic Alaska linked to warming and increased shrub habitat. PLOS One 11:e0152636.

Taylor, R.H., and P.R. Wilson. 1990. Recent increase and southern expansion of Adelie penguin populations in the Ross Sea, Antarctica, related to climatic warming. New Zealand Journal of Ecology 14:25–29.

Teplitsky, C., and V. Millien. 2014. Climate warming and Bergmann's rule through time: Is there any evidence? Evolutionary Applications 7:156–168.

Thackeray, S.J., T.H. Sparks, M. Frederiksen, S. Burthe, P.J. Bacon, J.R. Bell, M.S. Botham, et al. 2010. Trophic level asynchrony in rates of phenological change for marine, freshwater and terrestrial environments. Global Change Biology 16:3304–3313.

Tierson, W.C., G.F. Mattfeld, R.W. Sage, and D.F. Behrend. 1985. Seasonal movements and home ranges of white-tailed deer in the Adirondacks. Journal of Wildlife Management 49:760–769.

Todd, B.D., D.E. Scott, J.H.K. Pechmann, and J.W. Gibbons. 2010. Climate change correlates with rapid delays and advancements in reproductive timing in an amphibian community. Proceedings of the Royal Society B: Biological Sciences 278:2191–2197.

Towns, L., A.E. Derocher, I. Stirling, and N.J. Lunn. 2010. Changes in land distribution of polar bears in western Hudson Bay. Arctic 63:206–212.

Üveges, B., K. Mahr, M. Szederkényi, V. Bókony, H. Hoi, and A. Hettyey. 2016. Experimental evidence for beneficial effects of projected climate change on hibernating amphibians. Scientific Reports 6:26754.

Visser, M.E., and C. Both. 2005. Shifts in phenology due to global climate change: The need for a yardstick. Proceedings of the Royal Society B: Biological Sciences 272:2561–2569.

Vors, L.S., and M.S. Boyce. 2009. Global declines of caribou and reindeer. Global Change Biology 15:2626–2633.

Weishampel, J.F., D.A. Bagley, and L.M. Ehrhart. 2004. Earlier nesting by loggerhead sea turtles following sea surface warming. Global Change Biology 10:1424–1427.

Whitney, K.D., and C.A. Gabler. 2008. Rapid evolution in introduced species, "invasive traits" and recipient communities: Challenges for predicting invasive potential. Diversity and Distributions 14:569–580.

Williams, B.K., J.D. Nichols, and M.J. Conroy. 2002. Analysis and management of animal populations: Modeling, estimation, and decision making. Academic Press, San Diego, CA, USA.

Yoccoz, N.G., J.D. Nichols, and T. Boulinier. 2001. Monitoring of biological diversity in space and time. Trends in Ecology and Evolution 16:446–453.

Zuckerberg, B., A.M. Woods, and W.F. Porter. 2009. Poleward shifts in breeding bird distributions in New York State. Global Change Biology 15:1866–1883.

7 — Global Energy Sprawl
Scale and Solutions

Joseph M. Kiesecker
David E. Naugle

Introduction

By 2050, global energy consumption is expected to grow by 65% and global electricity demand is projected to nearly double (US Energy Information Administration 2017). This development will spur economic growth and improve quality of life, with 1.2 billion more people gaining access to electricity by 2030 (Casillas et al. 2010, Ouedrago 2013). But the growth of energy development and accompanying losses in wildlife will be enormous if we continue to extract energy in the same way that we do now. Glimpsing into our energy future, we are going to use more of every type of energy to fuel a growing and more affluent world population (Kiesecker and Naugle 2017). Climate change remains a serious concern and is characterized by many as the defining conservation challenge of this century. Efforts to reduce greenhouse gas emissions, the bulk of which are from energy-related activities, will strongly shape the future of energy development.

The world is shifting from traditional carbon energy sources to a mix of renewable energies (International Energy Agency 2016). Concomitant with this shift are the firsthand trade-offs that accompany it; namely, the additional human infrastructure associated with renewable sources (McDonald et al. 2009, Wise et al. 2009, Dale et al. 2011). Without careful planning, we could trade one crisis, climate change, for another, land use change and its associated conflicts (Kiesecker and Naugle 2017; Fig. 7.1).

This means that regardless of the energy development path we choose—business as usual or renewable resources instead of fossil fuels—the resulting energy footprint will continue to challenge us (Kiesecker and Naugle 2017). Now we are at a crossroads. We already have solid scientific evidence that shows the cumulative effects from energy development to the environment (Naugle 2011, Jones et al. 2015). New tools to mitigate those effects are available, but we have yet to realize their benefits. And we will not fully realize those benefits until decision makers make best practices the most common practices. Unfortunately, this issue has not received the attention it deserves.

In this chapter on global energy development and wildlife, we synthesize recent work on energy sprawl solutions (Kiesecker and Naugle 2017). First, we present an overview of global energy patterns and development scenarios that makes clear the scale of the challenge of sustainable development. Next we highlight case studies from around the world showing best-practice approaches to energy development that could be replicated. Finally, we close with a simple

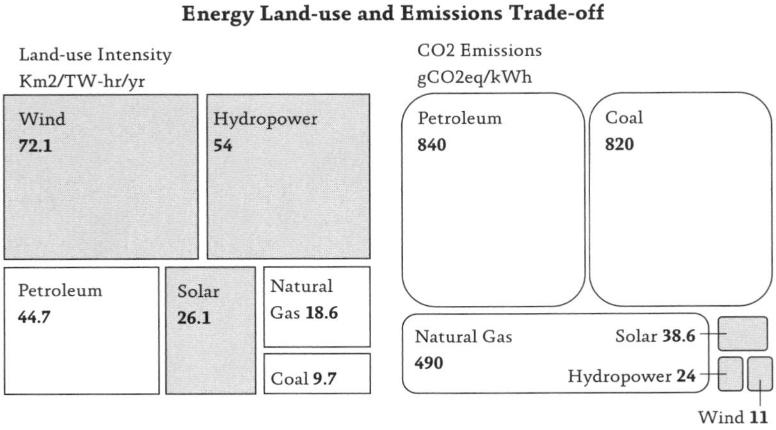

Energy Land-use and Emissions Trade-off

Land-use Intensity
Km2/TW-hr/yr

CO2 Emissions
gCO2eq/kWh

Wind **72.1**	Hydropower **54**

Petroleum **44.7**	Solar **26.1**	Natural Gas **18.6**
		Coal **9.7**

Petroleum **840**	Coal **820**

Natural Gas **490**   Solar **38.6**
Hydropower **24**
Wind **11**

Figure 7.1   Comparison of the energy footprints in terms of land required in kilometers squared per terawatt hours per year (km2/terawatt-hr/yr) and CO_2 emissions in grams of carbon dioxide equivalent per kilowatt hours (gCO_2eq/kWh). Land footprint adapted from McDonald et al. 2009, and CO_2 emissions data adapted from Schlömer et al. 2014. Land footprints and CO_2 emissions can vary greatly for different fossil fuel and renewable energy sources, but for simplicity, they were averaged across energy types. Given the difficulty of displaying graphically in comparison with other forms of energy, we intentionally excluded biofuels in the figure, given their large land-use intensity, which ranged from ~280 km2/terawatt-hr/yr for ethanol made from sugarcane to ~900 f km2/terawatt-hr/yr for biodiesel made from soy.

blueprint for turning best practice into common practice for future developments.

Energy Sprawl and the Global Geographies of Risk

To understand how and where future energy development might happen, we must first understand our inventory of potential energy resources. The single clear and consistent message about the amount of land we will need to devote to energy development on earth is that it is a lot; so much so that many scientists and most of the general public fail to comprehend the sheer magnitude of change that is upon us.

All energy sources have effects that create wastes during their production cycles, use nonrenewable resources during manufacturing or production, and take up space that converts land use to generate energy. Not surprisingly, different types of energy vary widely in their land footprint. Many of the renewable energy sectors and also newly established unconventional petroleum sources that are desperately needed to combat climate change have very large spatial footprints (McDonald et al. 2009; see Fig. 7.1). As human populations grow and demand for energy soars in developing countries, something will have to give. The larger energy footprint will inevitably lead to trade-offs with land use and greater risks to wildlife. Here we do not review specific impacts that future energy development will create but instead rely on syntheses from others who have compiled data regarding these impacts (Naugle 2011, Jones et al. 2015). In general, as energy development advances, we can expect negative effects on wildlife from habitat loss and fragmentation; direct mortality (from turbines, roads, powerlines); noise pollution (from drill rigs, construction); light pollution (substations, flares); spread of invasive species; loss of carbon stocks; increased impervious surfaces and resulting changes in hydrology; and altered water consumption (from dust suppression and fracking operations). Armed with this general understanding of the negative effects of energy development, we have a starting point from which we can manage these effects in a way that can meet demands but also reduce negative consequences for wildlife. That said, research on the consequences of land-based energy development is not necessarily proportionate to existing energy reserves or capacity. For example, Jones et al. (2015) found that most studies that examined wildlife effects were conducted in North America and Europe, whereas the new development frontiers are in developing countries, especially in the tropics (Oakleaf et al. 2015, 2017; Fig. 7.2).

Assessing Cumulative Risk

If we hope to achieve a sustainable balance between development and conservation, we need to proactively identify lands at risk of conversion and strategically plan to mitigate future effects. But striking this balance is possible only if we first understand where and how future development may occur. In this first-of-its-kind analysis, Oakleaf et al. combined nine potential development threats to identify where current natural lands are at future risk of conversion or modification (Oakleaf et al. 2015, 2017). First, spatial patterns of expected energy threats were aggregated for conventional and unconventional oil and gas, coal, solar, wind, hydropower, and biofuels; these were then merged with non-energy threats from mining, urbanization, and agricultural expansion to produce a global cumulative development threat map. Then, high cumulative development threat was overlaid with current natural areas to identify natural areas at future risk. The analysis tells us that our future world could look wildly different than it does today. In all, 20% of the world's remaining natural lands could be developed by midcentury, negating most or all of the wildlife values they cur-

rently support. South America's conversion of natural land to working land is predicted to double, while Africa's is projected to triple. Already-developed regions will not be immune to further development; people and wildlife living in Europe, Central America, and Southeast Asia will cope with a land-use intensity they have never before experienced. Future development could push half the world's biomes to more than 50% converted, and all biomes except boreal forest and tundra could lose over 25% of their natural lands (Oakleaf et al. 2015). And only 5% of natural lands that are at the highest risk of development are currently under legal protection. That 5% number is really a key finding, because it tells us two important things: how vulnerable these critical places are, and how we have to think about new solutions beyond traditional protection. These are sobering statistics, but this is really an opportunity to get ahead of this curve, to bring conservation science to the decision-making forefront, and to break new ground for wildlife conservation as a central part of future development.

Global Energy Patterns

Energy expansion is likely to be globally dispersed, but there are sector-specific geographic patterns (Oakleaf et al. 2017; Fig. 7.3). To a large degree, conventional oil and gas and coal will consist largely of intensification in regions that are currently high producers. Over two-thirds of the world's coal reserves and 75% of the world's coal production occurs in the United States, Russia, China, Australia, and India, and these five top-producing countries are likely to continue as the major coal producers. Although coal production and use is expected to decline in response to global commitments to reduce carbon dioxide emissions, unless countries make aggressive shifts, projections suggest coal will remain a staple of energy use for several decades. Like coal production, oil and gas development will mainly consist of intensification of the basins already producing oil or gas. Offshore oil and gas may be one exception, where

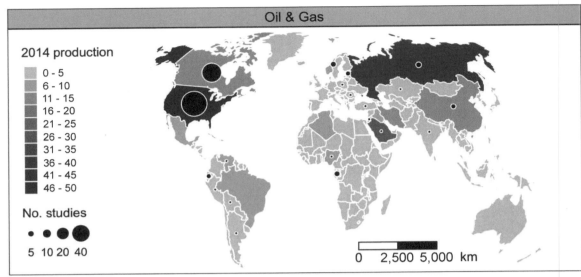

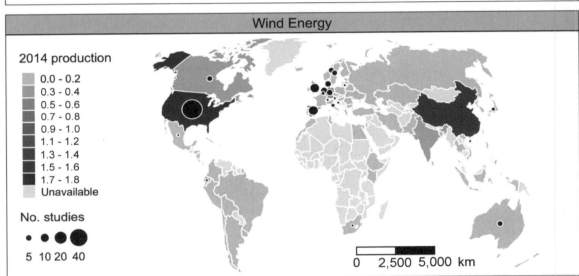

Figure 7.2   The number of published, peer-reviewed studies on the effects of oil, gas, and wind energy on selected indicators of biodiversity and ecosystem services by country, relative to 2014 annual production in quads (1 quad = 1 × 1015 British Thermal Units). Energy production data are from US Energy Information Administration (https://www.eia.gov/beta/international/). Figure adapted by J. Smith from Jones et al. 2015

resources remain unexploited or are in early stages of development.

The energy types that will likely have the largest expansion into new areas are renewable energy and unconventional oil and gas. The combination of technological advancements, including horizontal drilling and hydraulic fracturing, has spurred a rapid in-

crease in unconventional petroleum production during the last decade. Most development has centered in the United States, with over a million fracking events (Patterson et al. 2017), but several other countries, including Argentina, Russia, Mexico, Australia, and China, will soon be in a similar position. With renewable energy (hydropower, solar, wind)

Figure 7.3   Examples of energy sprawl. Upper left: View of coal extraction in the Cesar Valley of Colombia, South America (© Joseph Kiesecker). Upper right: Aerial photo of the sunpower facility in Rosamond, California, USA, with wind farm in the background (© Dave Lauridsen for The Nature Conservancy). Lower left: Aerial photograph showing natural gas well site and fracking operation with related road and pipeline infrastructure in northeastern Pennsylvania, USA, forest (Mark Godfrey © The Nature Conservancy). Lower middle: Aerial view of the Elk River Wind Project near the small town of Beaumont, in the southern Flint Hills region of Kansas, USA (© Jim Richardson). Lower right: A natural gas drilling platform at the mouth of Mobile Bay, Alabama, USA, with an extension of Dauphin Island with brown pelicans in the foreground (© Carlton Ward Jr.).

estimated to surpass coal as the largest supplier of electricity by 2040, it collectively has the greatest potential for expansion (International Energy Agency 2016), but siting will be a challenge given the land-use intensity of all renewable energy sectors (Kiesecker et al. 2011a). Not surprising, the same countries in southeast Asia, Africa, and South America that happen to contain many of the longest free-flowing or undammed rivers are most threatened by future hydropower development. Solar power, consisting of two main types of technologies—concentrating solar power and photovoltaics—currently makes up less than 1% of global electricity demand, but it is the fastest growing renewable electricity sector, projected to see a thirtyfold increase in capacity by 2040. A large portion of the globe is suited for solar development, with much of Africa, the Middle East, India, Mexico, portions of Brazil and Chile, and the southwestern United States having high potential and providing ample opportunity to site development in low-impact areas. Wind power generation currently meets only 4% of the global electricity demand, but it has more than doubled in the last 5 years and is projected to see a tenfold increase in capacity by 2040. United States, China, and Eastern Europe are currently experiencing the greatest growth in wind development, but South America, East Africa, New Zealand, and Japan also have high potential for future wind development. With rising energy demands and volatile oil prices, many countries have sought out alternative liquid fuels such as biofuels. Production of biofuels produced largely from corn or sugarcane or also from soybean or palm oil production is projected to double or triple in areas by 2040. Land conversion is expected to be highest in tropical South America and Africa and in select areas in Southeast Asia (Indonesia and Malaysia). With these types of global patterns in mind, we turn our attention to case studies that illustrate how negative effects of energy production have been successfully reduced.

Case Studies for Turning Best Practice into Common Practice

Given the human benefits of development, massive energy expansion and production is inevitable. Undoubtedly reduced consumption and increased energy efficiency will need to be part of the solution, but society must quickly recognize and subsequently embrace the inevitability of accelerated development as the new normal. To achieve better outcomes for wildlife and conservation, we need a better approach to planning energy developments. Without a better approach, the story will remain the same: a company quietly acquires land rights, plans its individual project, and afterward complies with wildlife laws and statutes in piecemeal fashion. The ensuing lawsuits slowly develop, deplete conservation and industry resources that could be put to better use, and most times do not safeguard the natural resources. Moreover, this piecemeal approach will likely slow production of renewable energy, which is critical for combatting climate change. Regional planning changes the old piecemeal method of development by inverting the timing and scale at which wildlife and other societal values are considered by industry (Kiesecker et al. 2010, 2011a, 2013). In this planning scheme, industry considers the potential effects of multiple projects before they have been individually planned and are already moving toward implementation. Regional planning looks at the bigger geographic picture, with individual projects planned collectively to avoid, reduce, and then mitigate undue negative effects. Regional rather than fragmented planning results in more effective conservation outcomes, reduces regulatory hurdles for industry, and provides cost savings to conservation and industry.

The case studies presented here advance next-generation planning concepts using best-practice examples from around the world and are intended to help readers envision how to successfully blend a landscape vision with a mitigation hierarchy for a sustainable future. The following case studies all thus: first, avoid or reduce effects on landscapes with irreplaceable wildlife or ecosystem service values; then ensure that effects are minimized and restored onsite using the best available technology; and, finally, offset residual impacts. Scenario modeling is an important component of regional planning because it allows decision makers to envision consequences of their development choices and proactively prevent problems before they occur. The following eight case studies represent the major ecosystems of the world that are experiencing development from every major energy sector. We cite the literature from which we synthesize each case study within their respective subheadings.

Solar Energy Zones in the United States (Cameron et al. 2012, 2017)

Rapid growth in installed solar power capacity poses challenges for land-use change. Policies and funding have successfully stimulated new projects, but permitting has delayed approvals, and insufficient attention has been paid to effects on wildlife and other natural resources. In 2012, the US Bureau of Land Management, in line with comprehensive assessments of the value of lands carried out by The Nature Conservancy, adopted a landscape approach to accelerate utility-scale solar energy development on public lands. The landscape plan, which identifies solar energy zones, applies to a six-state region and assesses the potential deployment of solar energy development over the next 20 years and its direct, indirect, and cumulative effects. Since then, 19 solar energy development zones have been identified, and large areas of the region have been designated as off-limits for development because it would not be the best use of public lands. This approach has encouraged cross-agency collaboration and has cut project permitting time from an average of 18–24 months down to 10. Reliable data on wildlife values has enabled better decisions about the siting, boundaries, and desirability of solar projects.

Regional Planning for Landscapes, Wildlife, and Culture in Mongolia (Heiner et al. 2013, Kiesecker et al. 2018)

Mongolia's vast landscapes are rich with coal, mineral, and wind energy resources. How this resource development takes place will greatly affect the economy, nomadic livelihoods, and environment of Mongolia. Potential for conflict over competing claims for land looms large, requiring that a delicate balance be struck between development and conservation. Recognizing this challenge, the government of Mongolia has supported landscape-scale planning for the entire country in support of sus-

tainable development and has enacted an enlightened planning framework into law. They have embraced proactive planning, which accounts for biological resources, ecosystem services, climate change, and projected development. This framework for identifying priority conservation areas and recommended practices at the outset is allowing the country to better minimize negative consequences and more effectively manage environmental and economic risks. Plans are helping guide project siting and mitigation and the establishment of new protected areas that support nomadic livelihoods and wildlife. The approach includes a commitment to protect 30% of the country. To date, real progress has been made on that commitment. Since 2012, the government has designated 150,000 square kilometers of new protected land, an area approximately the size of Nepal.

Developing Oil Resources in Venezuela's Marine Ecosystems (Klein et al. 2017)

Venezuela is one of the most biologically diverse countries in the world, and its southern Caribbean basin is a center of extraordinary marine biodiversity. It is also the source of the country's plentiful and economically vital offshore oil resources. Risks to marine ecosystems from oil production are well known, but comprehensive planning, as frequently used in terrestrial developments, has rarely been seen in plans for the development of marine environments. Working with Venezuela's national oil company, The Nature Conservancy and partners helped develop conservation-based standards and practices for inclusion in permits for each oil and gas lease. This partnership shows how collaboration among conservation organizations, petroleum developers, and environmental regulators can work. In this case and others, it is vital to identify priority conservation areas and recommend good practices at the start. Where development must proceed, likelihood of negative ecological consequences can be minimized by sticking to guidelines, policies, and contractual requirements. By helping energy companies engage all parties in avoiding risk from the outset, decisions can be made while there is still room to maneuver, so that ecological values are not irretrievably lost and livelihoods are not negatively affected.

Power of Rivers: Securing Sustainable Energy and Healthy Rivers (Opperman et al. 2011, 2015, Opperman 2017)

Hydropower, the world's largest, most mature, and most reliable source of low-carbon renewable energy, will need a massive overhaul on how projects manage social and environmental performance. Experience has shown that poorly planned hydroprojects can have enormous negative environmental and social effects. Future forecasts also highlight that 70% of new hydropower expansion is expected to occur in river basins that support the highest diversity of fish species and high fish productivity. Without success in reducing environmental and social conflicts, hydropower may not reach its potential contribution to sustainable energy systems. Examples around the world show that system-scale approaches to planning and management, rather than a project-by-project approach, can help solve challenges associated with hydropower development (Winemiller et al. 2016). A systemic approach allows government planners and regulators to explore, compare, and choose the best alternative to meet multiple objectives across a full river basin (Ziv et al. 2012). In China, the United States, Brazil, and Mexico, analysis has demonstrated how changes to river basin management could have reduced negative effects on fish species and human populations while having minimal effect on power generation capacity. Such system-scale approaches can also avoid project cancellations, delays, and cost overruns. Rather than assessing individual sequential projects, system planning and modeling can help governments identify a set of infrastructure investments that deliver broader benefits to both people and nature.

Conserving Sage Grouse in the United States and Canada (US Fish and Wildlife Service 2017)

Because of threats from energy development, cultivation, and catastrophic wildfire, invasive annual grasses, and conifer invasion, the greater sage-grouse (*Centrocercus urophasianus*) of western North America and southern Canada was considered a candidate species for listing under the US Endangered Species Act (ESA). Against this backdrop of precipitous decline, a proactive, coordinated, voluntary, and incentive-based effort provided a clear roadmap for landscape-scale conservation of the species and its ecosystem. Strategic conservation features prominently in this story because user groups and state and federal governments wished to avoid ESA listing and worked together to reduce threats. Regional planning and spatial prioritization provided the foundation for partners to target where to focus conservation efforts, balancing energy needs and habitat protection. As the primary land managers, the federal Bureau of Land Management and the United States Forest Service completed the bulk of the public land policy protections. But significant investment was also made by the Natural Resources Conservation Service–led Sage Grouse Initiative to conserve and enhance habitat within 1,500 private working ranches encompassing an area equivalent to twice the size of Yellowstone National Park. The sage-grouse story to date illustrates how a proactive, coordinated, landscape-scale conservation approach has the potential to maximize biodiversity conservation and maintain energy development. And it provides a real-world model for including wildlife conservation across all development planning.

Colombian Communities: Accounting for Ecosystem Services in Energy Development (Tallis et al. 2011)

In countries across Latin America, the competing demands of energy growth and environmental protections are often considered in the context of a third challenge: indigenous peoples and local community rights. Historically in this region, energy development has not always respected indigenous peoples' rights, and indigenous groups have in some cases borne the bulk of environmental loss and associated human consequences. These factors combine to create a complicated context for decision making. In Colombia, The Nature Conservancy has worked with the government to encourage the incorporation of ecosystem service concepts into environmental impact assessments and mitigation action relating to mining projects, developments that are important contributors to energy resourcing and economic well-being. Using the concept of servicesheds, areas that provide a specific ecosystem service benefit to a specific group of people, stakeholders were able to reveal potentially harmful loss of ecosystem services in advance of their harm (Tallis and Polasky 2009). This concept allowed land planners, developers, and policy makers to consider alternatives that let energy development progress rapidly without creating damaging inequalities. The serviceshed method provided a replicable, robust way to inform energy development, to identify how proposed development may affect people, and to determine whether possible ecosystem service losses create or worsen inequalities (Tallis et al. 2015).

Biofuels of Brazil: Sustaining Production Expansion and Environmental Quality (Kennedy et al. 2016a)

Biofuels have been embraced as a promising alternative to oil, because in principle they can reduce carbon emissions, enhance domestic energy security, and revitalize rural economies. But biofuel feedstock production can use up to a thousand times more land per unit energy than natural gas, oil, coal, wind, hydropower, or solar. By 2040, global biofuel production may double or triple, requiring an additional 44 to 118 million hectares of land, expansion that would exceed a land area equivalent to the state of Texas.

How this future land conversion occurs will have a major impact on wildlife, ecosystems, and the climate, and on challenges such as food production (Kennedy et al. 2016b). Kennedy et al. (2016a,b) investigated the potential for targeted land-use planning to achieve commodity production and environmental goals in the Brazilian Cerrado, the world's most diverse tropical savannah. They applied spatial optimization techniques to map marginal service values, assess economic and environmental trade-offs, and find efficient land-use patterns. The results demonstrate that biofuels production and environmental goals can be met with landscape planning, even in a biodiversity and agricultural hotspot. Legal, voluntary, and market mechanisms, if effectively implemented and widely adopted by agricultural producers in affected landscapes, hold promise of delivering large-scale conservation outcomes in the expanding biofuels sector.

Proactive Planning for Shale Gas in South Africa (Scholes et al. 2016)

The South African government has made high-level commitments to shale gas exploration and, pending successful exploration that yields viable hydrocarbon deposits, the government will likely consider development of those resources at a significant scale. In 2015, to place the South African Department of Environmental Affairs and other regulatory bodies in a position to make decisions in a timely and responsible fashion, the Council for Scientific and Industrial Research, in partnership with the National Biodiversity Institute, initiated a proactive planning approach to identify development areas while minimizing negative environmental, social, and economic impacts. The planning process traverses Eastern, Northern, and Western Cape provinces and includes 27 local municipalities encompassing 171,811 square kilometers. While this is still a work in progress, it is novel because of both the scale and timing of the assessment. The large scale of the assessment will provide flexibility to assess trade-offs in various alternative development designs and coordinate infrastructure development across multiple developers. By conducting this assessment well in advance of development proposals from industry, regulators can better design concessions that ensure environmental and social needs are on equal standing with economic values.

Three Simple Solutions for Curbing the Energy Footprint

We close by introducing tools in hand that, if applied broadly, can put us on a path to sustainability. Here are three things we can do today to better balance global development and wildlife conservation.

Proactive Landscape-Scale Mitigation

Given the scale of potential future energy development, society needs to adapt the way we plan for, regulate, and mitigate negative impacts (Kiesecker and Naugle 2017). Legally protected areas might not be able to steer development away from sensitive areas. Land-use planners must improve existing tools and create new approaches to address pending negative consequences. Environmental Impact Assessments (EIAs) are one of the chief tools currently used to mitigate impacts from energy development. EIAs, used with impact mitigation, are a systematic process for examining the environmental consequences of planned developments (McKenney and Kiesecker 2010, Villarroya et al. 2014). These tools also emphasize prediction and prevention of environmental damage through the application of the mitigation hierarchy: avoid, minimize, restore, or offset. Mitigation tools are conventionally implemented through a narrow spatial lens, however, at project or site levels that often result in uncoordinated, piecemeal mitigation that fails to deliver conservation outcomes at relevant ecological scales. On the basis of our assessment, we propose a shift in regulatory oversight with an eye toward regional-

scale, cumulative impact assessments and proactive mitigation planning that better accounts for future development threats from multiple sectors.

As development encroaches into more remote and previously undisturbed wildlife habitats, corporations, governments, development banks, and civil society groups must collaborate to avoid and minimize future negative consequences on remaining natural areas. To change future trajectories, we propose that environmental licensing, impact mitigation, and financing should target where development could affect significant proportions of natural areas. Mitigation requirements should include procedures for proactively evaluating the compatibility of proposed development with conservation goals to determine when negative effects should be avoided and when development can proceed. Given the expansive scale of expected impacts from a variety of sectors, developers will need to compensate for residual effects using biodiversity offsets.

Biodiversity offsets, also known as set-asides, compensatory habitat, or mitigation banks, can maintain or enhance environmental assets in situations where development is moving forward despite negative impacts. The same regional plans that can help guide appropriate sites for development can also be used to guide the siting of offsets so that the most ecologically important and at-risk areas are secured.

Accept Trade-Offs from Renewables

The spatial area required to generate energy from renewables exceeds that of traditional carbon sources. Now that we have acknowledged this, we can address it. Likewise, we hope this chapter enables readers to shed their denial or guilt about elevated consequences resulting from society's shift to renewable energy. Unless this shift happens, we will never reduce the far greater effects of climate change. It is bad news that wind and solar require more land than fossil fuels, but one of the great advantages of most renewable sources is that we can put them just about anywhere. Some places may be sunnier or windier than other places, but renewables are not as tightly linked to a specific patch of ground as fossil fuels, and there are more than enough good places for siting renewables from which to pick and choose. For example, an analysis we completed shows that the United States has 14 times the amount of degraded land needed to meet the goal of 20% of electricity generated by wind energy established by the US Department of Energy (Kiesecker et al. 2011b).

Policy and Enabling Conditions

Achieving sustainability hinges in part on our ability to break down long-held regulatory and financial barriers to enable master planning. Society needs more examples of pathways that help facilitate this transition. This should include a library of practices and policies that countries can adapt or insert into national-level policy. We encourage more cases studies showing how multi-objective planning across sectors can provide social and economic benefits to developing countries rich in natural resources. The environmental community should also help simplify the transition to renewable energy, especially for countries willing to make necessary changes. To help facilitate the proliferation of renewable energy, the conservation and academic communities will need to alter their current approach. The standard conservation approach has been to use data on species and ecological systems to identify the best places for new nature reserves and protected areas. But if we are to simultaneously address climate change and protect wildlife, we will need to be just as data intensive and analytical about anticipating where development will occur (Kiesecker et al. 2018). In other words, at the same time we site protected areas we should also provide guidance on the siting of renewable energy, removing one of the hurdles that might slow development.

Summary

Development, energy development in particular, will continue to influence land-use change and imperil wildlife populations around the world. On the positive side, 1.2 billion more people in developing countries are expected to gain access to electricity by 2030 thanks to infrastructure projects (International Energy Agency 2013). Conservationists should not stand in the way of securing these basic needs, but should instead seek pragmatic ways of achieving them while also maintaining wildlife and associated conservation values (Kiesecker and Naugle 2017). Ideas and case studies we have presented here suggest that, in many places around the world, opportunities remain for development (wind farms, solar fields, roads, etc.) amid opportunities for protected areas—as long as enough foresight and data are applied to identify what areas are no-go zones and what areas are zoned for development. We recognize that there is an alternative narrative for conservation in which economic growth and development are seen as a curse and thus rejected altogether (Kingsnorth 2013). And although to some, the idea that conservation and development are incompatible is appealing (Kingsnorth 2013), that position is not compatible with success in the face of the coming energy, food, and material demands required to meet the population-growth projections of our twenty-first-century world. We also recognize that this pragmatic approach is not without its risks. For example we do not know what level of development, if crossed, will cause irreversible loss of biodiversity. Compromises also requires trade-offs that allow development to move forward in some places while conservation gains are accrued in different places. How do we track local gains and losses in a way that ensures that no wildlife species ends up with a net loss across the entire landscape? Answering these questions will require a mix of modeling and detailed studies that carefully uncovers species and ecosystem responses to development that can be guided with structured decision making (Lyons et al. 2008, Game et al. 2013).

LITERATURE CITED

Cameron, D.R., B.S. Cohen, and S.A. Morrison. 2012. An approach to enhance the conservation-compatibility of solar energy development. PLOS One 7:e38437.

Cameron, D.R., L. Crane, S.S. Parker, and J.M. Randall. 2017. Solar energy development and regional conservation planning. Pages 67–75 in J.M. Kiesecker and D.E. Naugle, editors. Energy sprawl solutions. Island Press, Washington, DC, USA.

Casillas, C.E., and D.M. Kammen. 2010. The energy-poverty-climate nexus. Science 330:1181–1182.

Dale, V.H., R.A. Efroymson, and K.L. Klineet. 2011. The land use-climate change-energy nexus. Landscape Ecology 26:755–773.

Game, E.T., P. Kareiva, and H.P. Possingham. 2013. Six common mistakes in conservation priority setting. Conservation Biology 27:480–485.

Heiner, M., Y. Bayarjargal, J. Kiesecker, D. Galbadrakh, N. Batsaikhan, G. Munkhzul, I. Odonchimeg, et al. 2013. Identifying conservation priorities in the face of future development: Applying Development by Design in the Mongolian Gobi. The Nature Conservancy. http://www.nature.org/media/smart-development /development-by-design-gobi-english.pdf.

International Energy Agency. 2013. World energy outlook 2013. https://www.iea.org/newsroom/news/2013 /november/world-energy-outlook-2013.html.

International Energy Agency. 2016. World energy outlook 2016. http://www.worldenergyoutlook.org /weo2016/.

Jones, N.F., L. Pejchar, and J.M. Kiesecker. 2015. The energy footprint: How oil, natural gas, and wind energy affect land for biodiversity and the flow of ecosystem services. BioScience 65:290–301.

Kennedy, C.M., P.L. Hawthorne, D.A. Miteva, L. Baumgarten, K. Sochi, M. Matsumoto, J.S. Evans, et al. 2016a. Optimizing land use decision-making to sustain Brazilian agricultural profits, biodiversity, and ecosystem services. Biological Conservation 204:221–230.

Kennedy, C.M., D.A. Miteva, L. Baumgarten, P.L. Hawthorne, K. Sochi, S. Polasky, J.R. Oakleaf, E.M. Uhlhorn, and J. Kiesecker. 2016b. Bigger is better: Improved nature conservation and economic returns from landscape-level mitigation. Science Advances 2:e1501021.

Kiesecker, J.M., and D.E. Naugle, editors. 2017. Energy sprawl solutions: Balancing global development and conservation. Island Press, Washington, DC, USA.

Kiesecker, J., H. Copeland, A. Pocewicz, and B. McKenney. 2010. Development by design: Blending landscape level planning with the mitigation hierarchy. Frontiers in Ecology and Environment 8:261–266.

Kiesecker, J.M., H.E. Copeland, B.A. McKenney, A. Pocewicz, and K.E. Doherty. 2011a. Energy by design: Making mitigation work for conservation and development. Pages 159–181 in D.E. Naugle, editor. Energy development and wildlife conservation in western North America. Island Press, Washington, DC, USA.

Kiesecker, J.M., J.S. Evans, J. Fargione, K. Doherty, K.R. Foresman, T.H. Kunz, D. Naugle, N.P. Nibblelink, and N.D. Niemuth. 2011b. Win-win for wind and wildlife: A vision to facilitate sustainable development. PLOS One 6:e17566.

Kiesecker, J.M., K. Sochi, M. Heiner, B. McKenney, J. Evans, and H. Copeland. 2013. Development by design: Using a revisionist history to guide a sustainable future. Pages 495–507 in S.A. Levin, editor. Encyclopedia of biodiversity, second edition. Academic Press, Waltham, Massachusetts, USA.

Kiesecker, J.M., K. Sochi, J. Evans, M. Heiner, C.M. Kennedy, and J.R. Oakleaf. 2018. Conservation in the real world: Pragmatism does not equal surrender. Pages 152–158 in P. Kareiva, M. Marvier, and B. Silliman, editors. Effective conservation science: Data not dogma. Oxford University Press, Oxford, UK.

Kingsnorth, P. 2013. Dark ecology: Searching for truth in a post-green world. Orion Magazine, January/February. http://www.orionmagazine.org/index.php/articles/article /7277.

Klein, E., J. J. Cardenas, R. Martinez, J.C. Gonzalez, J. Papadakis, K. Sochi, and J. M. Kiesecker. 2017. Planning for offshore oil. Pages 77–87 in J.M. Kiesecker and D.E. Naugle, editors. Energy sprawl solutions: Balancing global development and conservation. Island Press, Washington, DC, USA.

Lyons, J.E., M.C. Runge, H.P. Laskowski, and W.L. Kendall. 2008. Monitoring in the context of structured decision-making and adaptive management. Journal of Wildlife Management 72:1683–1692.

McDonald, R.I., J. Fargione, J. Kiesecker, W.M. Miller, and J. Powell. 2009. Energy sprawl or energy efficiency: Climate policy impacts on natural habitat for the United States of America. PLOS One 4:e0006802.

McKenney, B.A., and J.M. Kiesecker. 2010. Policy development for biodiversity offsets: A review of offset frameworks. Environmental Management 45:165–176.

Naugle, D.E., editor. 2011. Energy development and wildlife conservation in western North America. Island Press, Washington, DC, USA.

Oakleaf, J.R., C.M. Kennedy, S. Baruch-Mordo, P.C. West, J.S. Gerber, L. Jarvis, and J. Kiesecker. 2015. A world at risk: Aggregating development trends to forecast global habitat conversion. PLOS One 10:e0138334.

Oakleaf, J., C.M. Kennedy, S. Baruch-Mordo, and J. Kiesecker. 2017. Geography of risk. Pages 9–19 in J.M. Kiesecker and D.E. Naugle, editors. Energy sprawl solutions. Island Press, Washington, DC, USA.

Opperman, J. 2017. Sustainable energy and healthy rivers. Pages 113–124 in J.M. Kiesecker and D.E. Naugle, editors. Energy sprawl solutions. Island Press, Washington, DC, USA.

Opperman, J., J. Royte, J. Banks, L.R. Day, and C. Apse. 2011. The Penobscot River, Maine, USA: A basin-scale approach to balancing power generation and ecosystem restoration. Ecology and Society 16:7.

Opperman, J., G. Grill, and J. Hartmann. 2015. The power of rivers: Finding balance between energy and conservation in hydropower development. Final report, The Nature Conservancy. https://www.nature.org/media/freshwater/power-of-rivers-report.pdf.

Ouedraogo, N.S. 2013. Energy consumption and economic growth: Evidence from the economic community of West African States (ECOWAS). Energy Economics 36:637–647.

Patterson, L.A., K.E. Konschnik, H. Wiseman, J. Fargione, K.O. Maloney, J. Kiesecker, J. Nicot, et al. 2017. Unconventional oil and gas spills: Risks, mitigation priorities, and state reporting requirements. Environmental Science and Technology 51:2563–2573.

Schlömer, S., T. Bruckner, L. Fulton, E. Hertwich, A. McKinnon, D. Perczyk, J. Roy, et al. 2014. Annex III: Technology-specific cost and performance parameters. Pages 1329–1356 in Climate change 2014: Mitigation of climate change, Working Group III contribution to the Fifth Assessment Report of the Intergovernmental Panel on Climate Change 2014. Cambridge University Press, Cambridge, UK.

Scholes, R., P. Lochner, G. Schreiner, L. Snyman-Van der Walt, and M. de Jager, editors. 2016. Shale gas development in the Central Karoo: A scientific assessment of the opportunities and risks. CSIR/IU/021MH/EXP/2016/003/A, ISBN 978-0-7988-5631-7.

Tallis, H., and S. Polasky. 2009. Mapping and valuing ecosystem services as an approach for conservation and natural-resource management. Annals of the New York Academy of Sciences 1162:265–283.

Tallis, H., S. Wolny, and J.S. Lozano. 2011. Including ecosystem services in mitigation. Report to the Colombian Ministry of the Environment, Mines and Territorial Development. Natural Capital Project: Stanford. http://www.naturalcapitalproject.org/wp-content/uploads/2017/05/Including-Ecosystem-Services-in-Mitigation.pdf.

Tallis, H., C.M. Kennedy, M. Ruckelshaus, J. Goldstein, and J.M. Kiesecker. 2015. Mitigation for one and all: An integrated framework for mitigation of development impacts on biodiversity and ecosystem services. Environmental Impact Assessment Review 55:21–34.

US Energy Information Administration. 2017. International energy outlook 2017. https://www.eia.gov/outlooks/aeo/.

US Fish and Wildlife Service. 2017. Greater sage-grouse 2015 Endangered Species Act finding. https://www.fws.gov/greaterSageGrouse/findings.php.

Villarroya, A., A.C. Barros, and J.M. Kiesecker. 2014. Policy development for environmental licensing and biodiversity offsets in Latin America. PLOS One 9:e107144.

Winemiller, K.O., P.B. McIntyre, L. Castello, E. Fluet-Chouinard, T. Giarrizzo, S. Nam, I.G. Baird, et al. 2016. Balancing hydropower and biodiversity in the Amazon, Congo, and Mekong. Science 351:128–129.

Wise, M., K. Calvin, A. Thomson, L. Clarke, B. Bond-Lamberty, R. Sands, S.J. Smith, A. Janetos, and J. Edmonds. 2009. Implications of limiting CO2 concentrations for land use and energy. Science 324:1183–1186.

Ziv, G., E. Baranb, S. Namc, I. Rodríguez-Iturbed, and S.A. Levin. 2012. Trading-off fish biodiversity, food security, and hydropower in the Mekong River Basin. Proceedings of the National Academy of Science of the United States of America 109:5609–5614.

8

SAMANTHA M. WISELY

Wildlife Disease Management in the Global Context

Introduction

Diseases have been a subject of wildlife management since the early twentieth century, when wildlife management focused on the conservation and production of game species (Leopold 1947). As the focus of management has broadened in the twenty-first century to include ecosystem and biodiversity management, macro- and micro-parasites may best be viewed as an intrinsic part of the biological diversity that composes ecosystems. By including an ecological view of pathogens and parasites in trophic cascades and biological networks of ecosystems (Hatcher and Dunn 2011), scientists can better understand the role that pathogens play in regulating ecosystem function and the services that ecosystems provide (De Vos et al. 2016).

Despite their natural role in the world, pathogens and the diseases they cause can have enormous effects on human and animal health, and action must sometimes be taken. In these cases, management attempts to reduce the biological or economic effects of disease. Integral to wildlife disease management is the management of human perception of disease risk. Cultural perceptions can create social conflict about how to manage disease. If risks are over- or underemphasized, resulting management actions can have adverse effects on sympatric species and the habitats in which they live. This chapter focuses on the evolving role wildlife management has taken in disease management (Table 8.1), why wildlife diseases are receiving increased attention, how the concept of wildlife management has evolved in the twenty-first century, and the contemporary methods used to manage wildlife diseases. Wildlife disease management is a global issue, and this chapter draws on examples of wildlife disease outbreaks and management actions from around the world to illustrate these concepts.

The Case for Wildlife Disease Management

Threats to Wildlife

When diseases threaten wildlife populations, domestic species, or humans, management action is typically taken. Although wildlife diseases can be a natural part of the ecosystem, epizootics in already endangered or threatened species can push a species to extinction. For example, avian malaria has been implicated in the extinction of 16 endemic bird species on the Hawaiian Islands, and pandemics of Ranavirus and chytrid fungus have been implicated

Table 8.1 Glossary of terms used in wildlife disease management and ecology

Term	Definition
Pathogen	A biological agent that causes clinical illness.
Disease	The clinical illness caused by a disorder of function or structure often brought on by a pathogen.
Pandemic	A global outbreak of disease.
Epizootic	A large-scale outbreak of an animal disease.
Endemic	An endemic host is a plant or animal that is locally adapted to a region. An endemic pathogen is a pathogen that is locally present.
Serology	A diagnostic assay often used to detect host antibodies to a particular pathogen.
qPCR	A diagnostic assay used to find specific sequences of DNA. Often used to identify the presence or absence of a host species or of a particular pathogen in a host as inferred by the presence of their nucleic acid.
Antibody	A protein produced by a host that chemically neutralizes a pathogen's antigenic site.
Reproductive rate	R_0, the basic reproductive number of a parasite, measured as the number of secondary infections arising from a primary infection. If R_0 is >1, the pathogen will spread in the host population.
Reservoir species	A host species in which a pathogen is endemic and is transmitted from individual to individual.
Keystone species	A species that has an outsized effect on the environment compared with its relative biomass in the ecosystem.
Zoonotic	Having a wildlife origin.
Vector	Typically an arthropod species that serves as a host of a pathogen and is capable of transmitting the pathogen to other species.

as disease stressors contributing to the global collapse of amphibian species (MacPhee and Greenwood 2013). Disease is the primary threat of extinction to Tasmanian devils (*Sarcophilus harrisii*) from an unusual infectious facial tumor (McCallum 2007). From an ecosystem management perspective, if the affected host is a keystone species, the pathogen and disease syndrome can endanger entire ecosystems by altering community composition and ecosystem function. For example, prairie dogs (*Cynomys* spp.) are considered a keystone species in the grasslands of central North America. They dig extensive burrow systems that create habitat for many rodent species and are a mainstay prey item of many mammalian and raptor predators. Sylvatic plague became established in the Great Plains in the mid-twentieth century and has caused high levels of mortality (85%–100%; Cully et al. 2010). Cascading effects occur where prairie dog colonies have been extirpated from the landscape. Declines in prairie dog populations cause decreases in rodent abundance and diversity (Collinge et al. 2008) and the collapse of a population of a main predator of prairie dogs, the

black-footed ferret (*Mustela nigripes*; Gage et al. 2005). Thus, this bacterial pathogen appears to cause a cascading effect on the vertebrate community composition in this grassland ecosystem.

In addition to managing disease in protected wildlife species, disease in game species that decreases either population size or quality of trophy specimens often triggers a management response. The rapid spread of chronic wasting disease (CWD) across the United States triggered many state wildlife agencies to close state borders to the importation of white-tailed deer in the hope of containing the spread (CWD Alliance, http://cwd-info.org/). In Europe, the precipitous decline of European rabbits caused by the emergence of two diseases, myxomatosis and rabbit hemorrhagic disease, in the last half of the twentieth century, triggered a slew of management activities including bag limits, a controversial live vaccine, translocations, and habitat management (Williams et al. 2007).

Ironically, some wildlife management practices, such as artificial feeding, exacerbate disease outbreaks but are difficult to curtail because of intense public support. Feedgrounds established in Wyoming at the beginning of the twentieth century relieve damage by elk (*Cervus canadensis*) to standing hay crops by providing winter forage, but have allowed brucellosis, pasteurellosis, necrotic stomatitis, and other diseases to persist in this population (Dean et al. 2004). Other forms of supplementary feeding, such as baiting stations that are used to increase hunting opportunities, are legal and popular in many parts of the United States, but have been suggested to proliferate CWD and bovine tuberculosis (Williamson 2000).

When wildlife is threatened with extinction from a disease outbreak, but that disease does not affect people or domestic animals, agencies that manage wildlife usually take the lead in managing the outbreak and the affected species. The International Union for Conservation of Nature (IUCN) provides international assistance on wildlife health and disease to the Convention on International Trade in Endangered Species of Wild Fauna and Flora (CITES), which, in addition to being an international agreement on wildlife trade, tackles pressing wildlife issues. For example, from 2015 to 2017, more than 200,000 Saiga antelope (*Saiga tatarica*) died from multiple wildlife epidemics (ProMED-mail 2017). Massive, undiagnosed die-offs from disease prompted the Wildlife Health Specialist Group of the IUCN to pen a white paper encouraging world governing bodies to allow timely movement of diagnostic specimens by fast-tracking or waiving endangered species (CITES) permits for diagnostic specimens so that diagnosis, treatment, and management can be fast-tracked (IUCN 2016). When global cooperation is not dictated, it is typically the national and provincial or state government agencies that lead management efforts on more localized wildlife disease outbreaks.

Threats to Livestock

Some wildlife diseases also affect domestic animals such as livestock and companion animals. Indeed, 79% of reportable livestock diseases are of wildlife origin (Miller et al. 2013). These disease outbreaks can have a huge economic burden on industry, government, and the public. For example, foot and mouth disease is caused by a virus that originated in wild suids and bovids in Africa. Globally, it has cost tens of billions of dollars to detect, treat, and control (Shwiff et al. 2016). Other diseases of livestock that originate in wildlife include rinderpest, bovine tuberculosis, African swine fever, and brucellosis (Kilpatrick et al. 2009). The H5N1 avian influenza virus has cost the poultry industry in Asia, Africa, and Europe US$10 billion through direct loss of animals or through massive culling to contain the spread of disease (FAO-OIE 2005, 2007). Migrating waterfowl are the reservoir species of avian influenza, which is transmitted to poultry around the world, causing severe pandemics periodically (Gilbert et al. 2006).

Many wildlife species act as reservoirs for pathogens transmitted into domestic species. Feedback loops can then be established, such that spillback from domestic animals to wildlife occurs, and these diseases become very difficult to contain or eradicate. Bovine tuberculosis is a global animal disease that is extremely difficult to control, given its propensity to co-circulate among domestic ruminants, wild ruminants, and other wildlife species, including European badgers (Gordejo and Vermeersch 2006). In the European Union, multiple countries have been managing for bovine tuberculosis in both domestic and wild animal populations for 40 years and have not achieved disease-free status (ECOA 2016). Management responses include enhanced biosecurity to prevent domestic animal and wildlife contact, vaccination, and culling of both domestic herds and wildlife (Schiller et al. 2010).

International organizations such as the Food and Agriculture Organization (FAO) of the United Nations and the World Organisation for Animal Health (OIE) often set the research and response agendas for global outbreaks or regional outbreaks of global importance, while national agriculture ministries provincial agricultural departments, and livestock in lustries work to contain local to regional disease issues.

Threats to Humans

Some of the most high-profile wildlife diseases are those that threaten human health. Worldwide, 43% of human infectious diseases originate in wildlife (Jones et al. 2008). Often the wildlife reservoir is a food item (hunting or farmed wildlife), and thus transmission occurs from direct handling of affected animals. Tularemia contracted from rabbits, brucellosis contracted from wild pigs, and HIV/AIDS from great ape bushmeat (Karesh and Noble 2009) have contributed to disease transmission from wildlife to humans. Oftentimes, the wildlife reservoir species transmits the pathogen to livestock herds, where it circulates and is eventually transmitted to humans. Bats have been implicated in transmitting multiple viruses to livestock or other food animals, which are then transmitted to and circulate in people. Examples include severe acute respiratory syndrome (SARS) virus transmission to palm civets (*Paradoxurus hermaphroditus*) in China (Li et al. 2005); Middle East respiratory syndrome (MERS) virus to camels in Saudi Arabia (Azhar et al. 2014); Nipah virus to domestic pigs in Malaysia (Montgomery et al. 2008); and Hendra virus to horses in Australia (Field et al. 2001). In each case the virus was transmitted from bats to livestock, where it circulated and exposed humans to the virus. In most of these cases, the virus then mutated to transmit directly from human to human. Domestic companion animals such as dogs and cats have also been implicated as intermediate host species in the transmission of disease from wildlife to humans. Rabies, leptospirosis, and toxoplasmosis have all been transmitted to humans from companion animals. The World Health Organization (WHO) of the United Nations is the primary international organization that directs and coordinates responses to human disease emergence globally. They coordinate with national and provincial public health ministries to coordinate responses within a country.

Global Change and the Link to Wildlife Diseases

Emerging infectious diseases (EIDs) are diseases that have increased in distribution or incidence and have never been described before or have recently re-emerged (Daszak et al. 2001). Since 1940 there has been a steady climb in the number of emerging infectious diseases that infect humans globally, and this increase cannot be explained merely by the increase in surveillance or diagnostic capacity (Jones et al. 2008). Forty-three percent of all infectious diseases of humans are of wildlife origin, and 75% of emerging infectious diseases have a wildlife origin. Zoonotic viruses and protozoa are the most likely to become EIDs, and zoonotic pathogens are twice as likely as non-zoonotic pathogens to become an EID (Taylor et al. 2001). There is no one cause for this increase in EIDs, but rather multiple, synergistic explanations for why wildlife diseases are emerging in wildlife, domestic animals, and humans. Given the rapid increase of zoonotic diseases, wildlife science is increasingly being used to understand these diseases, and wildlife management is often used to control its emergence. We explore the reasons for the proliferation of wildlife diseases below.

The Intrinsic Nature of Pathogens

The nature of some pathogens makes them amenable to emergence. Some pathogens have low host specificity and as such threaten entire faunal guilds with loss of biodiversity. The entire global amphibian fauna is threatened with extinction from the dermal fungus (*Batrachochytrium dendrobatidis*; Bd, chytrid fungus). This fungus thickens the skin of amphibians, which impedes the exchange of water, air, and electrolytes. It has been found on every continent except Antarctica (Olson et al. 2013). Similarly, another emerging fungal disease, white-nosed syndrome, has caused a tenfold decrease in the abundance of hibernating bats and threatens this group with extinction (Frick et al. 2015). Although the fungus is also present in Europe, the pathology and high mortality associated with American bats has not been observed in Europe (Puechmaille et al. 2011), but it has been proposed that the observed pattern of increased resistance in Europe is due to historical exposure to the fungus and elimination of susceptible species prior to the discovery of white-nosed syndrome in North America.

Globalization

In addition to the intrinsic features of pathogens, extrinsic influences also cause the emergence of pathogens. Globalization has increased air travel, trade, and the transportation of goods, which increases the frequency and expands the distribution of infected humans, animals, and arthropod vectors (Fig. 8.1). Air travel rapidly moves people from one continent to another and has facilitated the human-to-human spread of such zoonotic diseases as SARS (Mangili and Genreau 2005) and influenza (Grais et al. 2004). The rapid spread of Zika virus throughout the Americas is attributed to air travel by infected humans (Turrini et al. 2016). Transportation of animals, either wildlife for the exotic pet trade or domestic pets, has also facilitated the emergence of wildlife diseases. Ranavirus is a water-borne pathogen of fish and amphibians that contributes to the global collapse of amphibians. The global aquaculture industry of bullfrogs has facilitated the rapid spread of this virus (Claytor et al. 2017), along with chytrid fungus. More than 100,000 tons of frog legs are shipped each year around the world (Gratwicke et al. 2010). Regional transportation of amphibians such as salamanders for fish bait has further spread the disease within countries (Jancovich et al. 2005, Picco et al. 2008). The exotic pet trade can also facilitate the spread of disease to humans. In 2003, three family members contracted monkeypox virus from their pet prairie dog. Epidemiological investigation revealed that this prairie dog had been housed with exotic rodents from Africa, which are known hosts to monkeypox virus and the likely source of the pathogen into the United States.

Exotic Invasive Species

When exotic species are introduced to a nonnative area and become established, they are alien invasive

Figure 8.1   For vector-borne diseases, surveillance includes surveying for arthropod vectors. Here, a technician is dragging for questing ticks in a communal cattle pasture in Eswatini, Africa. Tick-borne diseases are a major threat to livestock health in Sub-Saharan Africa and affect the livelihoods of people reliant on financial and social capital from their herds. Kimberly Ledger

species. Twenty-five of the 100 worst alien invaders are linked to diseases in native wildlife (Hatcher et al. 2006, 2012). Many exotic invasive species are intentionally released as unwanted pets (Burmese pythons, *Python bivittatus*, and Gambian rats, *Cricetomys gambianus*, in Florida), as exotic game (nilgai, *Boselaphus tragocamelus*, in Texas), or as biocontrol (mongoose in the Caribbean islands). But often, the transportation of pathogen hosts and vectors is unintentional. These stowaways can be disease-vectoring mosquitos or other arthropods, or fish and amphibians in ballast water, or rodents in the cargo holds of ships (Hing et al. 2016).

When invasive species are reservoir hosts to pathogens, they create pathogen pollution in the native habitat. These exotic pathogens can have devastating effects on naïve populations of native wildlife that have never been exposed to these exotic pathogens and thus are highly susceptible. For example, in Great Britain, poxvirus is carried by invasive gray squirrels (*Sciurus carolinensis*) and transmitted to the endangered red squirrel (*Sciurus vulgaris*). The disease caused by the squirrel poxvirus has contributed to the precipitous decline of red squirrels in Great Britain. Although the virus appears to have no detrimental effects on gray squirrels, it is nearly always lethal in red squirrels. In North America and Australia, wild pigs are an exotic invasive species that are increasing in abundance and spreading in distribution. They are host to multiple exotic pathogens such as pseudorabies virus and *Brucella suis*, and are a potential reservoir for several foreign animal diseases. African swine fever, classical swine fever, and hoof and mouth disease are not currently in the United States or Australia. If wild pigs became reservoirs for any of these diseases, it would devastate the swine and/or beef industries in these countries (Pedersen et al. 2013). Pseudorabies, a herpesvirus of swine carried by wild pigs in the United States, is the third leading cause of death in the endangered Florida panther (*Puma concolor coryi*; Glass et al. 1994).

Habitat Degradation

Habitat alterations and land-use change have also been linked to the emergence of wildlife diseases. For example, declines in water quality due to the increase in nutrient load can increase pathogen abundance (Johnson et al. 2010). Vector-borne pathogens such as West Nile virus and *Plasmodium* (the causative agent of malaria) can increase because nutrient-rich environments tend to favor larval development of many arthropod vectors. Amphibians may also be more susceptible to parasite infections because of an increase in intermediate hosts and parasites. Nematode infections of wading birds have also been higher in nutrient-rich environments (Johnson et al. 2010). Oftentimes, large-scale land alterations bring humans into closer contact with wildlife and the diseases they carry. Road construction in African tropical forests is thought to have increased hunting pressure and therefore human contact with wildlife. Two human zoonoses, HIV/AIDS and Ebola virus, are thought to have originated in wildlife but were transmitted to humans who hunted great apes and bats, respectively. These pathogens quickly evolved to be capable of human-to-human transmission. In South America, the clear-cutting of rain forests for cattle pastures has dramatically increased the abundance of vampire bats (Desmodontinae), a vector of rabies throughout the Americas (Jones et al. 2013). Smaller-scale changes to land use can also affect disease emergence. Urban composting and point-source feeding of wildlife can increase disease transmission via increased contact rates and population density in urban landscapes (Murray et al. 2016).

This increased contact between humans and wildlife has not just led to increased human diseases; human diseases have also affected wildlife. Leprosy in armadillos (Dasypodidae) originated with humans in North America. As armadillo populations have grown, transmission is now from armadillos to humans (Sharma et al. 2015). Ecotourism, research, and hunting are responsible for outbreaks of disease in great apes (Epstein and Price 2009). In humans, paramyxovirus causes measles, and herpes simplex virus 1 causes cold sores and genital herpes. Both pathogens have been observed in great ape populations that are in close contact with human populations and cause much more severe disease in these nonhuman primate species (Epstein and Price 2009).

Global Climate Change

Climate change has been implicated in the latitudinal and altitudinal expansion of numerous vector-borne diseases. This progression has been particularly well documented in Europe, where bluetongue virus (causing hemorrhagic disease in ruminants including deer) and its vector, the midge (Culicoides spp.), have increased their northward distribution by 800 km (Purse et al. 2005). Similarly, ixodid ticks (Ixodes spp.) in North America have expanded into Canada, bringing with them Lyme disease (Ogden et al. 2008). In addition to expanding the distribution of vectors, increased temperatures can also decrease the generation time of arthropod vectors and increase the bite rate, both of which can increase the spread of a pathogen (Siraj et al. 2017). Increased temperatures can also increase the efficiency of viruses and other pathogens to grow inside the vector species (Gage et al. 2008). But climate change does not just change temperature; increases in extreme weather events and changes to precipitation have been documented to influence not only arthropod vectors, but also the distribution of other reservoir species, including rodents and bats (Mills et al. 2010). For example, increased rainfall increases seed production, which in turn increases the abundance of rodents, facilitating the increase of hantavirus in people in the United States, Africa, and Europe (Klempa 2009). Changes to climate can also alter vertebrate host responses to pathogens by increasing physiological stress and lowering immune response, which increases host susceptibility and ultimately increases pathogen prevalence (Hing et al. 2016). It is hypothesized that climate-related decreases in food

availability of fruit bats (Megachiroptera) in Australia increased physiological stress, decreased immune function, and increased viral shedding rates of the Hendra virus. This cascading effect is likely an important influence of ongoing outbreaks of the disease in horses and humans (Plowright 2008).

Evolving Wildlife Disease Management Strategies

In the mid-1980s, wildlife management practitioners called for a science-based approach to management (Romesburg 1981, McNab 1983). Today, effective wildlife management is based on science in its decision-making processes, implementation, and evaluation. Wildlife disease management has undergone a similar epiphany more recently (Joseph et al. 2013). Any wildlife disease is a complex system of interacting biotic and abiotic agents that can interact in deterministic and stochastic ways. Thus, outcomes are not always reliably predictable, yet managers must be able to define and ultimately accept certain levels of uncertainty to implement management plans. Using decision-making frameworks to manage diseases requires a four-step system: define the problem, enumerate options, implement a chosen strategy, and evaluate outcomes so that the strategy can be continued, modified, or terminated (Orton et al. 2011). This framework should also advance scientific understanding of disease ecology.

Complicating this process are the numerous agencies often involved in wildlife disease management. This is particularly true if the disease affects human or livestock health or if the breadth of the outbreak is transboundary. International, national, and regional agencies may all need to coordinate to respond to an event. Recognition that public health officials, medical doctors, veterinarians, and wildlife managers may each view a wildlife disease issue through a very different professional lens was the impetus for One Health, an initiative that works to find common ground and interface among these disciplines (Zinsstag et al. 2011). This more holistic approach to the identification and management of diseases continues to evolve and is an active scientific discipline unto itself (Fig. 8.2; Davis et al. 2017).

Despite the evolving strategies and disciplines, the end goal can be only one of four options: prevention, control, eradication, or doing nothing. Prevention assumes that some baseline knowledge of the system exists, and a proactive approach to preventing the spread or emergence of a disease is the goal. Biosecurity at borders that seeks to repel infected hosts from entering an area is an example of a management strategy that might achieve the goal of prevention. Control of a pathogen requires the recognition that a pathogen exists in the managed system and that eradication is not feasible or practical. Control implies that the pathogen is kept at a current

or diminished level of prevalence. Eradication seeks to eliminate a pathogen from the defined landscape. This can be a costly or impossible enterprise for most systems unless the geographic boundaries are small or extremely well defined (insular systems). Each of these goals requires disease surveillance to determine whether these goals are being met. Finally, the do-nothing approach may be the most desirable when the pathogen is sublethal, or when chronic, system-wide infections dominate a community.

Once a goal has been set, designing the path forward requires some baseline knowledge of the system (defining the problem). Then strategies can be defined and options weighed. Consideration of which option to take includes evaluation of the efficacy of the method, the cost of the method, and acceptability of the proposed method to the general public. Methods generally target one or more players in the epidemiological system: the disease agent, the host, the vector, or the environment. The discussion that follows outlines basic strategies, based on the target of management.

Management of the Disease Agent

Some management strategies are based on manipulating or managing the pathological agent (Wobeser 2002, Blancou et al. 2009). The pathogen may be amenable to control by reducing its reproductive rate, such that prevalence declines over time (Hatcher and Dunn 2011, for a full epidemiological treatment of reproductive rate and management of wildlife diseases). For some directly transmitted pathogens with stable antigenic regions on which the host immune system can act, this may be amenable. Worldwide, vaccination has been used as a strategy to control rabies, either through oral bait vaccines, as is the case in the United States (Fig. 8.3), or via trap, vaccinate, and release campaigns in Australia, Africa, and Europe. Vaccination has been used or considered for use to control bovine tuberculosis in opossums in New Zealand (Skinner et al. 2005) and wild boar in Iberia (Beltrán-Beck et al. 2012), sylvatic

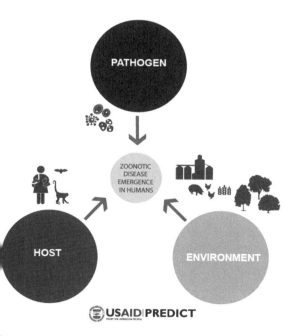

Figure 8.2   An example of One Health in action, the USAID Predict program recognizes the importance of incorporating information about the environment, host, and pathogen ecology to understand the causes and predict outbreaks of infectious diseases globally. The project uses pathogen surveillance, virus discovery, and global health capacity-building as tools to achieve the overall goal using a One Health platform to address and mitigate global outbreaks of infectious diseases.

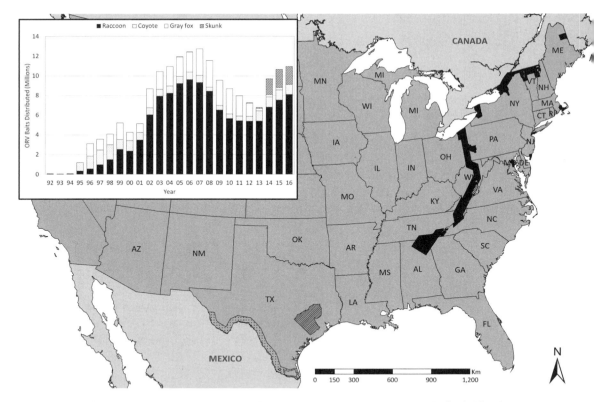

Figure 8.3   Oral rabies vaccines presented in baits have been used to prevent the spread of wildlife rabies into naïve wildlife populations and to reduce rabies transmission in enzootic areas in the United States since 1992. The graph presents the number of oral vaccines delivered by species target since 1992. The map indicates the different oral rabies vaccination (ORV) zones in the United States in 2016. Raccoon rabies is enzootic along the eastern United States. Aerial distribution of vaccine baits along the Appalachian Mountains has created an ORV zone that has prevented the appreciable spread of raccoon rabies west. Along the Texas-Mexico border, an ORV zone has virtually eliminated canine rabies in foxes and coyotes. The experimental skunk ORV zone in Texas ceased in 2017.

plague in prairie dogs (Rocke et al. 2010), and brucellosis in elk in Wyoming (Roffe et al. 2004). Vaccination of wildlife has had varying degrees of success at managing wildlife diseases. Efficacy in mounting a long-lasting immune response, the cost and ability to deliver the vaccine, and the ability to vaccinate enough hosts to lower the reproductive rate of the pathogen are some of the hurdles that must be overcome using this strategy.

Managing the Host

As with managing disease agents, some host-management strategies may be based on reducing the number of infected hosts to reduce the reproductive rate. Management practices have included reducing overall population density, reducing the infected population, isolating uninfected individuals from infected ones if geographic isolation is possible, or artificially selecting for resistant animals (McCallum 2008). By changing the host range, density, or contact with susceptible hosts, these management strategies work to decrease the transmission of pathogens from infected to susceptible individuals (De Vos et al. 2016).

Understanding the specific host ecology of a system is imperative for effective management. An example that illustrates the importance of understanding social behavior of hosts comes from the bovine tuberculosis system in Great Britain. The bacterial pathogen circulates in a number of wildlife hosts but is prevalent in European badgers (*Meles meles*). A management strategy of culling badgers was implemented but ultimately found to be ineffective. It was later determined that the social nature of badgers (they live in social groups) was disrupted by culling, such that individuals ranged more widely and interacted more frequently with other social groups when members of their group had been culled. This change in behavior increased the transmission potential of tuberculosis in badgers and, ultimately, in cattle (McDonald et al. 2008).

Managing the Environment

Still other management strategies may be based on modifying the environment to reduce vector or host habitat to break or reduce the cycle of transmission (Ward et al. 2009). For pathogens such as the causative agent of anthrax (*Bacillus anthracis*), disinfecting watering holes in Southern Africa can reduce the pathogen load in the environment (Berry 1993). For arthropod vectors that have aquatic larvae, such as mosquitos, controlling the amount of standing water can reduce mosquito abundance, and application of insecticide has been practiced for over 100 years around the world (Lacey and Lacey 1990). Host populations can also be manipulated to reduce transmission rates by modifying the environment. Food sources that congregate animals can amplify transmission rates because of increased contact rates. For example, supplemental feeding of deer in Michigan has increased contact rates, and deer populations have dramatically increased. Supplemental feeding supports sick deer that might otherwise succumb more quickly to the disease. For these reasons, bovine tuberculosis has become endemic in white-tailed deer in this region, and cattlemen have lost tuberculosis-free status, resulting in an export ban. While banning supplemental feeding of deer could reduce transmission, the political clout of the hunt-

ing community outweighs the clout of the cattle industry in this region, creating conflict among different stakeholders (Hicking 2002).

Managing Human Perceptions

Oftentimes, management of wildlife diseases is based on human perception of the disease or of the host. The fear of loss of revenue from big game tourism, the threat of a ban on livestock trade, or the political clout of certain constituencies can lead to decisions about wildlife management that do not reduce disease threat, or ignore it completely. Foot and mouth disease in Southern Africa provides an example of how different protected areas manage for this endemic viral disease based on the priorities of managing agencies. The disease is endemic in native hoofstock (buffalo, *Syncerus caffer*, in particular, but also impala, *Aepyceros melampus*) and causes animal losses in cattle. Because Kruger National Park in South Africa is marketed as a "big five" (i.e., buffalo; elephant, *Loxodonta africana*; rhinoceros, *Ceratotherium simum* or *Diceros bicornis*; leopard, *Panthera pardus*; lion, *P. leo*) game-viewing preserve, removing buffalo is not an option. To reduce transmission to buffalo, upward of 15,000 head of cattle have been culled from surrounding ranches during outbreaks. By contrast, in Etosha National Park in Namibia, the park is not permitted to have buffalo, as a way of preventing foot and mouth disease in local cattle (De Vos et al. 2016). Changing human demographics can alter how the public perceives and ultimately accepts wildlife management for diseases. In the United States and parts of Europe, urban dwellers are moving to rural areas for increased quality of life (Starrs 1995). Their perception of wildlife is often conservation oriented, with a focus on care of individual wildlife. Within this demographic, culling may not be an acceptable form of wildlife disease control. This attitude may stand in contrast to long-standing rural agricultural communities that may view wildlife as competitors or disease risks (Peterson et al. 2006).

Contemporary Methods for Wildlife Disease Management

Technological advances have increased the tools in the toolbox available to wildlife disease managers. Here we explore techniques within three broad methods commonly used in wildlife disease management: prediction, surveillance, and pathogen reduction.

Prediction

Increased computing power, bioinformatics, and the scale of data collection have allowed modeling to play a larger role in the decision-making process. Modeling can assist managers by predicting alternate consequences from different management strategies (Russell et al. 2017), the spread of a pathogen based on climatic data (Altizer et al. 2013), or the cost of alternative management scenarios (Fenichel et al. 2010).

Transmission models are frequently employed to understand basic epidemiological behavior of pathogens. Compartmental models of disease transmission incorporate disease states (susceptible, infected, recovered) of individuals to understand population- and community-level effects of the basic reproductive rate of the pathogen (Hatcher and Dunn 2016). For example, the control of canine distemper virus (CDV) in African lions has been modeled to understand community-level effects. In this model, the increase in lion fitness and abundance increased competitive pressure on cheetahs (*Acinonyx jubatus*), which threatened their population with extinction (Chauvenet et al. 2012).

In addition to compartmental models, network models of transmission identify animal social networks such that contact rates can be estimated and transmission rates inferred. These transmission models help managers understand how pathogens are spread through populations, which can influence decisions on interventions. For example, individual variation in susceptibility and infectiousness in host populations can create super-spreaders of pathogens. Identification and selective removal of these host individuals may greatly curb pathogen transmission (Lloyd-Smith et al. 2005). Network models have been used to predict disease spread and vaccination strategies within groups of chimpanzees (*Pan troglodytes*; Rushmore et al. 2014) and rabies spread in raccoons (Reynolds et al. 2015), and many additional species have recently been modeled (Craft 2015).

Correlative models of biotic and abiotic parameters that predict disease emergence have influenced much of the spatially explicit predictive modeling. These models have been particularly beneficial for modeling the vectors of arthropod-borne diseases that are dependent on climatic conditions for emergence and spread. Multiple models have predicted the geographic spread of ticks (McPherson et al. 2017), mosquitos (Tagaris et al. 2017), and midges (Samy and Peterson 2016) in relation to global change.

Some models do not parameterize epidemiological entities but focus on economic parameters associated with diseases. Bioeconomic models parameterize cost-benefit analyses of alternate control strategies. In Central and South America, vampire bats are a reservoir of rabies that frequently infects cattle when the bats feed on the cattle. Shwiff et al. (2016) showed that there was an economic benefit to vaccinating cattle: for every US$1 spent on vaccine, a producer saved US$6 that would have gone to rabies treatment or animal loss from rabies. Furthermore, they showed that vaccination was more efficient than bat culling at preventing losses.

Surveillance

Surveillance is a critical tool for empirically assessing the epidemiological behavior of a disease, determining wildlife disease management goals, and assessing the efficacy of management strategies to achieve those goals. Monitoring for infectious diseases is not a new method of disease management, but new tools are available that increase our knowledge of pathogens and their evolution and epidemiology.

Molecular surveillance is a basic tool for understanding the disease state of individuals and the prevalence of disease in populations. Commonly used strategies include

1. Surveillance for host antibodies. Serology has been a cornerstone of molecular surveillance for nearly a century. Serology typically detects antibodies produced by the host in response to the antigens of the pathogen. Although surveillance using this method can provide evidence of transmission on a population level, it defines only whether an individual was exposed to the pathogen, not whether it is infected or infective (Gilbert et al. 2013).
2. Surveillance for the pathogen. Amplification of DNA using priming sequences specific to a known pathogen (polymerase chain reaction, PCR) or next-generation sequencing of previously undescribed pathogens, can identify the presence or absence of the pathogen's nucleic acid sequence. Often these data are used to infer the presence of the pathogen, but, particularly for viral and some bacterial agents, the nucleic acid can be present without the pathogen being viable. Thus caution should be used in interpreting results. When this method is coupled with pathogen isolation and cell culture, a more definitive result can be inferred.
3. Phylodynamics and molecular epidemiology. These techniques allow for the reconstruction of past epidemiological events by using the nucleic acid or amino acid sequence information of the pathogen. The sequence is linked to transmission by the assumption that more genetically similar pathogens have a more recent common ancestor in a host. Thus, the more closely related the genotypes of the pathogen, the more likely that direct transmission from one host to another occurred. Understanding the spread and transmission of a pathogen can help implement management strategies for future outbreaks. This method was re-

cently employed in Great Britain, where it was inferred that Ranavirus in frogs has spread from lake to lake via human movement of animals (Price et al. 2016).

Transboundary surveillance is an important concept to consider when managing wildlife diseases, which can occur at and spread to large geographic scales. Because the distribution of wildlife hosts and arthropod vectors and the pathogens they carry does not follow geopolitical boundaries, outbreaks of wildlife diseases and biosecurity responses require joint efforts that are often advised or organized by international bodies such as the United Nations. For example, global surveillance for avian influenza has been conducted through the WHO's Global Influenza Surveillance and Response system for more than 50 years (http://www.who.int/influenza/gisrs_laboratory/en/). International surveillance is not limited to official governing bodies. The International Society for Infectious Diseases maintains the Program for Monitoring Emerging Diseases (ProMED, https://www.promedmail.org/aboutus/), which is an internet reporting system describing new cases of infectious diseases in plants, animals, and humans. There are no restrictions on submissions, but submissions are moderated with follow-up investigations.

Disease sentinel species provide surveillance systems for pathogens of importance in species other than the ones being surveyed (Halliday et al. 2007). Wildlife have the potential to be used as sentinels for diseases of wildlife, livestock, and humans (Neo and Tan 2017). Because medical and veterinary ethics and the right to privacy preclude most surveillance of domestic animals and people, wildlife hosts may provide surveillance capabilities when other methods are not feasible. Good sentinels of disease include wide-ranging, omnivorous, and long-lived species that are subject to pathogens similar to the species of interest. Good sentinels also live in close association with the environment or species of interest. For example, invasive small Indian mongooses (*Herpestes auropunctatus*) inhabit many Caribbean islands where the incidence of human gastrointestinal illness is high. Mongooses live in close association with human habitations and are considered a human commensal species that frequents open sewers and trash piles, where they can acquire and carry human pathogenic *Salmonella*. Because these animals are frequently culled as part of invasive species control efforts, gut microbial samples can be acquired and screened for common human pathogenic gastrointestinal microbes from a large number of animals (Miller et al. 2013). These data may provide epidemiological information necessary for outbreaks of infectious intestinal diseases.

Pathogen Reduction

Wildlife management to manipulate the population size of susceptible hosts remains one of the most frequently used tools for disease management. By decreasing the number of susceptible individuals in the population, a pathogen is prevented from increasing its rate of transmission and eventually is either held at a constant prevalence or eliminated from the host population. Multiple methods exist for reducing the number of susceptible individuals in a population.

1. Vaccination. Vaccines can be an effective method of management, but may be too costly or inefficient for many disease systems. In order for vaccination to work, managers must be able to vaccinate a sufficient portion of the population to prevent pathogen spread and induce herd immunity. If the target population is an elusive wildlife species, this can be a daunting task. In addition, vaccines must produce in individuals an immune response sufficient in size and duration that a population-level response can occur. Finally, many wildlife vaccines do not have a large enough market for vaccine manufacturers to commit resources to developing or manufacturing the product.

If a vaccine is in use, oftentimes vaccinated animals cannot be distinguished serologically from animals that have been exposed to and are possible carriers of a pathogen. In these cases, surveillance becomes difficult to execute. If the vaccinated population is livestock, the inability to distinguish between exposed and vaccinated animals can impede the industry by blocking sales and exportation because of false positives associated with vaccinated animals (Horan et al. 2008).

Although not a management method for most disease systems, vaccination can be an effective tool. The size and scope of the United States Department of Agriculture (USDA) Animal and Plant Health Inspection Service (APHIS) Wildlife Services rabies program illustrates the commitment necessary to reduce rabies virus transmission in the United States (see Fig. 8.3). This multipronged approach has been successful for many reasons, including the financial commitment of the US government, the financial commitment of the vaccine manufacturer, and the goodwill and coordination of state, provincial, and international agencies.

2. Animal control. Culling can reduce the number of susceptible individuals in a population. Culling to eradicate or greatly reduce population numbers is often most acceptable when the host species is an alien invasive species. The United States currently has a nationwide campaign to control wild pig populations, in part because of the threat they pose to livestock trade should a foreign animal disease be introduced into wild pigs (Pedersen et al. 2013). Culling is most often nonselective, meaning it is unknown whether the culled animal harbors the pathogen or not. This method is less efficient than targeted culling and often costly (Lanfranchi et al. 2003). Culling can disrupt the social system of highly social animals, leading to greater transmission rates among host individuals.

3. Fertility control. This method uses either temporary or permanent methods of sterilization to decrease host population abundance or curb sexual activity and behavior of host species. This method may be favorable over culling animals because it is less disruptive to behavior and more socially accept-

ble to some members of the public. The method can be effective if it can reduce contact by reducing sexual behavior, but sterilization can take many years to realize a decrease in population density and transmission. Sterilization techniques must be chosen thoughtfully. For example, targeting the gonadotropic hormones eliminates breeding behavior, but vaccines that sterilize via zona pellucida are associated with increased breeding behavior (Killian et al. 2007). Empirical studies of the efficacy of fertility control to reduce diseases are lacking, but a survey of sterilized and intact street dogs in India showed a lower prevalence of diseases in dogs from cities that had fertility control of dogs versus those that did not (Yoak et al. 2014).

Summary

Management of wildlife diseases is biologically complex. The interconnectedness of pathogens within an ecosystem can have cascading effects at multiple trophic levels. For example, pathogens of predators can change prey dynamics. On Isle Royale, Michigan, USA, canine parvovirus substantially reduced wolf (*Canis lupus*) populations, which never fully returned to previous densities. This release of predators initially increased the density of moose (*Alces alces*) on the island, but the moose population growth rate eventually became regulated more by climatic and vegetative conditions than by top-down dynamics from wolf predators (Wilmers et al. 2006). Thus, disease epidemics and management responses to them can change the dynamics of entire ecosystems and initiate community-level effects. These potential consequences should be considered in the planning phases of disease management.

The complexity of human emotions toward wildlife diseases also influences wildlife disease management. Human reaction to pathogens and the diseases they cause is often visceral and dogmatic, which can heighten the sense of urgency when managing wildlife diseases. The lay public may overreact to the discovery of a wildlife pathogen, or there may be conflicting viewpoints about how to manage a pathogen if a commodity is involved. More than any other aspect of wildlife management, the management of wildlife diseases requires not only interagency communication and cooperation but also open and transparent communication with the public to successfully implement surveillance and response planning.

In our increasingly complex and globalized society, coordination often requires multinational cooperation and commitment. Within nations, each ministry is typically designated to set policy within the separate regulatory realms of animal health, public health, and wildlife health, yet effective management of disease requires a coordinated commitment from each regulatory viewpoint. Distrust and failure to cooperate slows the management process and allows disease outbreaks to grow and therefore cost more money, time, and human capital to control.

LITERATURE CITED

Altizer, S., R.S. Ostfeld, P.T. Johnson, S. Kutz, and C.D. Harvell. 2013. Climate change and infectious diseases: From evidence to a predictive framework. Science 341:514–519.

Azhar, E.I., S.A. El-Kafrawy, S.A. Farraj, A.M. Hassan, M.S. Al-Saeed, A.M. Hashem, and T.A. Madani. 2014. Evidence for camel-to-human transmission of MERS coronavirus. New England Journal of Medicine 370:2499–2505.

Beltrán-Beck, B., C. Ballesteros, J. Vicente, J. De la Fuente, and C. Gortázar. 2012. Progress in oral vaccination against tuberculosis in its main wildlife reservoir in Iberia, the Eurasian wild boar. Veterinary Medicine International 2012:978501.

Berry, H.H. 1993. Surveillance and control of anthrax and rabies in wild herbivores and carnivores in Namibia. Revue Scientifique et Technique-Office International Des Epizooties 12:137–146.

Blancou, J. 2009. Options for the control of disease 1: Targeting the infectious or parasitic agent. Pages 97–120 *in* R.J. Delahay, G.C. Smith, and M.R. Hutchings, editors. Management of disease in wild mammals. Springer, Berlin, Germany.

Chauvenet, A.L.M., S.M. Durant, R. Hilborn, and N. Pettorelli. 2012. Correction: Unintended consequences of conservation actions: Managing disease in complex ecosystems. PLOS One 7: e28671.

Claytor, S.C., K. Subramaniam, N. Landrau-Giovannetti, V.G. Chinchar, M.J. Gray, D.L. Miller, C. Mavian, M. Salemi, S. Wisely, and T.B. Waltzek. 2017. Ranavirus phylogenomics: Signatures of recombination and inversions among bullfrog ranaculture isolates. Virology 511:330–343.

Collinge, S.K., C. Ray, and J.F. Cully Jr. 2008. Effects of disease on keystone species, dominant species, and their communities. Pages 129–144 in R.S. Ostfeld, F. Keesing, and V.T. Eviner, editors. Infectious disease ecology: Effects of ecosystems on disease and of disease on ecosystems. Princeton University Press, Princeton, New Jersey, USA.

Craft, M.E. 2015. Infectious disease transmission and contact networks in wildlife and livestock. Philosophical Transactions of the Royal Society B: Biological Sciences 370:20140107.

Cully, F.J. Jr., T.L. Johnson, S.K. Collinge, and C. Ray. 2010. Disease limits populations: Plague and black-tailed prairie dogs. Vector-Borne and Zoonotic Diseases 10:7–15.

Daszak, P., A.A. Cunningham, and A.D. Hyatt. 2001. Anthropogenic environmental change and the emergence of infectious diseases in wildlife. Acta Tropica 78:103–116.

Davis, M.F., S.C. Rankin, J.M. Schurer, S. Cole, L. Conti, P. Rabinowitz, and C.E.R. Group. 2017. Checklist for one health epidemiological reporting of evidence (COHERE). One Health 4:14–21.

Dean, R., M. Gocke, B. Holz, S. Kilpatrick, T. Kreeger, B. Scurlock, S. Smith, E.T. Thorne, and S. Werbelow. 2004. Elk feedgrounds in Wyoming. Wyoming Game and Fish Department, Cheyenne, Wyoming, USA.

De Vos, A., G.S. Cumming, D. Cumming, J.M. Ament, J. Baum, H.S. Clements, J.D. Grewar, K. Maciejewski, and C. Moore. 2016. Pathogens, disease, and the social-ecological resilience of protected areas. Ecology and Society 21:20.

ECOA (European Court of Auditors). 2016. Eradication, control and monitoring programmes to contain animal diseases. Special report no. 6. Publications Office of the European Union, Luxembourg.

Epstein, J.H., and J.T. Price. 2009. The significant but understudied impact of pathogen transmission from humans to animals. Mount Sinai Journal of Medicine 76:448–455.

FAO-OIE (Food and Agriculture Organization-World Organisation for Animal Health). 2005. A global strategy for the progressive control of highly pathogenic avian influenza (HPAI). Food and Agricultural Organization of the United Nations, Rome, Italy, and World Organisation for Animal Health, Paris, France.

FAO-OIE (Food and Agriculture Organization-World Organisation for Animal Health). 2007. The global strategy for prevention and control of H5N1 highly pathogenic avian influenza (HPAI). Food and Agricultural Organization of the United Nations, Rome, Italy, and World Organisation for Animal Health, Paris, France.

Fenichel, E.P., R.D. Horan, and G.J. Hickling. 2010. Management of infectious wildlife diseases: Bridging conventional and bioeconomic approaches. Ecological Applications 20:903–914.

Field, H., P. Young, J.M. Yob, J. Mills, L. Hall, and J. Mackenzie. 2001. The natural history of Hendra and Nipah viruses. Microbes and Infection 3:307–314.

Frick, W.F., S.J. Puechmaille, J.R. Hoyt, B.A. Nickel, K.E. Langwig, J.T. Foster, K.E. Barlow, et al. 2015. Disease alters macroecological patterns of North American bats. Global Ecology and Biogeography 24:741–749.

Gage, K.L., and M.Y. Kosoy. 2005. Natural history of plague: Perspectives from more than a century of research. Annual Review of Entomology 50:505–528.

Gage, K.L., T.R. Burkot, R.J. Eisen, and E.B. Hayes. 2008. Climate and vectorborne diseases. American Journal of Preventive Medicine 35:436–450.

Gilbert, A.T., A.R. Fooks, D.T.S. Hayman, D.L. Horton, T. Müller, R. Plowright, A.J. Peel, et al. 2013. Deciphering serology to understand the ecology of infectious diseases in wildlife. EcoHealth 10:298–313.

Gilbert, M., X. Xiao, J. Domenech, J. Lubroth, V. Martin, and J. Slingenbergh. 2006. Anatidae migration in the western Palearctic and spread of highly pathogenic avian influenza H5N1 virus. Emerging Infectious Diseases 12:1650.

Glass, C.M., R.G. McLean, J.B. Katz, D.S. Maehr, C.B. Cropp, L.J. Kirk, A.J. McKeiman, and J.F. Evermann. 1994. Isolation of pseudorabies (Aujeszky's disease) virus from a Florida panther. Journal of Wildlife Diseases 30:180–184.

Gordejo, F.R., and J.P. Vermeersch. 2006. Towards eradication of bovine tuberculosis in the European Union. Veterinary Microbiology 112:101–109.

Grais, R.F., J.H. Ellis, A. Kress, and G.E. Glass. 2004. Modeling the spread of annual influenza epidemics in the US: The potential role of air travel. Health Care Management Science 7:127–134.

Gratwicke, B., M.J. Evans, P.T. Jenkins, M.D. Kusrini, R.D. Moore, J. Sevin, and D.E. Wildt. 2010. Is the international frog legs trade a potential vector for deadly amphibian pathogens? Frontiers in Ecology and the Environment 8:438–442.

Halliday, J.E.B., A.L. Meredith, D.L. Knobel, D.J. Shaw, B.M.C. Bronsvoort, and S. Cleaveland. 2007. A framework for evaluating animals as sentinels for infectious disease surveillance. Journal of the Royal Society Interface 4:973–984.

Hatcher, M.J., and A.M. Dunn. 2011. Parasites in ecological communities: From interactions to ecosystems. Cambridge University Press, Cambridge, UK.

Hatcher, M.J., J.T. Dick, and A.M. Dunn. 2006. How parasites affect interactions between competitors and predators. Ecology Letters 9:1253–1271.

Hatcher, M.J., J.T. Dick, and A.M. Dunn. 2012. Diverse effects of parasites in ecosystems: Linking interdependent processes. Frontiers in Ecology and the Environment 10:186–194.

Hicking, G.J. 2002. Dynamics of bovine tuberculosis in wild white-tailed deer in Michigan, Michigan Department of Natural Resources Wildlife Division Report No. 3363. http://digitalcommons.unl.edu/michbovinetb/27.

Hing, S., E.J. Narayan, R.A. Thompson, and S.S. Godfrey. 2016. The relationship between physiological stress and wildlife disease: Consequences for health and conservation. Wildlife Research 43:51–60.

Horan, R.D., C.A. Wolf, E.P. Fenichel, and K.H. Mathews Jr. 2008. Joint management of wildlife and livestock disease. Environmental Resource Economics 41:47–70.

IUCN Species Survival Commission Wildlife Health Specialist Group. 2016. CITES and timely movement of emergency diagnostic specimens for conservation purposes. http://www.iucn-whsg.org/node/1585.

Jancovich, J.K., E.W. Davidson, N. Parameswaran, J. Mao, V.G. Chinchar, J.P. Collins, B.L. Jacobs, and A. Storfer. 2005. Evidence for emergence of an amphibian iridoviral disease because of human-related spread. Molecular Ecology 14:213–214.

Johnson, P.T., A.R. Townsend, C.C. Cleveland, P.M. Glibert, R.W. Howarth, V.J. McKenzie, E. Rejmankova, and W.H. Ward. 2010. Linking environmental nutrient enrichment and disease emergence in humans and wildlife. Ecological Applications 20:16–29.

Jones, B.A., D. Grace, R. Kock, S. Alonso, J. Rushton, M.Y. Said, D. McKeever, et al. 2013. Zoonosis emergence linked to agricultural intensification and environmental change. Proceedings of the National Academy of Sciences of the United States of America 110:8399–8404.

Jones, K.E., G.P. Nikkita, M.A. Levy, A. Storeygard, D. Balk, J.L. Gittleman, and P. Daszak. 2008. Global trends in emerging infectious diseases. Nature 451: 990–993.

Joseph, M.B., J.R. Mihaljevic, D.L. Arellano, J.G. Kueneman, D.L. Preston, P.C. Cross, and T.J. Johnson. 2013. Taming wildlife disease: Bridging the gap between science and management. Journal of Applied Ecology 50:702–712.

Karesh, W.B., and E. Noble. 2009. The bushmeat trade: Increased opportunities for transmission of zoonotic disease. Mount Sinai Journal of Medicine 76:429–434.

Killian, G., K. Fagerstone, T. Kreeger, L. Miller, and J. Rhyan. 2007. Management strategies for addressing wildlife disease transmission: The case for fertility control. Wildlife Damage Management Conference 12:265–271.

Kilpatrick, A.M., C.M. Gillin, and P. Daszak. 2009. Wildlife-livestock conflict: The risk of pathogen transmission from bison to cattle outside Yellowstone National Park. Journal of Applied Ecology 46:476–485.

Klempa, B. 2009. Hantaviruses and climate change. Clinical Microbiology and Infections 15:518–523.

Lacey, L.A., and C.M. Lacey. 1990. The medical importance of riceland mosquitoes and their control using alternatives to chemical insecticides. Journal of the American Mosquito Control Association 2:1–93.

Lanfranchi, P., E. Ferroglio, G. Poglayen, and V. Guberti. 2003. Wildlife vaccination, conservation, and public health. Veterinary Research Communications 27:567–574.

Leopold, A. 1947. The land ethic. A Sand County almanac and sketches here and there. Oxford University Press, Oxford, UK.

Li, W., Z. Shi, M. Yu, W. Ren, C. Smith, J.H. Epstein, H. Wang, et al. 2005. Bats are natural reservoirs of SARS-like coronaviruses. Science 310:676–679.

Lloyd-Smith, J.O., P.C. Cross, C.J. Briggs, M. Daugherty, W.M. Getz, J. Letto, M.S. Sanchez, A.B. Smith, and A. Swei. 2005. Should we expect population thresholds for wildlife disease? Trends in Ecology and Evolution 20:511–519.

Macnab, J. 1983. Wildlife management as scientific experimentation. Wildlife Society Bulletin 11:397–401.

MacPhee, R.D., and A.D. Greenwood. 2013. Infectious disease, endangerment, and extinction. International Journal of Evolutionary Biology 2013:571939.

Mangili, A., and M.A. Gendreau. 2005. Transmission of infectious diseases during commercial air travel. Lancet 365:989–996.

McCallum, H. 2008. Tasmanian devil facial tumour disease: Lessons for conservation biology. Trends in Ecology and Evolution 23:631–637.

McDonald, R.A., R.J. Delahay, S.P. Carter, G.C. Smith, and C.L. Cheeseman. 2008. Perturbing implications of wildlife ecology for disease control. Trends in Ecology and Evolution 23:53–56.

McPherson, M., A. García-García, F.J. Cuesta-Valero, H. Beltrami, P. Hansen-Ketchum, D. MacDougall, and N.H. Ogden. 2017. Expansion of the Lyme disease vector Ixodes scapularis in Canada inferred from CMIP5 climate projections. Environmental Health Perspectives 125: 057008.

Miller, R.S., M.L. Farnsworth, and J.L. Malmberg. 2013. Diseases at the livestock-wildlife interface: Status, challenges, and opportunities in the United States. Preventive Veterinary Medicine 110:119–132.

Mills, J.N., K.L. Gage, and A.S. Khan. 2010. Potential influence of climate change on vector-borne and zoonotic diseases: A review and proposed research plan. Environmental Health Perspectives 118:1507.

Montgomery, J.M., M.J. Hossain, E. Gurley, D.S. Carroll, A. Croisier, E. Bertherat, N. Asgari, et al. 2008. Risk factors for Nipah virus encephalitis in Bangladesh. Emerging Infectious Diseases 14:1526–1532.

Murray, M.H., D.J. Becker, R.J. Hall, and S.M. Hernandez. 2016a. Wildlife health and supplemental feeding: A review and management recommendations. Biological Conservation 204:163–174.

Murray, M.H., J. Hill, P. Whyte, and C.C.S. Clair. 2016b. Urban compost attracts coyotes, contains toxins, and may promote disease in urban-adapted wildlife. EcoHealth 13:285–292.

Neo, J.P.S., and B.H. Tan. 2017. The use of animals as a surveillance tool for monitoring environmental health hazards, human health hazards and bioterrorism. Veterinary Microbiology 203:40–48.

Ogden, N.H., L. St-Onge, I.K. Barker, S. Brazeau, M. Bigras-Poulin, D.F. Charron, C.M. Francis, et al. 2008. Risk maps for range expansion of the Lyme disease vector, Ixodes scapularis, in Canada now and with climate change. International Journal of Health Geographics 7:24.

Olson, D.H., D.M. Aanensen, K.L. Ronnenberg, C.I. Powell, S.F. Walker, J. Bielby, T.W. Garner, G. Weaver, and M.C. Fisher. 2013. Mapping the global emergence of Batrachochytrium dendrobatidis, the amphibian chytrid fungus. PLOS One 8:e56802.

Orton, L., F. Lloyd-Williams, D. Taylor-Robinson, M. O'Flaherty, and S. Capewell. 2011. The use of research evidence in public health decision making processes. Systematic review. PLOS One 6:e21704.

Pedersen, K., S.N. Bevins, J.A. Baroch, J.C. Cumbee Jr., S.C. Chandler, B.S. Woodruff, T.T. Bigelow, and T.J. DeLiberto. 2013. Pseudorabies in feral swine in the United States 2009–2012. Journal of Wildlife Diseases 49:709–713.

Peterson, M.N., A.G. Mertig, and J. Liu. 2006. Effects of zoonotic disease attributes on public attitudes towards wildlife management. Journal of Wildlife Management 70:1746–1753.

Picco, A.M., and J.P. Collins. 2008. Amphibian commerce as a likely source of pathogen pollution. Conservation Biology 22:1582–1589.

Plowright, R.K., H.E. Field, C. Smith, A. Divljan, C. Palmer, G. Tabor, P. Daszak, and J.E. Foley. 2008. Reproduction and nutritional stress are risk factors for Hendra virus infection in little red flying foxes (Pteropus scapulatus). Proceedings of the Royal Society of London B: Biological Sciences 275:861–869.

Price, S.J., T.W.J. Gamer, A.A. Cunningham, T.E.S. Langton, and R.A. Nichols. 2016. Reconstructing the emergence of a lethal infectious disease of wildlife supports a key role for spread through translocations by humans. Proceedings of the Royal Society of London B: Biological Sciences 283:20160952.

ProMED-mail. 2017. Peste des petits ruminants—Mongolia (03): (hovd) saiga antelope. Archive No. 20170309.4889954. Accessed 28 December 2017.

Puechmaille, S.J., W.F. Frick, T.H. Kunz, P.A. Racey, C.C. Voigt, G. Wibbelt, and E.C. Teeling. 2001. White-nose syndrome: Is this emerging disease a threat to European bats? Trends in Ecology and Evolution 26:570–576.

Purse, B.V., P.S. Mellor, D.J. Rogers, A.R. Samuel, P.P. Mertens, and M. Baylis. 2005. Climate change and the recent emergence of bluetongue in Europe. Nature Review Microbiology 3:171–181.

Rocke, T.E., N. Pussini, S.R. Smith, J. Williamson, B. Powell, and J.E. Osorio. 2010. Consumption of baits containing raccoon pox–based plague vaccines protects black-tailed prairie dogs (Cynomys ludovicianus). Vector-Borne and Zoonotic Diseases 10:53–58.

Roffe, T.J., L.C. Jones, K. Coffin, M.L. Drew, S.J. Sweeney, S.D. Hagius, P.H. Elzer, and D. Davis. 2004. Efficacy of single calfhood vaccination of elk with Brucella abortus strain 19. Journal of Wildlife Management 68:830–836.

Romesburg, H.C. 1981. Wildlife science: Gaining reliable information. Journal of Wildlife Management 45:293–313.

Rushmore, J., D. Caillaud, R.J. Hall, R.M. Stumpf, L.A. Meyers, and S. Altizer. 2014. Network-based vaccination improves prospects for disease control in wild chimpanzees. Journal of the Royal Society Interface 11:20140349.

Russell, R.E., R.A. Katz, K.L. Richgels, D.P. Walsh, and E.H. Grant. 2017. A framework for modeling emerging diseases to inform management. Emerging Infectious Diseases 23:1.

Samy, A.M., and A.T. Peterson. 2016. Climate change influences on the global potential distribution of bluetongue virus. PLOS One 11:pe0150489.

Schiller, I., B. Oesch, H.M. Vordermeier, M.V. Palmer, B.N. Harris, K.A. Orloski, B.M. Buddle, T.C. Thacker, K.P. Lyashchenko, and W.R. Waters. 2010. Bovine tuberculosis: A review of current and emerging diagnostic techniques in view of their relevance for disease control and eradication. Transboundary and Emerging Diseases 57:205–220.

Sharma, R., P. Singh, W.J. Loughry, J.M. Lockhart, W.B. Inman, M.S. Duthie, M.T. Pena, et al. 2015. Zoonotic leprosy in the southeastern United States. Emerging Infectious Diseases 21:2127.

Shwiff, S.A., S.J. Sweeney, J.L. Elser, F.S. Miller, M.L. Farnsworth, P. Nol, S.S. Shwiff, and A.M. Anderson. 2016. A benefit-cost analysis decision framework for mitigation of disease transmission at the wildlife-livestock interface. Human-Wildlife Interactions 10:91–102.

Siraj, A.S., R.J. Oidtman, J.H. Huber, M.U. Kraemer, O.J. Brady, M.A. Johansson, and T.A. Perkins. 2017. Temperature modulates dengue virus epidemic growth rates through its effects on reproduction numbers and generation intervals. PLOS Neglected Tropical Diseases 11: e0005797.

Skinner, M.A., D.L. Keen, N.A. Parlane, K.L. Hamel, G.F. Yates, and B.M. Buddle. 2005. Improving protective efficacy of BCG vaccination for wildlife against bovine tuberculosis. Research in Veterinary Science 78:231–236.

Starrs, P.F. 1995. Conflict and change on the landscapes of the arid American west. Pages 271–285 in E.M. Castle, editor. The changing American countryside: Rural people and places. University Press of Kansas, Lawrence, Kansas, USA.

Tagaris, E., R.E.P. Sotiropoulous, A. Sotiropoulos, I. Spanos, P. Milonas, and A. Michaelakis. 2017. Climate change impact on the establishment of the invasive mosquito species (IMS). Pages 689–694 in T. Karacostas, A. Bais, and P.T. Nastos, editors. Perspectives on atmospheric sciences. Springer Atmospheric Sciences. Springer, Berlin, Germany.

Taylor, L.H., S.M. Latham, and M.E.J. Woolhouse. 2001. Risk factors for human disease emergence. Philosophical Transactions of the Royal Society London B: Biological Sciences 356:983–989.

Turrini, F., S. Ghezzi, I. Pagani, G. Poli, and E. Vicenzi. 2016. Zika virus: A re-emerging pathogen with rapidly evolving public health implications. New Microbiology 39:86–90.

Ward, A.I. 2009. Options for the control of disease 3: Targeting the environment. Pages 147–168 in R.J. Delahay, G.C. Smith, and M.R. Hutchings, editors. Management of disease in wild mammals. Springer, Berlin, Germany.

Williamson, S.J. 2000. Feeding wildlife . . . just say no. Wildlife Management Institute, Washington, DC, USA.

Wilmers, C.C., E. Post, R.O. Peterson, and J.A. Vucetich. 2006. Predator disease outbreak modulates top-down, bottom-up and climatic effects on herbivore population dynamics. Ecology Letters 9:383–389.

Wobeser, G. 2002. Disease management strategies for wildlife. Revue Scientifique et Technique-Office International des Epizooties 21:159–178.

Yoak, A.J., J.F. Reece, S.D. Gehrt, and I.M. Hamilton. 2014. Disease control through fertility control: Secondary benefits of animal birth control in Indian street dogs. Preventive Veterinary Medicine 113:152–156.

Zinsstag, J., E. Schelling, D. Waltner-Toews, and M. Tanner. 2011. From "one medicine" to "one health" and systemic approaches to health and well-being. Preventive Veterinary Medicine 101:148–156.

9 — The Effects of Wildlife-Based Ecotourism

WALT ANDERSON
MARISSA C. G. ALTMANN

Introduction

"Ecotourists save the world" (Brodowsky 2010: xi). Who can argue with saving the world? If ecotourism is the answer, then surely, we need to understand the specifics, so we can be part of the solution. Brodowsky's (2010) brief introduction to her book mentions that many species are at risk of extinction. "On a lighter note, it's not too late." Just join one of the more than 300 diverse volunteer opportunities featured in the book, and you "can make a difference for our species, our planet, and ultimately future generations" (Brodowsky 2010: xi). These activities (volunteering in wildlife rehabilitation centers, assisting with scientific research, helping with trail maintenance) may require paying thousands of dollars for participation costs and project donations. Do all of these examples truly constitute ecotourism?

The International Ecotourism Society (TIES) defines ecotourism as "responsible travel to natural areas that conserves the environment, sustains the well-being of the local people and involves interpretation and education" (TIES 2017). Using this definition, we focus specifically on wildlife-based ecotourism (WBE), in which positive or negative effects on wildlife are as important as other ecotourism criteria (Fig. 9.1).

Our objectives are to assess the growth of the ecotourism industry; examine the costs and benefits of ecotourism with respect to social, economic, and environmental metrics; and assess how humans manage the business of wildlife-based ecotourism. The literature on ecotourism is enormous and growing rapidly. We make no pretense of covering the field comprehensively. We feel that identifying strengths and opportunities can help the reader better understand how tourism has the potential to be a major tool for wildlife conservation. Like any tool, it must be wielded with foresight, intelligence, and humility, and it is important to know when travel purported to be ecotourism may cause more harm than good.

What Is Ecotourism?

"Ecotourism" is unlike other eco-labels that consumers might be familiar with, such as "organic" and "fair trade," in that the term itself is neither regulated by a certification body nor defined by a strict set of approved standards. There are no trademarks, audits, or mechanisms in place to ensure that all businesses claiming to be ecotours or eco-lodges are, in fact, operating in line with the TIES definition. Whereas TIES has been involved in the ecotourism industry

Figure 9.1   (A) Birding in Arizona: Internationally, birding is big business that can, if done responsibly, support local communities and protect wildlife habitats (Walt Anderson). (B) Sea kayaking in Alaska, sea lions: Small-scale ecotourism in Alaska contrasts dramatically with giant cruise ships doing conventional mass tourism (Walt Anderson). (C) Ecotourists in Amazonia: The scale of Amazonia is so vast that ecotourism operations often need to diversify to be financially viable year-round. For example, they may have to host fishing or general tourists at times, even if they prefer ecotourism (Walt Anderson). (D) Safari elephant watching: East African safaris, at their best, support community-based conservation that places high value on live wildlife. International demand for wildlife products and policies that disenfranchise local people remain serious issues (Walt Anderson).

for more than 20 years, many other organizations and experts have published their own definitions, guidelines, and certifications based on ecotourism; none are the official authority on the matter (World Tourism Organization 2002a, Medina 2005).

Without sufficient regulation within the ecotourism sector, travelers are left to choose from a confusing marketplace that is essentially greenwashed (in which tourism operators are promoting ecotourism as a marketing tool, whereas their practices oppose the core sustainability principles) (Self et al. 2010). The word ecotourism, in practice, means little as an indicator of what a business may or may not achieve for the environment or communities (Me-

letis and Campbell 2008). The general philosophy behind ecotourism has become an important travel marketing tool, regardless of whether ecotourism-labeled destinations truly benefit ecosystems and communities.

Ecotourism's root concept of travel for positive change has evolved into other related forms of tourism that are more transparent and accountable. For example, the Global Sustainable Tourism Council (GSTC) establishes and manages global standards for sustainable tourism, which it defines as tourism that "takes full account of its current and future economic, social and environmental impacts, addressing the needs of visitors, the industry, the environment and host communities" (GSTC 2017) (additional certification programs are discussed later in this chapter).

Despite offshoot certification programs and labels like sustainable tourism, ecotourism itself remains a large, unregulated market; the term ecotourism is unlikely to ever become auditable because of the massive administrative and financial costs that a formal regulatory program would require. Nonetheless, ecotourism remains an important guiding philosophy for how tourism can minimize negative effects and maximize positive conservation outcomes for people and planet alike.

Growth of the Ecotourism Industry

Ecotourism can be thought of as a form of alternative tourism that contrasts with conventional mass tourism, which often results in the Disneyfication of destinations and caters to travelers who prefer ease of access and the comforts of home (Orams 2001, Torres 2002: 91). Aspirational consumers, 40% of the global public and growing, are more likely to seek destinations that are authentic, with genuine benefits and respect for the communities and environments (Globescan/BBMG 2016). Consumers increasingly have higher expectations for products and services that minimize environmental effects and provide benefits to a cause, and tourism is no exception (WTTC 2015, CREST 2016).

The potential effects of these preferences are significant. In 2015, travel and tourism generated 9.8% of global gross domestic product (GDP), equivalent to US$7.2 trillion (CREST 2016). In 2016, international tourist arrivals grew by 2.9% to exceed US$1.2 billion (CREST 2017). The economic contribution of tourism and travel is expected to grow by 4% annually over the next decade, reaching over US$10.9 trillion, or 10.8% of global GDP, by 2026 (WTTC 2016). The industry contributed over 280 million jobs, or 9.5% of global employment, in 2015 (WTTC 2015).

Ecotourism was estimated to grow to 25% of global travel by 2016 and to generate US$470 billion in revenue annually (CREST 2015). Between 2014 and 2017, 60% of all leisure travelers in the United States participated in sustainable tourism (CREST 2017). Tourism, including ecotourism, is growing most quickly in emerging economies, where arrivals are expected to increase twice as quickly compared with advanced economies between 2010 and 2030 (CREST 2016). This is significant, considering that many biodiversity hotspots are located in developing countries. The 140,000 protected areas worldwide receive at least eight billion visitors annually, contributing >US$600 billion to national economies, though only about US$10 billion is spent to protect and manage these areas, a fraction of what is needed (Balmford et al. 2015).

Ecotourism is often associated with small, remote businesses, but many conventional mass tourism hotel and resort chains are incorporating waste reduction, energy savings, and community benefits as part of their corporate social responsibility strategies (WTTC 2015). Travelers may choose conventional mass tourism out of convenience, ease of booking, or lack of awareness, but with increased information sharing via social media and travel review websites like TripAdvisor, tourists are now more likely to be guided toward smaller-scale ecotourism destinations. Simultaneously, destinations are increasingly incorporating ecotourism as part of their conservation strategies.

Traditional conservation paradigms have favored strict protected areas rather than sustainable use (Terborgh et al. 2002, Salafsky 2011). After World War II, concern grew for the welfare of communities that were being excluded from protected areas (Terborgh et al. 2002). This shift resulted in efforts to ensure that conservation management works with, rather than against, people (Terborgh et al. 2002). Consequently, the conservation community has begun to acknowledge the potential for ecotourism to support this strategy for biodiversity protection.

The shift in conservation practice from natural resource objectives to the prioritization of human livelihoods has been criticized, perhaps most succinctly as a "focus on human welfare as a goal (as opposed to a means)" (Salafsky 2011: 4). Nevertheless, sustainable development has become entrenched within the global conservation framework (Terborgh et al. 2002, Salafsky 2011). As such, ecotourism has become a tool that can support livelihoods and responsible practices, leading to positive conservation outcomes (Geolleague et al. 2015). Ecotourism has been incorporated in international policy, such as the Convention on Biological Diversity (Secretariat of the CBD 2004, CBD 2017), and has become a part of the programmatic work of conservation organizations worldwide (The Nature Conservancy, World Wildlife Fund, Wildlife Conservation Society). Multiple high-level events have aimed to raise awareness of these contributions on the global stage (WTO 2002b, 2017). These meetings have also situated ecotourism within at least 15 sustainable development goals that frame development priorities for UN member states and civil society to address between 2015 and 2030 (WTO 2015). Ecotourism has been identified as most closely related to the goals addressing economic growth (Goal 8), consumption and production (Goal 12), and use of marine resources (Goal 14); however, life on land (Goal 15), sustainable communities (Goal 11), and the remaining sustainable development goals are all closely linked (WTO 2015).

The Importance of Wildlife for Ecotourism

Despite the ecological, economic, and cultural value of biodiversity worldwide, the earth is facing rapid rates of species extinction, with habitat loss the dominant threat (Vié et al. 2008). Based on the International Union for Conservation of Nature (IUCN) Red List Index, 22% of mammal species are globally threatened or extinct, along with 13.6% of birds, 27%

of warm-water reef-building corals, and 32.4% of amphibians (Vié et al. 2008). Habitat loss is the greatest threat to terrestrial vertebrates, followed by pollution, intrinsic factors, and other human-linked risks (Vié et al. 2008). Some threats to biodiversity are causing rapid declines, such as disease, invasive species, intensive deforestation, and sudden coral-bleaching events. The Red List Index shows that mobile taxa, such as birds and dragonflies, are less threatened; their losses appear to be more chronic over the course of multiple decades (Vié et al. 2008).

For WBE to be economically and environmentally sustainable, tourism businesses must directly contribute to the conservation of the biodiversity on which they depend (Roe et al. 1997). The definition and principles offered by TIES expect that ecotourism "conserves the environment" by reducing environmental and cultural effects and producing "direct financial benefits for conservation" (TIES 2017). Ecotourism, however, is generally not expected to contribute to in situ conservation actions (Isaacs 2000, TIES 2017). There are many disparate definitions of ecotourism, with few precise standards for conservation outcomes (IUCN 2016).

The role of ecotourism in conservation may be evolving through the advent of science-based standards, guidelines, and widespread reconceptualization of ecotourism. For example, in 2016 the IUCN formally requested implementation of many of these actions at the World Conservation Congress, including an updated definition and support for auditing and certification schemes to help prevent negative human effects on species (IUCN 2016).

Charismatic megafauna (popular, large species including predators, primates, and marine mammals) are often used as flagship species (marketing tools for conservation actions that are appealing to the public; Simberloff 1998, Leader-Williams and Dublin 2000,). Flagship species like marine mammals, sea turtles (Chelonioidea), and others greatly influence the tourism sector; according to travel agents, animal-related activities are the second-most popular ecotourism attractions, surpassed only by visitation to historic sites (CREST 2015). Travelers often choose destinations based on the presence of flagship species, creating tourism economies in geographies that might otherwise rely on extractive industries.

Conservation focused on flagship species may be a compromise between conservation marketing and sound ecological management, as a flagship species does not necessarily need to be an indicator (reflecting the ecological health of the ecosystem), umbrella (protecting which will also protect other species), or keystone species (playing a critical role in conserving an ecosystem). Rather, flagship species are chosen based on their potential to garner public support (Simberloff 1998, Leader-Williams and Dublin 2000). A flagship species–based management approach may collapse if flagship species disappear or if management regimes for two flagship species in one area contradict each other (Simberloff 1998). If appropriate flagship species are selected (such as species that are also keystones), however, they can play a crucial role in linking science-based wildlife management strategies with socioeconomic support (Simberloff 1998, Leader-Williams and Dublin 2000, Eckert and Hemphill 2005, Catlin et al. 2013).

Improperly managed ecotourism can result in negative effects for flagship species, thereby reducing the sustainability of such projects. For any WBE project to be successful as a conservation strategy, it must directly support the conservation of wildlife by mitigating threats, providing tangible economic benefits for local communities, being sustainable, and not producing any direct negative effects that might cause wildlife declines and ecosystem degradation.

Examples of Wildlife-Based Ecotourism

Australia is arguably a world leader in ecotourism (Weaver 2001, Jenkins and Wearing 2003). Wildlife-based ecotourism emphasis tends to be along the east coast, particularly the Great Barrier Reef, but this site is highly vulnerable to climate change. Aus-

tralia has shown leadership in having a National Ecotourism Strategy and in co-managing popular parks like Uluru and Kakadu with the Aborigines. Mon Repos, in Queensland, is arguably one of the world's most well-regulated and successful sites for sea turtle ecotourism, a testament to decades of adaptive management and thorough monitoring (Tisdell and Wilson 2002). Australian national policies are vulnerable to the whims of political parties (Weaver 2001), however, and indigenous tourism ventures "remain peripheral to the ecotourism industry" (Zeppel 2003: 71).

What we understand as the community benefits of tourism in developing countries often do exist in developed nations. Regardless of the potential for the socioecological benefits of this sector in the United States, there is no US national ecotourism strategy; policies are left up to individual states, unlike in Canada, where a national policy exists (Fennel 2001).

We do not equate visitation to the many national parks in the United States with WBE, though it does exist in pockets and is increasing with time (e.g., Yellowstone National Park; Simoni 2013). In the Sutter Buttes mountains of California, an experiment begun in 1976 established contracts with landowners to allow high-quality interpretive day trips into the privately owned mountain range (Anderson 2004). A group called the Sutter Buttes Naturalists evolved into the Middle Mountain Foundation, which continues interpretive hikes to this day and which has spun off a conservation land trust. Working with a positive spirit of constructive collaboration, this partnership exemplifies how creative solutions can be generated with little up-front investment.

Of course, some ecotours focus on flagship species. One of the most prominent subjects of WBE has been the mountain gorilla (Gorilla beringei). Tourism has rebounded in Rwanda, Africa, following the horrific civil war and genocide, and tourists now pay a premium of US$1,500/person for an hour of visitation with habituated gorillas, which has resulted in major financial returns to local communities

(Rwanda Development Board 2017). More than 400 community projects have been supported by gorilla tourism revenues in the past 12 years, and gorilla numbers have increased in response to protection (Rwanda Development Board 2017). Visitors throughout mountain gorilla habitat are encouraged to take the Gorilla Friendly Pledge, a commitment for trekkers to take measures to protect mountain gorillas from disease transmission, disturbance, and other tourism-related impacts (International Gorilla Conservation Programme 2017). The Gorilla Friendly Pledge is just one aspect of the trademarked Gorilla Friendly Tourism initiative.

Today there are many players in the ecotourism arena (Weaver 2001, Anderson 2017). Our hope is that available knowledge and management tools help maximize the positive outcomes of WBE, such as the ecological benefits described next.

Ecological Benefits of Wildlife-Based Ecotourism

Ecotourism has grown to be characterized by visible sustainability actions within the tourism sector, such as the use of local and even recycled materials; architecture designed to exist in balance with the environment; food grown and harvested on-site or locally; responsible water and energy use; and more. These green practices may even command premium pricing for environmentally conscious clients, as described above. But the potential ecological contributions of WBE stretch far beyond these activities.

Wildlife-based ecotourism can support wildlife conservation through financial contributions for conservation activities, practical contributions (conservation management, monitoring, research), socioeconomic incentives, and education (Higginbottom et al. 2001). It has generally been assumed that the conservation benefits of WBE outweigh the net costs caused by the direct effects of tourism itself or the exploitative practices that could occur in the absence of ecotourism economies. Positive effects have been reported across a variety of species, although only a portion involves conservation benefit apart from financial or educational gain. Wildlife-based ecotourism can be successful in mitigating threats to wildlife, such as fishing, ranching, agriculture, and development; the presence of divers has no serious detrimental effects at a carefully managed northern red snapper (*Lutjanus* spp.) and Nassau grouper (*Epinephelus striatus*) aggregation site in Belize, where the net economic benefits of WBE are larger than they would be for fishing (Heyman et al. 2010). Evidence for WBE leading to concrete conservation effects is ambiguous at best; however, researchers reported that WBE only leads to direct conservation outcomes (population stability or growth) in a small portion of case studies (Krüger 2005, Altmann 2016). An analysis of 251 WBE case studies by Krüger (2005) reported that 63% of cases were ecologically sustainable (tourism will not pose a risk to the site or species in the foreseeable future), but only 17.6% made a positive contribution to conservation.

One of the clearest indicators of conservation resulting from WBE is population stability or increase, such as the growth in elephant seal (*Mirounga* spp.) colonies (Le Boeuf and Campagna 2013), attributed to five management practices: (1) restricting visitor numbers and access; (2) monitoring effects on wildlife and habitat; (3) encouraging research; (4) using trained volunteer guides; and (5) requiring independent oversight of tour operators. Population stability or increase can also be attributed to the creation of protected wildlife reserves as part of WBE strategy. For example, the establishment of the Wechiau Community Hippo Sanctuary in Ghana, Africa, has stabilized local hippopotamus (*Hippopotamus amphibius*) populations, abated threats to biodiversity, and supported community livelihoods for more than a decade (Sheppard et al. 2010). Mossaz et al. (2015: 116) report "many mechanisms by which tourism makes substantial and significant contributions to conservation of African big cats" (lion, *Panthera leo*; leopard, *P. pardus*; cheetah, *Acinonyx jubatus*). The protection of Moutohora (Whale Island) Wildlife Management Reserve, a hotel-managed marine reserve in Vietnam, has led to increases in fish density, average size, and numbers while engaging tourists and the local community in various environmental initiatives (Svensson et al. 2009). Even if not officially designated as sanctuaries or reserves, WBE sites can also act as refugia in mosaic landscapes, as indicated by species richness and composition compared with strictly protected areas (Salvador et al. 2011).

The presence of WBE may result in other positive outcomes for wildlife populations. Tourism activity can reduce predation of threatened species by native and nonnative predators (Tisdell and Wilson 2002, Leighton et al. 2010). It may also reduce poaching if the presence of tourists acts as a deterrent (Tisdell and Wilson 2002, Jachmann et al. 2011).

Ecological Costs of Wildlife-Based Ecotourism

Even though carefully managed WBE should prevent significant negative effects resulting from WBE, researchers have documented a wide range of behavioral, physiological, and demographic effects of WBE on in situ populations of over 160 species (Green and Higginbottom 2001, Krüger 2005, Altmann 2016). The varied case studies documenting these effects suggest they are influenced by species characteristics and the human activities involved (Altmann 2016). Blumstein et al. (2017) provide a comprehensive assessment of behavioral and ecological effects of WBE.

Many case studies report negative effects of WBE on birds. For example, the number and distribution of southern giant petrel (*Macronectes giganteus*) nests decreased at a tourism site in the Antarctic, where behavioral and physiological effects were also noted (Pfeiffer and Peter 2004). Coetzee and Chown (2016) reviewed WBE effects on other antarctic species and reported negative physiological and population responses. Lower fledging weight in yellow-eyed penguin (*Megadyptes antipodes*) chicks in New Zealand occurred at sites with tourist activity, which could indicate decreased survival (McClung et al.

2003). Juvenile hoatzins (*Opisthocomus hoazin*) at tourist-exposed nests in Ecuador demonstrated increased mortality, lower body mass, and stronger hormonal response to experimental stress than juveniles at non-tourist sites (Müllner et al. 2004). A lack of adequate management and enforcement in response to increased ecotourism in coastal Patagonia apparently resulted in an increase in damage to eggs and nests of multiple seabird species (Yorio et al. 2002). A review of IUCN assessment data reported that 63 species of birds are threatened at least in part as a result of tourism activities (Steven et al. 2013).

Wildlife-based ecotourism activities also affect marine species. Hueter and Tyminski (2012) described the risks posed by the growing whale shark tourism (*Rhincodon typus*) market to the conservation of this species. A meta-analysis of studies on whale-watching disturbance reported common disruptions of activity budget, path characteristics, traveling, and resting behaviors (Senigaglia et al. 2016). The amount of stress-induced corticosterone levels in marine iguanas (*Amblyrhynchus cristatus*) in the Galápagos was reduced at sites heavily exposed to tourism, which may inhibit acute corticosterone release (Romero and Wikelski 2002).

Some activities linked to WBE, such as provisional feeding, have profound negative effects on wildlife behavior and physiology (Semeniuk et al. 2009, Brena et al. 2015). Even though truly responsible WBE would not permit provisioning, careful management and enforcement are required to ensure that such activities do not take place at WBE sites (Orams 2002, Brena et al. 2015).

Multiple authors recommend population-specific longitudinal studies to accurately document WBE effects on wildlife (Bejder et al. 2006, Senigaglia et al. 2016). Out of the 208 WBE case studies analyzed by Altmann (2016), 69 studies were based on data sets spanning 1 year or less, and only 20 were based on data sets spanning 10 years or more. More longitudinal research on the topic of WBE clearly is needed to accurately represent the effects occurring on populations at these sites.

Wildlife-based ecotourism researchers tend to identify certain types of physiological or behavioral impacts as negligible or positive when they do not directly affect population size. One common example is habituation (when wildlife tolerates the presence humans without displaying signs of stress or avoidance; McKinney 2014). Tourism-related habituation may result in increased poaching by lowering stress response (Ménard et al. 2014); however, it can also provide access to poachers (Cardiff et al. 2009) and can also potentially increase the market demand of some species for the pet trade (Meng et al. 2014). Habituation may also lead to increased human-wildlife conflict (Bejder et al. 2006, Zbinden et al. 2007, Ellenberg et al. 2009, Marino and Johnson et al. 2012, McKinney 2014, Webb and McCoy 2014).

Authors have expressed concern that habituation findings may be a result of limited research methodologies and may actually reflect seasonal variance in behavior (harbor seals; Andersen et al. 2012), chronic physiological stress phenomenon (black grouse, *Lyrurus tetrix*; Arlettaz et al. 2015), or changes to population structure (Indo-Pacific bottlenose dolphins, *Tursiops aduncus*; Bejder et al. 2006a). The above authors were only able to contextualize the findings of their research with the use of varied methods or longer-term data sets. Analytic methods may also mask or distort the actual effects of WBE on wildlife (Williams and Ashe 2007, Weinrich and Corbelli 2009). These findings are of concern, given the widespread classification of many WBE sites as no or low effect, and emphasize the need for longitudinal research that also considers individual variation in behavioral and physiological response (Altmann 2016, Blumstein et al. 2017).

Because of greater awareness of WBE effects, researchers have called for it to be reconceptualized as a consumptive use of biodiversity (Cater and Cater 2007, Meletis and Campbell 2007, Altmann 2016). The positive effects of WBE may offset any associated costs (Higginbottom et al. 2001), but it is important to follow the precautionary principle when assessing the often-limited data sets collected at WBE sites

worldwide. Once the potential effects of WBE are acknowledged, enterprises can then implement effective adaptive management for the benefit of wildlife, communities, and industry.

Socioeconomic Benefits

A major rationale for ecotourism is the potential for significant socioeconomic benefits. These include "foreign exchange earnings, economic development and diversification, distribution of income to local economies/communities, tendency for ecotourists to spend more and stay longer, generation of income for conservation and reserve/park management, increased employment opportunities, local infrastructure development," and promotion of cultural and historical values and traditions (Jenkins and Wearing 2003: 214–215). Potential employment opportunities include on-site service providers, tour guides, sustainability managers within the travel industry, and protected-area administrators (Anderson 2017).

Bushell and McCool (2007) describe benefits to protected areas and associated communities that are closely related to the overall survival of species in locations where tourism occurs: (1) income for protected areas and conservation work; (2) sustainable use of natural and cultural heritage; (3) practice linked to conventions and guidelines; (4) attachment to heritage; (5) stewardship ethics among the public; (6) collaboration with local stakeholders and industry; (7) support for local and indigenous community development; and (8) contributions to civil society and to heritage and respect for others. For example, the positive or negative perspectives that nearby residents had toward Masoala National Park in Madagascar, Africa, were influenced by actual or potential benefits (Ormsby and Mannle 2006).

Wildlife-based ecotourism is most successful when well integrated to the local community through on-the-ground support and generation of alternative livelihood opportunities to turn people away from unsustainable consumptive wildlife practices. Local community involvement has been identified repeatedly as a critical component of WBE and sustainable tourism and conservation-development projects (Brandon et al. 1998, Krüger 2005, Ostrom 2009, Waylen et al. 2010, Mossaz et al. 2015).

Socioeconomic Costs

Ecotourism can contribute to local training and employment, worker welfare, heritage preservation, and well-being of visitors, among other benefits (Bushell and McCool 2007), but whether ecotourism consistently delivers these benefits has been subject to debate. The financial benefits resulting from WBE projects may be unequal or inadequate in providing alternative livelihood opportunities when compared with more detrimental practices (Bookbinder et al. 1998, Isaacs 2000, Lindberg 2001, Buckley 2016). Tourism employment as a limited resource may create conflict and reduce incentives for communities to participate in conservation action.

Adequate enforcement of regulations is another concern. Many communities face rapid tourism growth, and tour operators may feel pressured to generate the most market share and revenue by overstepping guidelines (such as minimum distance requirements for tourists or prohibitions on provisional feeding), which may be detrimental to socioecological systems over time (Ostrom 2009). Communities may also lack enforcement capacity.

Ecotourism has been criticized as a tool of globalization and neocolonialism, as tourism and sustainability reflect an unequal distribution of power from developed to developing countries (Mowforth and Munt 2009). Ecotourism has been called elitist, patronistic, and commodifying and has been attacked as a neoliberal strategy resulting in gains for conservationists and globalists while disenfranchising and objectifying communities and landscapes (Cater 2006). Whether the benefits outweigh the drawbacks is largely a reflection of the project planning process and a variety of site-specific characteristics. The work of Ostrom (2009), Salafsky (2011), the IUCN TAPAS subgroup on Communities and Heri-

tage, and others can provide insight into how ecotourism can be best orchestrated to maximize positive effects for human communities, who are integral to the success of WBE.

Hunting and Wildlife-Based Ecotourism

Hunting tourism can provide a meaningful participant experience for travelers and economic benefits to local communities, and help maintain viable wildlife populations, often through habitat protection and restoration (Loveridge et al. 2006, Lindsey et al. 2007, Gressier 2014). In North America, sport hunters spent more than US$40 billion in 2012; some of this revenue supported organizations that purchase habitat (Ducks Unlimited, Rocky Mountain Elk Foundation; Arnett and Southwick 2015). Wildlife-based hunting tourism in Africa also produces substantial economic returns (Dalal-Clayton 1991, Lindsey et al. 2007). In places where local communities have been displaced or marginalized by the creation of protected areas, hunting can be one component of community-based natural resource management programs that are designed to reduce human-wildlife conflict and provide economic returns to communities (Newsome et al. 2005).

Lindsey et al. (2007) examined the importance of trophy hunting in 23 countries, where land managed for trophy hunting exceeds that of strictly protected areas. Outcomes were mixed: some communities and wildlife populations have benefited, while in others, the combined factors of local and national corruption, human-wildlife conflict, inadequate sharing and governance policies, overexploitation, and more have limited the effectiveness of hunting tourism in contributing to conservation (Lindsey et al. 2007, Sinclair 2008). Trophy hunting can also have behavioral and demographic consequences for hunted and nonhunted species (artificial selection that results in declining trophy quality and other genetic changes, stress [landscape of fear], population sinks that can even affect nearby protected areas, local extirpation; Muposhi et al. 2017). Lions, a key species for hunting and photographic tourism, are declining rapidly in Africa, in part because of trophy hunting practices, but there are proven ways to reverse that trend (Brink et al. 2016).

Trophy hunting is a controversial subject (Gressier 2014, Di Minin et al. 2016). Elimination of trophy hunting would likely have serious negative consequences for conservation in Africa, but proper management (net conservation benefit, biological sustainability, socioeconomic cultural benefit, adaptive management, accountable and effective governance, animal welfare) could help trophy hunting achieve its conservation potential (Loveridge et al. 2006, Di Minin et al. 2016). In short, the challenges and recommendations regarding hunting tourism are similar to those facing other forms of WBE, as described elsewhere in this chapter.

Although hunting tourism and ecotourism are generally treated as two distinct management strategies, it has been argued that they differ only in the type of activities involved and their respective categorization as consumptive and nonconsumptive (Novelli et al. 2006). According to TIES, authentic ecotourism is nonconsumptive or nonextractive, but as discussed in the above section, the many documented effects of WBE support deeper consideration of this categorization (Meletis and Campbell 2007, TIES 2017). Hunting tourism will always be distinct from other wildlife tourism activities because of the controversies over hunting. Given the greater recognition of the effects of all forms of ecotourism on wildlife, however, perhaps more clarity will emerge on the ethics of both strategies and their respective roles in wildlife conservation.

Managing Wildlife-Based Ecotourism

To a large extent, all ecotourism is local; one must consider a site's social, economic, and environmental attributes, as no one size fits all. There are literally thousands of sites or projects that claim to follow ecotourism principles. Claims are cheap, and

exaggerations or wishful thinking often compete with honest marketing. Despite the potential conservation benefits that could arise from genuine WBE, even the most well-intentioned enterprises may simply lack the capacity to ensure that these projects are planned with careful consideration of environmental, social, and economic factors.

Honey warns that "When poorly planned, unregulated, and overhyped, ecotourism light, like mass tourism or even traditional nature tourism, can bring only marginal financial benefits but serious environmental and social consequences" (2008: 69). The buyer-beware admonition is insufficient. A serious ecotraveler should do everything possible to choose providers and destinations that conform to ecotourism principles, as must other industry stakeholders to prevent exploitation covered up by greenwashing. Most travelers are unable to distinguish and differentiate claims about destinations that provide genuine positive benefits for communities and ecosystems versus those that do not. For example, there are two primary tour operators in the Burunge Wildlife Management Area near Tarangire National Park in Tanzania, Africa. One high-end company portrays itself as a caring ecotourism operator with anti-poaching squads and ties to the local community; interviews with locals show an entirely different story (Bluwstein 2017). Visitors to its camps may feel they are supporting conservation when, in fact, their comforts come at significant costs to local communities; conversely, a smaller-scale operation nearby provides high-quality ecotourism experiences at a fraction of the cost while working collaboratively with locals (Bluwstein 2017).

Any tourism-based economy is inherently unpredictable, as visitation and revenue are dependent on external factors. Global markets must lead to a steady stream of tourists who can afford leisure travel. Tourism can be influenced by disease outbreaks, political turmoil, and environmental changes, such as those caused by climate change, which may alter the desirability of WBE destinations. For example, in Madagascar, a hurricane devastated parts of the island in 2000, and political turmoil peaked in 2002. The net result was a sharp drop in tourism and thus local support for national park conservation (Ormsby and Mannle 2006). While recognizing the dependence on market and environmental factors, Steven et al. (2013) reported that tourism revenue to protected areas contributes significantly to conservation of IUCN-listed threatened bird species. They recommend promoting specialist markets such as avitourism on private and public lands, because many threatened species cannot be protected by management of protected areas alone. Unpredictability can be addressed to some extent by diversification.

Wildlife-based ecotourism destinations are most resilient when they integrate supplemental economic opportunities. For example, agricultural goods or handicrafts could be produced for WBE destinations but could also be marketed to nearby cities or even internationally to generate revenue. Diversification models are increasingly used to reduce the risk of inadvertently creating overdependence on WBE, particularly when threatened species are concerned (e.g., gorillas).

Education is also a key component of ecotourism for tourists and communities (Buckley 2009, TIES 2017). Considered a vital aspect of any wildlife conservation strategy, education integrates a process of information provision, communication, and capacity building (Fien et al. 2001). The objective is to shape educational outputs (interpretive materials) and outcomes (enhanced knowledge of and commitment to conservation, increased capacity to effect conservation) to long-term impacts (Fien et al. 2001). Education is the strongest predictor of environmental concern in most people (Smith-Sebasto 1995). Based on surveys in southern Costa Rica, Stem et al. (2003) reported that indirect benefits of community participation in ecotourism projects (training, ideas exchange) have higher correlations with conservation awareness than those of direct employment benefits.

Well-executed experiential education, where experience is followed by focused reflection, can enhance awareness building, attitude formation, and empowerment in tourists (Ewert 1996, AEE 2017). One can argue that place-based learning without travel can lead to commitment to conserving nature (Russell 1994), but transformative travel experiences can certainly inspire, inform, and motivate a person to be an active environmental steward back home. For example, a meta-analysis of experiences with whales, dolphins, and sea turtles demonstrated increased knowledge, empathy, and desire to engage in future marine conservation actions (Zeppel 2008). High-quality ecotourism experiences in the Galápagos improved understanding of resource conservation issues and developed deeper commitments to conservation (Powell and Ham 2008, Buckley 2009).

Certification can maximize the conservation effect of ecotourism; it also helps tourists choose destinations that are committed to social and ecological responsibility (WTO 2002a). Certification programs (GSTC), Blue Flag, Rainforest Alliance) can provide useful guidance and incentive for the application of environmental best practices at tourism destinations and can help educate tourists and communities (WTO 2002a, Blackman et al. 2014). Many of these programs strive to mitigate wildlife-related effects alongside a wide variety of sociocultural and environmental sustainability components, often expecting enterprises to act as jacks-of-all-trades (GSTC 2017, Rainforest Alliance 2017). Such standards are often too general to adequately advise enterprises operating in areas where threatened and endangered species, and potentially human-wildlife conflict or wildlife exploitation, may be present.

Wildlife conservation is a challenging field requiring context- and species-specific expertise with regard to ecology and behavior; more targeted guidance is needed to ensure that WBE businesses are meeting their biodiversity targets. In response, the Wildlife Friendly Enterprise Network has developed a certification program for Wildlife Friendly Tourism: "travel that maximizes opportunities for travelers, communities, and businesses to not only engage tourists as partners in conservation but to advance the on-the-ground conservation of Key Species while minimizing negative impacts of tourists and tourism infrastructure on wildlife" (http://www.wildlife friendly.org). Wildlife Friendly certification provides a framework of expert-driven best practices and training materials for tourism businesses and tourist interactions that can be applied at a global scale but adapted for species and geographies as needed. In the Philippines, the community-run Sabang Mangrove Paddle Boat Tour Guide Association was awarded Wildlife Friendly certification for educating guests and surrounding communities and for protecting and planting mangroves (Fig. 9.2). This site is home to many species, including the Palawan water monitor lizard (*Varanus palawanensis*), a species endemic to the Philippines that was described in 2010 (Koch et al. 2010).

Gorilla Friendly Tourism, mentioned above, is a partnership between the Wildlife Friendly Enterprise Network and the International Gorilla Conservation Programme, a coalition of Fauna and Flora International and World Wildlife Fund. Gorilla Friendly is a first-of-its-kind species-focused tourism certification, based on the IUCN Best Practice Guidelines

Figure 9.2   The community-run Sabang Mangrove Paddle Boat Tour Guide Association (Puerto Princesa, Palawan, Philippines) was awarded Wildlife Friendly certification for its dedication to biodiversity conservation and environmental education. Julie Stein, Wildlife Friendly Enterprise Network

for Great Ape Tourism (http://www.gorillafriendly .org). This model is now being replicated with Sea Turtle Friendly Tourism, which was piloted in the Philippines in 2017. The Sea Turtle Friendly standards were developed in consultation with the Wider Caribbean Sea Turtle Conservation Network and other marine turtle experts.

Wildlife-based ecotourism has typically been focused on small-scale tourism, despite the potential for larger players to minimize negative effects and maximize positive conservation outcomes (IUCN 2008). The field of ecotourism is fragmented, with many stakeholders competing for limited funding and market share (Isaacs 2000). Participation in industry groups and capacity-building opportunities are financially or technologically inaccessible for many ecotourism enterprises. We believe that, rather than continue supporting the small percentage of ecotourism destinations that have the means to participate in these opportunities, or fit a narrow understanding of what ecotourism looks like, the entire tourism industry should engage in on-the-ground conservation action that demonstrates tangible benefits for communities, wildlife, and ecosystems. Certification programs and information-sharing networks can provide helpful guidance to assist a wide range of tourism businesses in working together to achieve these goals.

Wildlife-Based Ecotourism and Research

Management of WBE, as for conservation in general, relies on having good information, hence the need for scientific research. One recurring problem with collecting and disseminating good scientific results is that researchers from developed countries in North America and Europe have tended to publish their results in journals or books that are not accessible in host countries (e.g., Mexico) or have simply not shared the information with the locals who would benefit most (Peterson et al. 2016). This can result in a delay or disconnect between wildlife conservation needs and conservation management actions. As a result, some Neotropical countries have strict regulations regarding scientific collecting and other research, as they have experienced scientific exploitation that has provided little direct return to them (Freile et al. 2014). Research productivity in-country, however, has recently expanded significantly in Brazil, Argentina, and Mexico, greatly enhancing bird-related tourism in these countries (Freile et al. 2014). Accessible research findings can enhance ecotour guide training and, ultimately, tourist experiences (Buckley 2009).

Another major challenge noted by some authors is the lack of a common currency by which to evaluate WBE costs and benefits (Blumstein et al. 2017). Those authors suggest that assigning science-based values to the ecological services associated with the conservation of ecosystems supported by WBE (pollination, watershed protection, waste decomposition, and spiritual, aesthetic, and cognitive benefits) may help provide a currency that can be used in assessing trade-offs between costs and benefits.

As stated by Heller and Hollocher, "Science is often a prime intermediary in the transformation of nature" (1998: 16). Science, in particular ecology and conservation biology, plays a special role in paving the way for ecotourism. We need good science, common sense, a willingness to work collaboratively to achieve shared objectives, flexibility, and humility to deal with the challenges of growing populations, changing climate, and recurring conflicts.

Summary

Ecotourism is a value-laden concept, as is conservation biology. There are strong assumptions that biodiversity, cultural diversity, and sustainability are good and desirable. One challenge for ecotourism is to recognize that people do not all share the same values (Kellert 1996). Relatively wealthy visitors from developed countries may have developed naturalistic, ecologistic, aesthetic, humanistic, and moralistic values that counter more utilitarian or even neg-

ativistic values in people more worried about daily survival and food (Kellert 1996). A challenge for ecotourism is to achieve a reasonable balance so that local communities and the wildlife resources on which WBE depends both benefit.

Values change over time. Whale harvesting on an unsustainable scale is gone, while whale-watching trips proliferate. Regulations like the Migratory Bird Treaty Act, the Endangered Species Act, and others reinforce the values we are developing. We have lost many wetlands and forests, but large areas are now protected, and ecosystems have been restored. Higher standards of living usually result in better environmental protections (despite political setbacks at times), and there is hope for improvements in developed and developing countries. In productive and willing partnerships with others, we must use our intelligence and resources to foster closer relationships with nature, relationships that will enable our descendants to live fulfilling lives in a biodiverse world.

There are criticisms of protected area conservation and sustainable use approaches, but it is better to view them as a symbiotic mutualism (Fennell and Weaver 2005, Sinclair and Dobson 2015). We cannot depend on protected areas alone to save the world's biodiversity, and social justice demands that we strive to support the welfare of people who live with and are affected by wildlife. Fennell and Weaver propose creating "an international network of protected area 'ecotouriums' that is designed to stimulate positive socio-economic change within local communities and maintain and improve the ecological health of protected areas" (2005: 373). We echo the words of Terborgh et al.: "Let no one be seduced into thinking that efforts to promote sustainable development will result coincidentally in the preservation of nature, because there is no necessary link between the two" (2002: 7).

We acknowledge WBE is inherently consumptive, even if not inherently destructive. If properly managed, natural resources persist with no permanent negative effects; that is, the resource can be enjoyed in perpetuity. This is in direct opposition to a slash-and-burn mentality that exploits a resource and moves on to another site when the damage becomes unbearable (when wildlife relocate or populations are permanently affected). Observation and documentation of key indicators (wildlife abundance, species diversity, animal behavior, physiology) and acceptable levels of change are necessary to objectively determine effects at a tourism site and allow site managers to practice adaptive management (Duffus and Dearden 1990, Catlin et al. 2011).

As we have discussed, wildlife-based ecotourism is a growing field in which practices are evolving as people recognize and try to mitigate costs while optimizing benefits. Wildlife-based ecotourism is just one tool in a comprehensive conservation strategy. It is our collective responsibility to use this tool wisely, to emphasize the good while eliminating the bad. Leopold wrote, "A thing is right when it tends to preserve the integrity, stability and beauty of the biotic community. It is wrong when it tends otherwise" (1949: 224). Let us do it right.

LITERATURE CITED

AEE (Association for Experiential Education). 2017. What is experiential education? http://www.aee.org/what-is-ee.

Altmann, M. 2016. WBE as sustainable conservation strategy. Thesis, Prescott College, Prescott, Arizona, USA.

Andersen, S.M., J. Teilmann, R. Dietz, N.M. Schmidt, and L.A. Miller. 2012. Behavioural responses of harbour seals to human-induced disturbances. Aquatic Conservation: Marine and Freshwater Ecosystems 22:113–121.

Anderson, W. 2004. Inland island: The Sutter Buttes. Natural Selection and Middle Mountain Foundation. Prescott, Arizona, USA.

Anderson, W. 2017. Ecotourism examples: Antarctica, Galapagos, and the Greater Serengeti ecosystem, and building a career in ecotourism. http://www.geolobo.com/?page_id=638.

Arlettaz, R., S. Nusslé, M. Baltic, P. Vogel, R. Palme, S. Jenni-Eiermann, P. Patthey, and M. Genoud. 2015. Disturbance of wildlife by outdoor winter recreation: Allostatic stress response and altered activity-energy budgets. Ecological Applications 25:1197–1212.

Arnett, E.B., and R. Southwick. 2015. Economic and social benefits of hunting in North America. International Journal of Environmental Studies 72:734–745.

Balmford, A., J.M.H. Green, M. Anderson, J. Beresford, C. Huang, R. Naidoo, M. Walpole, and A. Manica. 2015. Walk on the wild side: Estimating the global magnitude of visits to protected areas. PLOS Biology 13: e1002074.

Bejder, L., A. Samuels, H. Whitehead, and N. Gales. 2006. Interpreting short-term behavioural responses to disturbance within a longitudinal perspective. Animal Behaviour 72:1149–1158.

Blackman, A., M.A. Naranjo, J. Robalino, F. Alpízar, and J. Rivera. 2014. Does tourism eco-certification pay? Costa Rica's blue flag program. World Development 58:41–52.

Blumstein, D.T., B. Geoffroy, D.S.M. Samiaand, and E. Bessa, editors. 2017. Ecotourism's promise and peril: A biological evaluation. Springer International, Cham, Switzerland.

Bluwstein, J. 2017. Creating ecotourism territories: Environmentalities in Tanzania's community-based conservation. Geoforum 83:101–113.

Bookbinder, M.P., E. Dinerstein, A. Rijal, and H. Cauley. 1998. Ecotourism's support of biodiversity conservation. Conservation Biology 12:1399–1404.

Brandon, K., K.H. Redford, and S. Sanderson, editors. 1998. Parks in peril: People, politics, and protected areas. Island Press, Washington, DC, USA.

Brena, P.F., J. Mourier, S. Planes, and E. Clua. 2015. Shark and ray provisioning: Functional insights into behavioral, ecological and physiological responses across multiple scales. Marine Ecology Progress Series 538:273–283.

Brink, H., R.J. Smith, K. Skinner, and N. Leader-Williams. 2016. Sustainability and long-term tenure: Lion trophy hunting in Tanzania. PLOS One 11: e0162610.

Brodowsky, P.K. 2010. Ecotourists save the world: The environmental volunteer's guide to more than 300 international adventures to conserve, preserve, and rehabilitate wildlife and habitats. Penguin Group, New York, USA.

Buckley, R. 2009. Ecotourism: Principles and practices. Center for Agriculture and Bioscience International, Oxon, UK.

Buckley, R.C., C. Morrison, and J.G. Castley. 2016. Net effects of ecotourism on threatened species survival. PLOS One 11:e0147988.

Bushell, R., and S.F. McCool. 2007. Tourism as a tool for conservation and support of protected areas: Setting the agenda. Pages 12–26 in R. Bushell and P. Eagles, editors. Tourism and protected areas: Benefits beyond boundaries. Centre for Agriculture and Biosciences International, Oxon, UK.

Cardiff, S.G., F.H. Ratrimomanarivo, G. Rembert, and S. M. Goodman. 2009. Hunting, disturbance and roost persistence of bats in caves at Ankarana, northern Madagascar. African Journal of Ecology 47:640–649.

Cater, C., and E. Cater. 2007. Marine ecotourism: Between the devil and the deep blue sea. Center for Agriculture and Bioscience International, Wallingford, Australia.

Cater, E. 2006. Ecotourism as a western construct. Journal of Ecotourism 5:23–39.

Catlin, J., R. Jones, and T. Jones. 2011. Revisiting Duffus and Dearden's wildlife tourism framework. Biological Conservation 144:1537–1544.

Catlin, J., M. Hughes, T. Jones, R. Jones, and R. Campbell. 2013. Valuing individual animals through tourism: Science or speculation? Biological Conservation 157:93–98.

CBD (Convention on Biological Diversity). 2017. COP decisions V/25, VI/24, VII/14. https://www.cbd.int/decisions/.

Coetzee, B.W., and S.L. Chown. 2016. A meta-analysis of human disturbance impacts on antarctic wildlife. Biological Reviews 9: 578–596.

CREST (Center for Responsible Travel). 2015. The case for responsible travel: Trends and statistics 2015. Washington, DC, USA.

CREST (Center for Responsible Travel). 2016. The case for responsible travel: Trends and statistics 2016. Washington, DC, USA.

CREST (Center for Responsible Travel). 2017. The case for responsible travel: Trends and statistics 2017. Washington, DC, USA.

Dalal-Clayton, D.B. 1991. Wildlife working for sustainable development. International Institute for Environment and Development, Sustainable Agriculture Programme, London, UK.

Di Minin, E., N. Leader-Williams, and C.J. Bradshaw. 2016. Banning trophy hunting will exacerbate biodiversity loss. Trends in Ecology and Evolution 31:99–102.

Duffus, D. A., and P. Dearden. 1990. Non-consumptive wildlife-oriented recreation: A conceptual framework. Biological Conservation 53:213–231.

Eckert, K.L., and A.H. Hemphill. 2005. Sea turtles as flagships for protection of the wider Caribbean region. Maritime Studies 3:119–143.

Ellenberg, U., T. Mattern, and P.J. Seddon. 2009. Habituation potential of yellow-eyed penguins depends on sex, character and previous experience with humans. Animal Behaviour 77:289–296.

Ewert, A. 1996. Experiential education and natural resource management. Journal of Experiential Education 19:29–33.

Fennell, D.A. 2001. Anglo-America. Pages 107–122 in D. Weaver, editor. The encyclopedia of ecotourism. Centre for Agriculture and Biosciences International, Oxon, UK.

Fennell, D., and D. Weaver. 2005. The ecotourium concept and tourism-conservation symbiosis. Journal of Sustainable Tourism 13:373–390.

Fien, J., W. Scott, and D. Tilbury. 2001. Education and conservation: Lessons from an evaluation. Environmental Education Research 7:379–395.

Freile, J.F., H.F. Greeney, and E. Bonaccorso. 2014. Current Neotropical ornithology: Research progress 1996–2011. Condor 116:84–96.

Geolleague, R., R. Harley, A. Koontz, S. Naval, and J. Stein. 2015. Payment for ecosystem services: Making environmental financing work, experiences from the Philippines. Relief International, Washington, DC, USA.

Globescan/BBMG. 2016. Five human aspirations and the future of brands. http://www.globescan.com/component/edocman/?view=document&id=262&Itemid=591.

Green, R., and K. Higginbottom. 2001. Negative effects of wildlife tourism on wildlife. Wildlife tourism research report 5. Cooperative Research Centre for Sustainable Tourism, Gold Coast, Australia.

Gressier, C. 2014. An elephant in the room: Okavango safari hunting as ecotourism? Ethnos: Journal of Anthropology 79:193–214.

GSTC (Global Sustainable Tourism Council). 2017. Glossary, Sustainable tourism. https://www.gstcouncil.org/gstc-criteria/glossary/.

Heller, B., and H. Hollocher. 1998. Ecotopias: The Disneyfication of Latin American and Caribbean nature in the age of globalization. Report prepared for the 1998 meeting of the Latin American Studies Association. Chicago, Illinois, USA.

Heyman, W.D., L.M. Carr, and P.S. Lobel. 2010. Diver ecotourism and disturbance to reef fish spawning aggregations: It is better to be disturbed than to be dead. Marine Ecology Progress Series 419:201–210.

Higginbottom, K., C. Northrope, and R. Green. 2001. Positive effects of wildlife tourism on wildlife. Wildlife tourism research report 6. Cooperative Research Centre for Sustainable Tourism, Gold Coast, Australia.

Honey, M. 2008. Ecotourism and sustainable development: Who owns paradise? Second edition. Island Press, Washington, DC, USA.

Hueter, R.E., and J.P. Tyminski. 2014. Issues and options for whale shark conservation in Gulf of Mexico and western Caribbean waters of the US, Mexico and Cuba. Mote Technical Report 1633:1–43.

International Gorilla Conservation Programme. 2017. Gorilla tourism. http://igcp.org/gorillas/tourism.

Isaacs, J. 2000. The limited potential of ecotourism to contribute to wildlife conservation. Wildlife Society Bulletin 28:61–69.

IUCN (International Union for Conservation of Nature). 2016. Improving standards in ecotourism. World Conservation Congress. 10 September. https://portals.iucn.org/congress/motion/065.

IUCN (International Union for Conservation of Nature). 2008. Biodiversity: My hotel in action: A guide to sustainable use of biological resources. Island Press, Gland, Switzerland.

Jachmann, H., J. Blanc, C. Nateg, C. Balangtaa, E. Debrah, F. Damma, E. Atta-Kusi., and A. Kipo. 2011. Protected area performance and tourism in Ghana. South African Journal of Wildlife Research 41:95–109.

Jenkins, J., and S. Wearing. 2003. Ecotourism and protected areas in Australia. Pages 205–233 in D.A. Fennell and R.K. Dowling. Ecotourism policy and planning. Centre for Agriculture and Biosciences International, Oxon, UK.

Kellert, S.R. 1996. The value of life: Biological diversity and human society. Island Press, Washington, DC, USA.

Koch, A., M. Gaulke, and W. Böhme. 2010. Unravelling the underestimated diversity of Philippine water monitor lizards (Squamata: Varanus salvator complex), with the description of two new species and a new subspecies. Zootaxa 2446:1–54.

Krüger, O. 2005. The role of ecotourism in conservation: Panacea or Pandora's box? Biodiversity and Conservation 14:579–600.

Leader-Williams, N., and H.T. Dublin. 2000. Charismatic megafauna as "flagship species." Pages 53–81 in A. Entwistle and N. Dunstone, editors. Has the Panda had its day? Future priorities for the conservation of mammal diversity. Cambridge University Press, Cambridge, UK.

Le Boeuf, B.J., and C. Campagna. 2013. Wildlife viewing spectacles: Best practices from elephant seal (Mirounga sp.) colonies. Aquatic Mammals 39:132.

Leighton, P.A., J.A. Horrocks, and D.L. Kramer. 2010. Conservation and the scarecrow effect: Can human activity benefit threatened species by displacing predators? Biological Conservation 143:2156–2163.

Leopold, A. 1949. A Sand County almanac and sketches here and there. Oxford University Press, New York, USA.

Lindberg, K. 2001. Economic impacts. Pages 363–367 in D. Weaver, editor. The encyclopedia of ecotourism. Centre for Agriculture and Biosciences International, Oxon, UK.

Lindsey, P.A., P.A. Roulet, and S.S. Romanach. 2007. Economic and conservation significance of the trophy hunting industry in sub-Saharan Africa. Biological Conservation 134:455–469.

Loveridge, A.J., J.C. Reynolds, and E.J. Milner-Gulland. 2006. Does sport hunting benefit conservation? Pages 224–240 in D.W. Macdonald and K. Service, editors. Key topics in conservation biology. Blackwell Publishing, Oxford, UK.

Marino, A., and A. Johnson 2012. Behavioural response of free-ranging guanacos (Lama guanicoe) to land-use change: Habituation to motorised vehicles in a recently created reserve. Wildlife Research 39:503–511.

McClung, M.R., P.J. Seddon, M. Massaro, and A.N. Setiawan. 2003. Nature-based tourism impacts on yellow-eyed penguins Megadyptes antipodes: Does unregulated visitor access affect fledging weight and juvenile survival? Biological Conservation 119:279–285.

McKinney, T. 2014. Species-specific responses to tourist interactions by white-faced capuchins (Cebus imitator) and mantled howlers (Alouatta palliata) in a Costa Rican Wildlife Refuge. International Journal of Primatology 35:573–589.

Medina, L.K. 2005. Ecotourism and certification: Confronting the principles and pragmatics of socially responsible tourism. Journal of Sustainable Tourism 13:281–295.

Meletis, Z.A., and L.M. Campbell. 2007. Call it consumption! Re-conceptualizing ecotourism as consumption and consumptive. Geography Compass 1:850.

Ménard, N., A. Foulquier, D. Vallet, M. Qarro, P. Le Gouar, and J.S. Pierre. 2014. How tourism and pastoralism influence population demographic changes in a threatened large mammal species. Animal Conservation 17:115–124.

Meng, X., A. Aryal, A. Tait, D. Raubenhiemer, J. Wu, Z. Ma, Y. Sheng, D. Li, F. Liu, F. Meng, and P. Wang. 2014. Population trends, distribution and conservation status of semi-domesticated reindeer (Rangifer tarandus) in China. Journal for Nature Conservation 22:539–546.

Mossaz, A., R.C. Buckley, and J.G. Castley. 2015. Ecotourism contributions to conservation of African big cats. Journal for Nature Conservation 28:112–118.

Mowforth, M., and I. Munt. 2009. Tourism and sustainability: Development, globalization and new tourism in the Third World. Third edition. Routledge, London, UK.

Müllner, A., K.E. Linsenmair, and M. Wikelski. 2004. Exposure to ecotourism reduces survival and affects stress response in hoatzin chicks (Opisthocomus hoazin). Biological Conservation 118:549–558.

Muposhi, V.K., E. Gandiwa, S.M. Makuza, and P. Bartels. 2017. Ecological, physiological, genetic trade-offs and socioeconomic implications of trophy hunting as a conservation tool. Journal of Plant and Animal Sciences 27:1–14.

Newsome, D., R.K. Dowling, and S.A. Moore. 2005. Wildlife tourism. Channel View Publications, Bristol, UK.

Novelli, M., J.I. Barnes, and M. Humavindu. 2006. The other side of the ecotourism coin: Consumptive tourism in Southern Africa. Journal of Ecotourism 5:62–79.

Orams, M.B. 2001. Types of ecotourism. Pages 23–36 in D.B. Weaver, editor. The encyclopedia of ecotourism.

Centre for Agriculture and Biosciences International, Oxon, UK.

Orams, M.B. 2002. Feeding wildlife as a tourism attraction: A review of issues and impacts. Tourism Management 23:281–293.

Ormsby, A., and K. Mannle. 2006. Ecotourism benefits and the role of local guides at Masoala National Park, Madagascar. Journal of Sustainable Tourism 14:271–287.

Ostrom, E. 2009. A general framework for analyzing sustainability of social-ecological systems. Science 325: 419–422.

Peterson, A.T., A.G. Navarro-Sigüenza, and A. Gordillo-Martínez. 2016. The development of ornithology in Mexico and the importance of access to scientific information. Archives of Natural History 43:294–304.

Pfeiffer, S., and H.U. Peter. 2004. Ecological studies toward the management of an antarctic tourist landing site (Penguin Island, South Shetland Islands). Polar Record 40:345–353.

Powell, R.B., and S.H. Ham. 2008. Can ecotourism interpretation really lead to pro-conservation knowledge, attitudes and behaviour? Evidence from the Galapagos Islands. Journal of Sustainable Tourism 16:467–489.

Rainforest Alliance. 2017. Find certified products. https://www.rainforest-alliance.org/find-certified.

Roe, D., N. Leader-Williams, and B. Dalal-Clayton. 1997. Take only photographs, leave only footprints: The environmental impacts of wildlife tourism. IIED Wildlife and Development Series 10:1–86.

Romero, L.M., and M. Wikelski. 2002. Exposure to tourism reduces stress-induced corticosterone levels in Galápagos marine iguanas. Biological Conservation 108:371–374.

Russell, C.L. 1994. Ecotourism as experiential environmental education? Journal of Experiential Education 17:16–22.

Rwanda Development Board. 2017. Increase of gorilla permit tariffs. http://www.rdb.rw/home/newsdetails/article/increase-of-gorilla-permit-tariffs.html.

Salafsky, N. 2011. Integrating development with conservation. Biological Conservation 144:973–978.

Salvador, S., M. Clavero, and R.L. Pitman. 2011. Large mammal species richness and habitat use in an upper Amazonian forest used for ecotourism. Mammalian Biology-Zeitschrift für Säugetierkunde 76:115–123.

Secretariat of the CBD (Convention on Biological Diversity). 2004. Guidelines on biodiversity and tourism development (CBD Guidelines). Secretariat of the Convention on Biological Diversity, Montreal, Canada.

Self, R.M., D.R. Self, and J. Bell-Haynes. 2010. Marketing tourism in the Galapagos Islands: Ecotourism or greenwashing? International Business and Economics Research Journal 9:111–126.

Semeniuk, C.A., S. Bourgeon, S.L. Smith, and K.D. Rothley. 2009. Hematological differences between stingrays at tourist and non-visited sites suggest physiological costs of wildlife tourism. Biological Conservation 142:1818–1829.

Senigaglia, V., F. Christiansen, L. Bejder, D. Gendron, D. Lundquist, D.P. Noren, A. Schaffar, et al. 2016. Meta-analyses of whale-watching impact studies: Comparisons of cetacean responses to disturbance. Marine Ecology Progress Series 542:251–263.

Sheppard, D.J., A. Moehrenschlager, J.M. Mcpherson, and J.J. Mason. 2010. Ten years of adaptive community-governed conservation: Evaluating biodiversity protection and poverty alleviation in a West African hippopotamus reserve. Environmental Conservation 37:270–282.

Simberloff, D. 1998. Flagships, umbrellas, and keystones: Is single-species management passé in the landscape era? Biological Conservation 83:247–257.

Simoni, S. 2013. Yellowstone National Park: A model to analyze an ecotourism destination. Agricultural Management / Lucrari Stiintifice Seria I, Management Agricol 15:197–202.

Sinclair, A.R.E. 2008. Integrating conservation in human and natural systems. Pages 471–495 in A.R.E. Sinclair, C. Packer, S.A.R. Mduma, and J.M. Fryxell, editors. Serengeti III: Human impacts on ecosystem dynamics. University of Chicago, Chicago, Illinois, USA.

Sinclair, A.R.E., and A. Dobson. 2015. Conservation in a human-dominated world. Pages 1–10 in A.R.E. Sinclair, K.L. Metzger, S.A.R. Mduma, and J.M. Fryxell, editors. Serengeti IV: Sustaining biodiversity in a coupled human-natural system. University of Chicago, Chicago, Illinois, USA.

Smith-Sebasto, N. 1995. The effects of an environmental studies course on selected variables related to environmentally responsible behavior. Journal of Experiential Education 26:30–34.

Stem, C.J., J.P. Lassoie, D.R. Lee, D.D. Deshler, and J.W. Schelhas. 2003. Community participation in ecotourism benefits: The link to conservation practices and perspectives. Society and Natural Resources 16:387–413.

Steven, R., J.G. Castley, and R. Buckley. 2013. Tourism revenue as a conservation tool for threatened birds in protected areas. PLOS One 8: e62598.

Svensson, P., L.D. Rodwell, and M.J. Attrill. 2009. Privately managed marine reserves as a mechanism for the conservation of coral reef ecosystems: A case study from Vietnam. AMBIO: A Journal of the Human Environment 38:72–78.

Terborgh, J., C. Van Schaik, L. Davenport, and M. Rao, editors. 2002. Making parks work: Strategies for preserving tropical nature. Island Press, Washington, DC, USA.

TIES (The International Ecotourism Society). 2017. What is ecotourism? http://www.ecotourism.org/what-is-ecotourism.

Tisdell, C., and C. Wilson. 2002. Economic, educational and conservation benefits of sea turtle based ecotourism: A study focused on Mon Repos. Wildlife Tourism Research Report Series 20:1–118.

Torres, R. 2002. Cancun's tourism development from a Fordist spectrum of analysis. Tourist Studies 2:87–116.

UNESCO. 2017. UNESCO and the International Year of Sustainable Tourism. http://en.unesco.org/iyst4d.

Vié, J.-C., C. Hilton-Taylor, and S.N. Stuart. 2008. Wildlife in a changing world: An analysis of the 2008 IUCN Red List of Threatened Species. International Union for Conservation of Nature, Gland, Switzerland.

Waylen, K., A. Fischer, P.J.K. McGowan, S.J. Thirgood, and E.J. Milner-Gulland. 2010. Effect of local cultural context on the success of community-based conservation interventions. Conservation Biology 24:1119–1129.

Weaver, D. 2001. Ecotourism. John Wiley and Sons Australia, Milton, Australia.

Webb, S.E., and M.B. McCoy. 2014. Ecotourism and primate habituation: Behavioral variation in two groups of white-faced capuchins (Cebus capucinus) from Costa Rica. Revista de Biología Tropical 62:909–918.

Weinrich, M., and C. Corbelli. 2009. Does whale watching in southern New England impact humpback whale (Megaptera novaeangliae) calf production or calf survival? Biological Conservation 142:2931–2940.

Williams, R., and E. Ashe. 2007. Killer whale evasive tactics vary with boat number. Journal of Zoology 272:390–397.

World Travel and Tourism Council. 2015. Travel and tourism 2015: Connecting global climate action. World Travel and Tourism Council, London, UK.

WTO (World Tourism Organization). 2002a. Voluntary initiatives for sustainable tourism: Worldwide inventory and comparative analysis of 104 eco-labels, awards and self-commitments. United Nations World Tourism Organization, Madrid. Spain.

WTO (World Tourism Organization). 2002b. World Ecotourism Summit final report. World Tourism Organization and UN Environment Programme, Madrid, Spain.

WTO (World Tourism Organization). 2015. Tourism and the sustainable development goals. World Tourism Organization, Madrid, Spain.

WTO (World Tourism Organization). 2017. Calendar of official #IY2017 events. http://www.tourism4development2017.org/news/calendar-official-iy2017-events-6th-september/.

WTTC (World Travel and Tourism Council). 2016. Travel and tourism economic impact 2016: World. London, UK.

Yorio, P., E. Frere, P. Gandini, and A. Schiavini. 2002. Tourism and recreation at seabird breeding sites in Patagonia, Argentina: Current concerns and future prospects. Bird Conservation International 11:231–245.

Zbinden, J.A., A. Aebischer, D. Margaritoulis, and R. Arlettaz. 2007. Insights into the management of sea turtle internesting area through satellite telemetry. Biological Conservation 137:157–162.

Zeppel, H. 2003. Sharing the country: Ecotourism policy and indigenous peoples in Australia. Pages 55–76 in D.A. Fennell and R.K. Dowling, editors. Ecotourism policy and planning. Centre for Agriculture and Biosciences International, Oxon, UK.

Zeppel, H. 2008. Education and conservation benefits of marine wildlife tours: Developing free-choice learning experiences. Journal of Environmental Education 39:3–18.

10 — Carnivores, Coexistence, and Conservation in the Anthropocene

DAVID CHRISTIANSON
MENNA JONES

Introduction

Perhaps no taxon evokes a stronger set of emotional responses in humans and a more complex suite of conservation challenges to biologists than carnivores. Of the known mammalian carnivores, most species are already extinct, about to go extinct, or steadily declining (Woodroffe 2000, Dalerum et al. 2009, Ripple et al. 2014). Located at the highest trophic levels, carnivore populations can persist only if their demographic rates can absorb natural and anthropogenic perturbations to consumers and producers in the ecosystems they occupy (Packer et al. 2005). Additionally, carnivores must also persist in the face of direct anthropogenic and environmental challenges to their own survival and reproduction (Creel et al. 2013). Many carnivore species do not appear to possess the adaptations that would ensure positive or even neutral population growth rates in the Anthropocene without conservation planning that recognizes these challenges (Dalerum et al. 2009). Additionally, people's attitudes and formal wildlife policies still commonly promote low carnivore densities or local extinction in many systems (Bergstrom 2017). Recent changes in perception and conservation priorities, however, have seen carnivores transition from de facto status as vermin toward more formally recognized and regulated taxa (Bergstrom 2017). Social, political, and legal challenges toward policies regulating carnivores or human-carnivore interaction that are now seen as malfeasant or openly hostile to carnivores, along with growing tolerance of carnivores and the development of coexistence practices, are underpinning conservation programs explicitly attempting to restore or grow carnivore populations (Bergstrom 2017). Our objectives in this chapter are to outline proximate and ultimate influences of global conservation issues for carnivores, including carnivore ecology, human perceptions, and economic considerations.

Global Conservation Challenges

Mammalian carnivores possess three general traits that strongly influence their relationship with humans. First, and most obviously, carnivores kill and eat other animals. Killing domestic or wild prey creates economic and emotional incentives for humans to retaliate (Kellert et al. 1996). Second, carnivores are generally wide-ranging, requiring more land per individual than comparably sized herbivores. Wide-ranging carnivores must therefore use landscapes that are also likely used by humans (Fig. 10.1; López-Bao et al. 2017). Third, many car-

Figure 10.1   A male lion in Kruger National Park, South Africa, shares the road with park visitors near the Malelane Gate. Surrounding agricultural lands can be seen in the distance. John Koprowski

nivores naturally persist at low densities, and their populations grow slowly, particularly in comparison with prey populations, because of the aforementioned large space requirement, low survival, low fecundity, and social constraints on association and breeding (Durant et al. 2017). Low densities predispose carnivores to localized extinction more readily than other species, which underscores the importance of metapopulation connectivity at large spatial scales for long-term persistence (Dalerum et al. 2009). These challenges generally apply to most carnivores, but they are exacerbated by body size (Carbone et al. 2007), and our discussion addresses large (>20 kg) and small (≤20 kg) carnivores separately.

Occupying the uppermost trophic levels of food webs, carnivores can have large effects on ecosystem functioning (Ripple et al. 2014). Examples abound of ecosystem restructuring following removal of native carnivores (Prugh et al. 2009), reintroduction of lost carnivores (Christianson and Creel 2014), or colonization by nonnative carnivores (Croll et al. 2005). Our understanding of the effects of carnivores on ecosystems has traditionally focused on direct, predatory interactions. These include killing of prey species and intraguild predation on other carnivores, with propagating or cascading consequences through

food webs (density-mediated trophic cascades and apparent competition). One often-cited example is the apparent recovery of deciduous trees, browsed by elk (*Cervus canadensis*), in the Greater Yellowstone Ecosystem following the reintroduction of gray wolves (*Canis lupus*) in 1995 and 1996 that has reduced elk density by more than half. Although it was originally argued that a behavioral response in elk to predation risk from wolves drove this trophic cascade—a behaviorally mediated trophic cascade (Ripple and Beschta 2004)—subsequent investigations challenged whether there has been any vegetation response or suggested that changes in elk density cannot be ruled out as an explanation for the response in Yellowstone vegetation, a density-mediated trophic cascade (Creel and Christianson 2009, Kauffman et al. 2010, Kimble et al. 2011, Winnie 2012). Additional research reported that elk do respond strongly to predation risk from wolves, but effects on elk (risk effects) and lower trophic levels are difficult to isolate because of high spatiotemporal heterogeneity in predation risk (Fortin et al. 2005, Winnie et al. 2006, Christianson and Creel 2010). The Yellowstone wolf restoration demonstrates that even specialized carnivores (elk are >90% of Yellowstone wolves' diet; Becker et al. 2009, Metz et al. 2012) can play complex roles in ecosystems, beyond the number of prey they remove. A growing appreciation for the ecological consequences of behavioral responses to predation risk from large carnivores is shedding light on entirely new pathways whereby carnivores can influence prey-population dynamics and shape ecosystem structure and function (Cherry et al. 2016).

Global Loss of Large Carnivores and Shifting Emphasis from Control to Protection

Despite being wide-ranging, long-lived, and generally robust, large carnivores prove relatively easy species to eradicate (Woodroffe 2000). Recent rapid declines in large carnivore populations have been influenced largely by anthropogenic factors, mostly intentional killing (Bergstrom 2017, Suutarinen and Kojola 2017). Systematic killing of carnivores perhaps began with the domestication of ungulates, when tools available for predator control were primitive. Today, livestock protection remains a primary reason for killing carnivores, along with fear, recreation, and protection of prey species, and many effective technologies now facilitate this killing (Carter et al. 2017). Projectile weapons and traps have long been effective at reducing or eliminating large carnivore populations. The use of poisons, explosives, and aircraft can greatly increase kill rates and facilitate population declines (Wagner and Conover 1999, Marquard-Petersen 2012). Predator bounties and legal and illegal markets for carnivore body parts can add incentives for killing large carnivores, yet substantial debate remains about whether bounties are effective at reducing undesired effects of predators (Bartel and Brunson 2003, Maria Fernandez and Ruiz de Azua 2010). In general, tolerance by local communities appears to be the primary factor influencing anthropogenic mortality rates for many carnivore species (Hazzah et al. 2009, Treves and Bruskotter 2014, Carter et al. 2017).

Carnivore effects (real or perceived) on native ungulates that are valued by local communities for meat, recreational hunting, or trophy hunting are often used to justify carnivore killing or carnivore reduction policies (Creel et al. 2013, 2015). Herbivore population dynamics are another important influence of carnivore population regulation (Vucetich et al. 2011). Functional and numerical response research suggests that large carnivore populations require a native ungulate population at least one order of magnitude larger in size (Dale et al. 1994, Sinclair and Krebs 2002, Becker et al. 2009, Vucetich et al. 2011). Thus, the space requirements for large carnivores often hinge on the space requirements of native herbivore populations, which can be migratory (Carbyn 1981, Valeix et al. 2012, López-Bao et al. 2017). Protecting the large areas used by herbivores and maintaining connectivity of seasonal herbivore ranges are challenging but may be particularly effective if the creation of such areas also reduces direct persecution of carnivores (Durant et al. 2015, López-Bao et al. 2017). Setting aside areas for reintroduction, recovery, or continued persistence of large carnivore populations and their prey is valued by many humans, but this objective is coming into conflict with increasing demand for livestock production and growing human populations in areas suitable for carnivores (Wittemyer et al. 2008, Durant et al. 2017).

Changing attitudes toward carnivores, which influence carnivore conservation programs, are well reflected in the ongoing saga of human-wolf interactions. Wolves lacked formal protection anywhere in the world outside tribal areas prior to 1969, and humans eagerly eradicated them from nearly all of Europe, many parts of Asia, and all the contiguous United States. The environmental movement ushered in the first legal protections for some wolf populations, preceded by seminal research that objectively examined wolf ecology and behavior, overturning long-held assumptions, defining their interactions with prey and humans in absolute terms, and facilitating more nuanced attitudes toward wolves and their role in ecosystems (Banks et al. 1967). Greater appreciation for the ecological functioning of wolves promoted recovery plans that included reintroductions and translocations. Gray wolf populations have been restored to six states in the United States and most large European countries and now persist in landscapes where they were aggressively targeted for eradication 100 years ago. Regulated trophy hunting of wolves has been initiated in some of these recovered populations.

Trophy Hunting of Large Carnivores

Like all aspects of carnivore conservation, regulated trophy hunting of large carnivores is highly controversial and has a complex history unique to each carnivore taxa. For example, regulated trophy hunting of lions (*Panthera leo*) has been an important industry for several decades, whereas regulated trophy hunting of wolves is a relatively recent development in the contiguous United States and Europe. Debate surrounding trophy hunting ranges from questioning the ethics of killing individual carnivores (Nelson et al. 2016) to whether or not revenues generated from trophy hunting net any conservation gain for carnivore populations (Brink et al. 2016) to how well specific hunting regulations minimize extinction risk (Begg et al. 2018). Large carnivores demand the highest premiums for hunting permits, which can represent a substantial portion of revenues for national conservation programs (Lindsey et al. 2006), but unanticipated population declines, low densities, and skewed demographics, even in areas with well-regulated trophy hunting, are not uncommon (Loveridge et al. 2007, Rosenblatt et al. 2014, Creel et al. 2015). Recent research suggests common paradigms of sustainable hunting developed for ungulates and waterfowl (compensatory hunting mortality) may not translate to large carnivores (Creel et al. 2015).

Employing trophy hunting as a management tool or tactic for supporting conservation is likely to become increasingly difficult with humans' continuing rapid disengagement from hunting generally; participation in hunting declined 16% in the United States from 2011 to 2016 (US Fish and Wildlife Service 2017), along with decreasing tolerance for consumptive or recreational uses of wildlife (Tichelar 2016) and increasing tolerance for large carnivores (Ericsson and Heberlein 2003). To manage conflicts, managers will need to facilitate coexistence by encouraging tolerance and coadaptation in both carnivores and people.

Conservation and Management of Smaller Carnivores

Smaller carnivores present many of the same conservation issues as large carnivores because they generally share the same three traits: they are predators, have large space requirements, and live at low density. Many species of small carnivore are threatened, particularly those that have specialist niche requirements, such as the black-footed ferret (*Mustela nigripes*) (Shoemaker et al. 2014), or are affected by alien invasive carnivores, such as the effects of feral American mink (*Mustela vison*) on the European polecat (*Mustela putorius*) (Barrientos 2015) and on transmitting disease to native carnivores (Sepúlveda et al. 2014). Mesocarnivores present some different management challenges, though, because smaller species typically live at higher densities, and more species are omnivorous, with broad and flexible dietary and, often, habitat niches (Fig. 10.2). Omnivorous species are more likely to adapt to anthropogenic landscapes. Landscape change is among the greatest challenges for wildlife in the Anthropocene. The cumulative human footprint is expanding and currently affects 75% of the surface area of the planet (Venter et al. 2016). There are winners and losers in how mammalian carnivore species cope with anthropogenic landscape change. Indeed, invasive predators that are a key threat to biodiversity globally are all mesocarnivores (Doherty et al. 2016).

Smaller carnivores frequently occur at higher population densities than do large carnivores because they eat smaller prey (Gittleman 1985). Body size scales with prey size in carnivores (Gittleman 1985,

Figure 10.2   Mesocarnivores like this black-backed jackal (*Canis mesomelas*) in Namibia may not have the charisma of the large carnivores but have great ecological importance, and populations often fluctuate in ecosystems based on the health of the large carnivore community. Amanda Veals

Jones 1997), in major part because the anatomical features proximal to subduing and killing prey limit the maximum size prey that a predator can kill (Jones 1997, Dayan and Simberloff 1998). In mammalian carnivores, these trophic structures are primarily the canine teeth (Dayan et al. 1989, Van Valkenburgh 1989, Jones 1997).

Small body size alone is not sufficient to buffer species against anthropogenic effects. The degree to which a carnivore is a specialist or a generalist in their diet and habitat requirements is a strong influencing factor in how resilient they are to anthropogenic challenges. Whereas most large mammalian predators are hypercarnivores (that is, most of their food is flesh of vertebrate prey), a much wider variation in dietary breadth exists among small and mesocarnivores.

Many species are hypercarnivorous, including species that are specialists on types of prey. Cats and some mustelids (families Felidae and Mustelidae) are the most specialized, their dentition lacking postcarnassial (the specialized meat-shearing molar tooth) molar teeth suited to eating plant or insect material (Van Valkenburgh 1988, 1989). The mustelids include species such as weasels and stoats that are predatory specialists on burrow-dwelling rodents, with long, thin bodies that enable them to hunt their prey in their underground refuges. These species will be more vulnerable to population decline. An extreme example of this is the black-footed ferret (*Mustela nigripes*), which has an obligate relationship with prairie dogs (*Cynomys* spp.), a burrow-dwelling rodent. Black-footed ferrets live within prairie dog burrows, and they specialize in eating them. With widespread persecution and decline in prairie dogs, black-footed ferrets very nearly went extinct (Shoemaker 2014).

Many of the smaller carnivores, however, are insectivore-carnivores or omnivores, and this dietary flexibility leads to greater resilience to anthropogenic landscape change. These include the canids (Canidae), raccoons (Procyonidae), civets and genets (Viverridae), mongooses (Herpestidae), and many of the mustelids, such as badgers and skunks (Mustelidae) (Van Valkenburgh 1988). Among the marsupial carnivores (Dasyuridae), the smaller species (<2kg) are all insectivore-carnivores (Jones 2003). Diet generalists are frequently habitat generalists because they can find food in a wider range of vegetation types.

The carnivore species that adapt well to anthropogenic landscapes are all mesocarnivores and are highly omnivorous (coyotes and raccoons; [Prange et al. 2004, Crimmins et al. 2015]). Their smaller body size means they are less likely to come into conflict with humans, through killing livestock or even biting or killing people. Omnivores are able to exploit the wider range of food sources that is available in anthropogenic landscapes. This includes small vertebrate prey such as rabbits, rodents, birds, and lizards, and the ability to switch to invertebrate prey when abundant and even seasonally available fruit. These species include red foxes (*Vulpes vulpes*) and coyotes (*Canis latrans*) in the Canidae; raccoons in the Procyonidae; and skunks, badgers, and stone martens (*Martes foina*; in Europe) in the Mustelidae (Nowak 1999, McDonald 2001). Note that these species are in the carnivore taxa identified as having generalist dentition and diets. One example of a highly successful larger mesocarnivore is the coyote (body mass = 7–20 kg) (Nowak 1999). The coyote's ability to adapt to human landscapes resides in its extreme omnivory (Newsome et al. 2015), unlike its larger congeners such as the wolf. Coyote populations have increased in recent decades across much of their North American range, facilitated by fragmentation of native vegetation for agriculture and by competitive release as wolves have been extirpated (Ripple et al. 2003, Merkle et al. 2009).

Some of these generalist mesocarnivore species have become invasive. Invasive species are defined as those that have been introduced by humans outside their native range and have become established, with individuals dispersing, surviving, and reproducing at multiple sites (Blackburn et al. 2011). Where potentially invasive, generalist mesocarnivores have been introduced to evolutionarily isolated land masses where native fauna have no coevolutionary experience with these predators, the results have been catastrophic (Woinarski et al. 2014). Exacerbating this problem, invasive carnivores are frequently co-introduced with invasive prey such as European rabbits (*Oryctolagus cuniculus*) and rodents (black rats, *Rattus rattus*) (Johnson 2006). These prey species have the highest intrinsic rates of increase among mammals and support very high densities of invasive predators, leading to spillover or hyperpredation on native prey that have lower reproductive rates and are not able to sustain population growth (Smith and Quin 1996, Courchamp et al. 2000). Nowhere is this better exemplified than in Australia, which has had more mammal extinctions and declines in the last 200 years than anywhere else in the world (Woinarski et al. 2014). These are attributed to predation by alien, invasive predators such as red foxes and feral cats, combined with loss of low, dense vegetation structure, caused by overgrazing and wildfire, that provides refuge from predators (Woinarski et al. 2014, 2015). All the native marsupial carnivore species have declined, with foxes and cats major contributory factors in widespread declines of all species of quolls (*Dasyurus* spp.; Jones 1997).

Red foxes and feral cats are ranked in the top 100 of the world's worst invasive alien species and are among the most widespread and destructive carnivore species globally (IUCN Global Invasive Species Database, http://www.iucngisd.org/gisd/species.php?sc=66). The red fox has expanded its native distribution northward over vast areas of the Canadian arctic in the last 100 years, possibly because of climate warming or food subsidies, and is threatening the smaller arctic fox (*Vulpes lagopus*) with which it competes (Berteaux et al. 2015). Feral cats present a major conservation threat globally (Doherty et al. 2016) and are responsible for widespread extinctions of island faunas (Medina et al. 2014).

Harnessing the Effects of Carnivores in Food Webs for Conservation

Effective conservation management for many small carnivore species, particularly habitat or diet specialists, requires detailed knowledge of the ecological interactions between the species and its environment, predators, competitors, and prey species. The issue of invasive species, however, requires a different approach. Traditional methods of control, such as trapping, shooting, and poisoning, are resource-intensive and need to be applied in perpetuity to protect smaller animals. In the case of feral cats, traditional methods of lethal control by poisoning are ineffective at large scales in open landscapes (outside fences and islands; Moseby and Hill 2011). New approaches are thus needed to reduce the effects of feral predators on other wildlife, including small carnivores.

Harnessing the power of natural ecological interactions in food webs is a cost-effective management tool that has broad applicability. The top-down structuring role that top carnivores have in suppressing mesocarnivores has widespread support globally (Soule 2010, Terborgh and Estes 2010), although its effects will vary with environmental heterogeneity, as discussed previously for trophic cascades involving wolves in Yellowstone National Park. A relatively small investment in restoring an apex predator can lead to continuous suppression of an invasive mesopredator and move management away from the ongoing resourcing required for traditional lethal control. Examples include the effects of wolves on coyotes and coyotes on red foxes (Newsome and Ripple 2015), dingoes (*Canis lupus dingo*) on foxes and cats, and Tasmanian devils (*Sarcophilus harrisii*) on cats (Hollings et al. 2014) in Australian ecosystems.

Reducing the population density of co-introduced invasive prey could also be effective as a management tool to reduce populations of invasive predators. The dramatic population decline of rabbits following introduction of rabbit haemorrhagic disease to Australia led to a reduction in populations of feral cats and foxes and the recovery of several species of small mammals (Pedler et al. 2016). The use of guardian dogs in conservation to protect threatened species, potentially including small predators, is another tool. Guardian dogs have been used for hundreds of years in Europe to protect livestock from wolf attack (Dalmasso et al. 2012). The dogs are bonded to livestock but function as an apex predator and either directly or indirectly, by creating a landscape of fear, exclude other predators, including, potentially, dingoes, foxes, and cats. Guardian dogs function as a surrogate top predator and can also protect wildlife (van

Bommel and Johnson 2016). This approach has great potential for protecting populations of native small carnivores.

Using food web approaches to managing invasive species has huge potential, particularly because at large scales outside of fenced reserves and islands, we will probably never be able to eradicate invasive predators (Wallach and Ramp 2015). The goal then is to reduce their impacts so that wildlife can coexist.

Coexistence between Carnivores and People: Human Dimensions of Predators

In general, many species of carnivores can persist in many ecosystems if anthropogenic mortality can be reduced or eliminated and if prey populations can be maintained at sufficient density (Balme et al. 2009). Exceptions among large carnivores include situations in which the carnivores pose a high risk to people or livestock, or the carnivores themselves are at risk of mortality from other species of carnivores. Where human populations encroach into tiger habitat in India, tigers pose a genuine risk of mortality for villagers. The persistence of species like wolves in North America requires not only sufficient wildlands outside formal protected areas (such as Yellowstone National Park) but also tolerance in rural communities and at the wildland-urban interface. Carnivore species such as African wild dogs (*Lycaon pictus*) and cheetahs (*Acinonyx jubatus*) that are highly susceptible to kleptoparasitism and even direct intraguild killings may also require areas with lower large carnivore densities or lower prey density to minimize interference competition with other large carnivores such as lions (Creel and Creel 2002), which may run counter to economic interests based on ecotourism or trophy hunting (Lindsey et al. 2006). Reducing anthropogenic mortality may be realized through protective policies to reverse immediate declines, but programs that focus on increasing human tolerance may be more effective in the long run (Carter et al. 2017).

Summary

Persistence in the face of habitat loss, prey depletion, disease, persecution, and direct consumption is the primary conservation issue carnivores face globally. Persistence of many carnivore populations is relatively tenuous owing to purely biological and ecological factors: slow population growth, sparse distribution, and their need to capture and kill other animals for food. These fundamental traits further contribute to reduced population growth when they intersect with anthropogenic nodes: fear, trophy hunting, livestock production, and land conversion. Tolerance of carnivores is generally increasing and plays a critical role in the design of carnivore conservation and management policies, but the influences of tolerance can be complex.

LITERATURE CITED

Balme, G.A., R. Slotow, and L.T.B. Hunter. 2009. Impact of conservation interventions on the dynamics and persistence of a persecuted leopard (*Panthera pardus*) population. Biological Conservation 142:2681–2690.

Banks, E., D. Pimlott, and B. Ginsburg. 1967. Ecology and behavior of wolf: Introductory statement. American Zoologist 7:221–222.

Barrientos, R. 2015. Adult sex-ratio distortion in the native European polecat is related to the expansion of the invasive American mink. Biological Conservation 186:28–34.

Bartel, R.A., and M.W. Brunson. 2003. Effects of Utah's coyote bounty program on harvester behavior. Wildlife Society Bulletin 31:736–743.

Becker, M.S., R.A. Garrott, P.J. White, R. Jaffe, J.J. Borkowski, C.N. Gower, and E.J. Bergman. 2009. Wolf kill rates: Predictably variable? Pages 339–369 *in* R.A. Garrott, P.J. White, and F.G.R. Watson, editors. Ecology of large mammals in central Yellowstone: Sixteen years of integrated field studies. Volume 3. Academic Press, San Diego, California, USA.

Begg, C.M., J.R.B. Miller, and K.S. Begg. 2018. Effective implementation of age restrictions increases selectivity of sport hunting of the African lion. Journal of Applied Ecology 55:139–146.

Bergstrom, B.J. 2017. Carnivore conservation: Shifting the paradigm from control to coexistence. Journal of Mammalogy 98:1–6.

Berteaux, D., D. Gallant, B.N. Sacks, and M.J. Statham. 2015. Red foxes (*Vulpes vulpes*) at their expanding front in the Canadian arctic have indigenous maternal ancestry. Polar Biology 38:913–917.

Blackburn, T. M., P. Pysek, S. Bacher, J.T. Carlton, R. P. Duncan, V. Jarosik, J.R.U. Wilson, and D.M. Richardson. 2011. A proposed unified framework for biological invasions. Trends in Ecology and Evolution 26:333–339.

Brink, H., R.J. Smith, K. Skinner, and N. Leader-Williams. 2016. Sustainability and long-term tenure: Lion trophy hunting in Tanzania. PLOS One 11:e0162610.

Carbone, C., A. Teacher, and J.M. Rowcliffe. 2007. The costs of carnivory. PLOS Biology 5:e22.

Carbyn, L.N. 1981. Territory displacement in a wolf population with abundant prey. Journal of Mammalogy 62:193–195.

Carter, N.H., J.V. López-Bao, J.T. Bruskotter, M. Gore, G. Chapron, A. Johnson, Y. Epstein, et al. 2017. A conceptual framework for understanding illegal killing of large carnivores. Ambio 46:251–264.

Cherry, M.J., R.J. Warren, and L.M. Conner. 2016. Fear, fire, and behaviorally mediated trophic cascades in a frequently burned savanna. Forest Ecology and Management 368:133–139.

Christianson, D., and S. Creel. 2010. A nutritionally mediated risk effect of wolves on elk. Ecology 91:1184–1191.

Christianson, D., and S. Creel. 2014. Ecosystem scale declines in elk recruitment and population growth with wolf colonization: A before-after-control-impact approach. PLOS One 9:e102330.

Courchamp, F., M. Langlais, and G. Sugihara. 2000. Rabbits killing birds: Modelling the hyperpredation process. Journal of Animal Ecology 69:154–164.

Creel, S., and D. Christianson. 2009. Wolf presence and increased willow consumption by Yellowstone elk: Implications for trophic cascades. Ecology 90:2454–2466.

Creel, S., and N.M. Creel. 2002. The African wild dog: Behavior, ecology, and conservation. Princeton University Press, Princeton, New Jersey, USA.

Creel, S., M.S. Becker, S.M. Durant, J. M'Soka, W. Matandiko, A.J. Dickman, D. Christianson, et al. 2013. Conserving large populations of lions: The argument for fences has holes. Ecology Letters 16:1413-e3.

Creel, S., M. Becker, D. Christianson, E. Dröge, N. Hammerschlag, M.W. Hayward, U. Karanth, et al. 2015. Questionable policy for large carnivore hunting. Science 350:1473–1475.

Crimmins, S.M., L.R. Walleser, D.R. Hertel, P.C. McKann, J.J. Rohweder, and W.E. Thogmartin. 2015. Relating mesocarnivore relative abundance to anthropogenic land-use with a hierarchical spatial count model. Ecography 39:524–532.

Croll, D.A., J.L. Maron, J.A. Estes, E.M. Danner, and G.V. Byrd. 2005. Introduced predators transform subarctic islands from grassland to tundra. Science 307:1959–1961.

Dale, B.W., L.G. Adams, and R.T. Bowyer. 1994. Functional response of wolves preying on barren-ground caribou in a multiple-prey ecosystem. Journal of Animal Ecology 63:644–652.

Dalerum, F., E.Z. Cameron, K. Kunkel, and M.J. Somers. 2009. Diversity and depletions in continental carnivore guilds: Implications for prioritizing global carnivore conservation. Biology Letters 5:35–38.

Dalmasso, S., U. Vesco, L. Orlando, A. Tropini, and C. Passalacqua. 2012. An integrated program to prevent, mitigate and compensate Wolf (Canis lupus) damage in the Piedmont region (northern Italy). Hystrix: Italian Journal of Mammalogy 23:54–61.

Dayan, T., and D. Simberloff. 1998. Size patterns among competitors: Ecological character displacement and character release in mammals, with special reference to island populations. Mammal Review 28:99–124.

Dayan, T., D. Simberloff, E. Tchernov, and Y. Yom-Tov. 1989. Inter- and intraspecific character displacement in mustelids. Ecology 70:1526–1539.

Doherty, T.S., A.S. Glen, D.G. Nimmo, E.G. Ritchie, and C.R. Dickman. 2016. Invasive predators and global biodiversity loss. Proceedings of the National Academy of Sciences of the United States of America 113:11261–11265.

Durant, S.M., M.S. Becker, S. Creel, S. Bashir, A.J. Dickman, R.C. Beudels-Jamar, L. Lichtenfeld, et al. 2015. Developing fencing policies for dryland ecosystems. Journal of Applied Ecology 52:544–551.

Durant, S.M., N. Mitchell, R. Groom, N. Pettorelli, A. Ipavec, A.P. Jacobson, R. Woodroffe, et al. 2017. The global decline of cheetah Acinonyx jubatus and what it means for conservation. Proceedings of the National Academy of Sciences 114:528–533.

Ericsson, G., and T.A. Heberlein. 2003. Attitudes of hunters, locals, and the general public in Sweden now that the wolves are back. Biological Conservation 111:149–159.

Fortin, D., H.L. Beyer, M.S. Boyce, D.W. Smith, T. Duchesne, and J. S. Mao. 2005. Wolves influence elk movements: Behavior shapes a trophic cascade in Yellowstone National Park. Ecology 86:1320–1330.

Gittleman, J. 1985. Carnivore body size: Ecological and taxonomic correlates. Oecologia 67:540–554.

Hazzah, L., M. Borgerhoff Mulder, and L. Frank. 2009. Lions and warriors: Social factors underlying declining African lion populations and the effect of incentive-based management in Kenya. Biological Conservation 142:2428–2437.

Hollings, T., M. Jones, N. Mooney, and H. Mccallum. 2014. Trophic cascades following the disease-induced decline of an apex predator, the Tasmanian devil. Conservation Biology 28:63–75.

Johnson, C.N. 2006. Australia's mammal extinctions: A 50,000 year history. Cambridge University Press, Port Melbourne, Victoria, Australia.

Jones, M. 1997. Character displacement in Australian Dasyurid carnivores: Size relationships and prey size patterns. Ecology 78:2569–2587.

Jones, M.E. 2003. Convergence in ecomorphology and guild structure among marsupial and placental carnivores. Pages 281–292 in M.E. Jones, C.R. Dickman, and M. Archer, editors. Predators with pouches: The biology of carnivorous marsupials. CSIRO Publishing, Melbourne, Australia.

Kauffman, M.J., J.F. Brodie, and E.S. Jules. 2010. Are wolves saving Yellowstone's aspen? A landscape-level test of a behaviourally-mediated trophic cascade. Ecology 91:2742–2755.

Kellert, S.R., M. Black, C.R. Rush, and A. J. Bath. 1996. Human culture and large carnivore conservation in North America. Conservation Biology 10:977–990.

Kimble, D.S., D.B. Tyers, J. Robison-Cox, and B.F. Sowell. 2011. Aspen recovery since wolf reintroduction on the northern Yellowstone winter range. Rangeland Ecology and Management 64:119–130.

Lindsey, P.A., R. Alexander, L.G. Frank, A. Mathieson, and S.S. Romanach. 2006. Potential of trophy hunting to create incentives for wildlife conservation in Africa where alternative wildlife-based land uses may not be viable. Animal Conservation 9:283–291.

López-Bao, J.V., J. Bruskotter, and G. Chapron. 2017. Finding space for large carnivores. Nature Ecology and Evolution 1:s41559-017-0140-017.

Loveridge, A.J., A.W. Searle, F. Murindagomo, and D.W. MacDonald. 2007. The impact of sport-hunting on the population dynamics of an African lion population in a protected area. Biological Conservation 134:548–558.

MacDonald, D.W., editor. 2001. The new encyclopaedia of mammals. Oxford University Press, Oxford, UK.

Maria Fernandez, J., and N. Ruiz de Azua. 2010. Historical dynamics of a declining wolf population: Persecution vs. prey reduction. European Journal of Wildlife Research 56:169–179.

Marquard-Petersen, U. 2012. Decline and extermination of an arctic wolf population in East Greenland, 1899–1939. Arctic 65:155–166.

Medina, F.M., E. Bonnaud, E. Vidal, and M. Nogales. 2014. Underlying impacts of invasive cats on islands: Not only a question of predation. Biodiversity and Conservation 23:327–342.

Merkle, J.A., D.R. Stahler, and D.W. Smith. 2009. Interference competition between gray wolves and coyotes in Yellowstone National Park. Canadian Journal of Zoology/Revue Canadienne De Zoologie 87:56–63.

Metz, M.C., D.W. Smith, J.A. Vucetich, D.R. Stahler, and R.O. Peterson. 2012. Seasonal patterns of predation for gray wolves in the multi-prey system of Yellowstone National Park. Journal of Animal Ecology 81:553–563.

Moseby, K.E., and B.M. Hill. 2011. The use of poison baits to control feral cats and red foxes in arid South Australia I: Aerial baiting trials. Wildlife Research 38:338–349.

Nelson, M.P., J.T. Bruskotter, J.A. Vucetich, and G. Chapron. 2016. Emotions and the ethics of consequence in conservation decisions: Lessons from Cecil the lion. Conservation Letters 9:302–306.

Newsome, T.M., and W.J. Ripple. 2015. A continental scale trophic cascade from wolves through coyotes to foxes. Journal of Animal Ecology 84:49–59.

Newsome, S.D., H.M. Garbe, E.C. Wilson, and S.D. Gehrt. 2015. Individual variation in anthropogenic resource use in an urban carnivore. Oecologia 178:115–128.

Nowak, R.M. 1999. Walker's mammals of the world. Johns Hopkins University Press, Baltimore, Maryland, USA.

Packer, C., R. Hilborn, A. Mosser, B. Kissui, M. Borner, G. Hopcraft, J. Wilmshurst, S. Mduma, and A.R.E. Sinclair. 2005. Ecological change, group territoriality, and population dynamics in Serengeti lions. Science 307:390–393.

Pedler, R.D., R. Brandle, J.L. Read, R. Southgate, P. Bird, and K.E. Moseby. 2016. Rabbit biocontrol and landscape-scale recovery of threatened desert mammals. Conservation Biology 30:774–782.

Prange, S., S.D. Gehrt, and E.P. Wiggers. 2004. Influences of anthropogenic resources on raccoon (Procyon lotor) movements and spatial distribution. Journal of Mammalogy 85:483–490.

Prugh, L.R., C.J. Stoner, C.W. Epps, W.T. Bean, W.J. Ripple, A.S. Laliberte, and J.S. Brashares. 2009. The rise of the mesopredator. BioScience 59:779–791.

Ripple, W.J., and R.L. Beschta. 2004. Wolves, elk, willows, and trophic cascades in the upper Gallatin Range of southwestern Montana, USA. Forest Ecology and Management 200:161–181.

Ripple, W.J., A.J. Wirsing, C.C. Wilmers, and M. Letnic. 2013. Widespread mesopredator effects after wolf extirpation. Biological Conservation 160:70–79.

Ripple, W.J., J.A. Estes, R.L. Beschta, C.C. Wilmers, E.G. Ritchie, M. Hebblewhite, J. Berger, et al. 2014. Status and ecological effects of the world's largest carnivores. Science 343:1241484.

Rosenblatt, E., M.S. Becker, S. Creel, E. Droge, T. Mweetwa, P.A. Schuette, F. Watson, J. Merkle, and H. Mwape. 2014. Detecting declines of apex carnivores and evaluating their causes: An example with Zambian lions. Biological Conservation 180:176–186.

Sepúlveda, M.A., R.S. Singer, E.A. Silva-Rodríguez, A. Eguren, P. Stowhas, and K. Pelican. 2014. Invasive American mink: Linking pathogen risk between domestic and endangered carnivores. EcoHealth 11:409–419.

Shoemaker, K.T., R.C. Lacy, M.L. Verant, B.W. Brook, T.M. Livieri, P.S. Miller, D.A. Fordham, and H. Resit Akçakaya. 2014. Effects of prey metapopulation structure on the viability of black-footed ferrets in plague-impacted landscapes: A metamodelling approach. Journal of Applied Ecology 51:735–745.

Sinclair, A.R.E., and C.J. Krebs. 2002. Complex numerical responses to top-down and bottom-up processes in vertebrate populations. Philosophical Transactions of the Royal Society of London B: Biological Sciences 357:1221–1231.

Smith, A.P., and D.G. Quin. 1996. Patterns and causes of extinctions and decline in Australian conilurine rodents. Biological Conservation 77:243–267.

Soulé, M.E. 2010. Conservation relevance of ecological cascades. Pages 337–352 in J. Terborgh and J. Estes, editors. Trophic cascades: Predators, prey, and the changing dynamics of nature. Island Press, Washington, DC, USA.

Suutarinen, J., and I. Kojola. 2017. Poaching regulates the legally hunted wolf population in Finland. Biological Conservation 215:11–18.

Terborgh, J., and J.A. Estes. 2010. Trophic cascades: Predators, prey and the changing dynamics of nature. Island Press, Washington, DC, USA.

Tichelar, M. 2016. The history of opposition to blood sports in twentieth century England: Hunting at Bay. Routledge, London, UK.

Treves, A., and J. Bruskotter. 2014. Tolerance for predatory wildlife. Science 344:476–477.

US Fish and Wildlife Service. 2017. 2016 National survey of fishing, hunting, and wildlife-associated recreation. US Fish and Wildlife Service, Washington, DC, USA.

Valeix, M., A.J. Loveridge, and D.W. MacDonald. 2012. Influence of prey dispersion on territory and group size of African lions: A test of the resource dispersion hypothesis. Ecology 93:2490–2496.

van Bommel, L., and C.N. Johnson. 2016. Livestock guardian dogs as surrogate top predators? How Maremma sheepdogs affect a wildlife community. Ecology and Evolution 6:6702–6711.

Van Valkenburgh, B.V. 1988. Trophic diversity in past and present guilds of large predatory mammals. Paleobiology 14:155–173.

Van Valkenburgh, B.V. 1989. Carnivore dental adaptations and diet: A study of trophic diversity within guilds. Pages 410–436 in J.L. Gittleman, editor. Carnivore behavior, ecology, and evolution. Springer, Boston, Massachusetts, USA. https://link.springer.com/chapter/10.1007/978-1-4613-0855-3_16.

Venter, O., E.W. Sanderson, A. Magrach, J.R. Allan, J. Beher, K.R. Jones, H.P. Possingham, et al. 2016. Sixteen years of change in the global terrestrial human footprint and implications for biodiversity conservation. Nature Communications 7:12558.

Vucetich, J.A., M. Hebblewhite, D.W. Smith, and R.O. Peterson. 2011. Predicting prey population dynamics from kill rate, predation rate and predator-prey ratios in three wolf-ungulate systems. Journal of Animal Ecology 80:1236–1245.

Wagner, K., and M. Conover. 1999. Effect of preventive coyote hunting on sheep losses to coyote predation. USDA National Wildlife Research Center, Staff Publications. http://digitalcommons.unl.edu/icwdm_usdanwrc/831.

Wallach, A.D., and D. Ramp. 2015. Let's give feral cats their citizenship. The Conversation e45165.

Winnie, J.A. 2012. Predation risk, elk, and aspen: Tests of a behaviorally mediated trophic cascade in the Greater Yellowstone Ecosystem. Ecology 93:2600–2614.

Winnie, J., D. Christianson, S. Creel, and B. Maxwell. 2006. Elk decision-making rules are simplified in the presence of wolves. Behavioral Ecology and Sociobiology 61:277–289.

Wittemyer, G., P. Elsen, W.T. Bean, A.C.O. Burton, and J.S. Brashares. 2008. Accelerated human population growth at protected area edges. Science 321:123–126.

Woinarski, J., A. Burbidge, and P. Harrison. 2014. The action plan for Australian mammals 2012. CSIRO Publishing, Melbourne, Australia.

Woinarski, J.C.Z., A.A. Burbidge, and P.L. Harrison. 2015. Ongoing unraveling of a continental fauna: Decline and extinction of Australian mammals since European settlement. Proceedings of the National Academy of Sciences of the United States of America 112:4531–4540.

Woodroffe, R. 2000. Predators and people: Using human densities to interpret declines of large carnivores. Animal Conservation 3:165–173.

11

RONALD R. SWAISGOOD
CARLOS RUIZ-MIRANDA

Moving Animals in the Right Direction

Making Conservation Translocation an Effective Tool

Introduction

Humans have been moving animals around on the landscape for thousands of years for a plethora of reasons, but today these translocations figure prominently in the conservation practitioner's toolbox (Seddon et al. 2012). Conservation translocation, the deliberate human-mediated movement of organisms on the landscape, is undertaken to fulfill a variety of conservation purposes, ranging from supplementation of small, isolated populations to repatriation of species once extinct in the wild to assisted migration, in which animals are relocated to suitable habitat outside the species' historical range to mitigate climate-mediated changes in habitat suitability.

"Reintroduction" and "translocation" are terms fraught with ambiguity, and they have had varying definitions. We follow recent precedent established by the International Union for Conservation of Nature (IUCN 2013), using "translocation" as the umbrella term to refer to human-mediated movements for conservation purposes and "reintroduction" to refer to these actions when they are used to reestablish a species to a part of its historical range from which it has been extirpated.

Translocation biology is in its heyday. Twenty years ago, the scientific approach to conducting translocations was in its infancy, and few agreed-upon standards and best practices were available. But the past two decades have seen a surge in the publication of peer-reviewed articles that collectively establish a strong baseline understanding for what works and what does not in translocations (Seddon et al. 2007). As a result, the success rates for translocations have increased substantially over the past two decades, and today more than half of attempts can be classified as fully successful (Seddon and Armstrong 2016). It is important for practitioners to understand that today the use of translocation as a conservation tool is a viable option, provided best practice standards are followed. As a starting point for guidance on best practices, including deciding whether translocation is the appropriate tool for addressing a conservation problem, consulting IUCN's Reintroduction Specialist Group guidelines is merited (IUCN 2013).

Once a decision is made that translocation is appropriate for the conservation circumstances, establishing a clear plan is essential (IUCN 2013). Elements of this plan include goal articulation, identification of measurable outcomes, monitoring, progress reviews, and an exit strategy. Previously, translocation research was largely descriptive and relied on a posteriori analysis of monitoring data, rather than question-driven hypothesis testing (Seddon et al. 2007). As required by adaptive management principles (Walters 1986), we can learn a great deal by doing translocations if a priori hypotheses from various alternative management actions are developed and tested. Translocations are in fact ideal platforms from which to conduct large-scale adaptive management experiments because they inherently involve human intervention.

A more difficult challenge is determining which questions to prioritize when developing a science-driven translocation program. Diverse approaches to this problem have been taken, probably reflecting real species and contextual differences and the interests and biases of the investigators. Armstrong and Seddon (2008) provide a framework of 10 key questions that can serve as a good starting point, and Batson et al. (2015) provide a more detailed list, but the devil will undoubtedly be in the detail. Practitioners and researchers formulating their management questions will need much more than general guidelines. In Batson and colleagues' terminology, we must devise a series of "tactics," each of which maps on to a specific goal-defined strategy, such as reducing post-release predation. If, during the planning stages, predation was determined to be a limiting factor, then several tactics, ranging from antipredator training to lethal control of predators at the release site, may be evaluated and prioritized. These decisions will of course be premised on a literature review evaluating past successes and failures of a tactic, and some sort of assessment of its taxonomic relevance.

In addition to these conservation translocations, there are a growing number of "mitigation translocations," which, although ostensibly motivated by conservation regulations, have benefited much less from the evidence-based best practices that have been developed in recent years. An important category of translocation that has not been given sufficient attention is mitigation-driven translocations, those undertaken to offset development impacts rather than to achieve a specific conservation advancement (Germano et al. 2015). Globally, the use of mitigation translocations is increasing rapidly, and its frequency far outstrips that for conservation translocations. These programs have great potential for conservation value but are plagued with a number of problems. At their core, these problems stem from two basic facts. First, they are undertaken by developers and their consultants to meet legal requirements to mitigate development impacts rather than as part of a larger strategic recovery goal. The result is often a piecemeal approach to conservation with no net benefit to the species or, worse, consultants doing the minimum to check the regulatory box. Second, mitigation translocations are essentially supply-driven rather than demand-driven, meaning that animals become available for translocation because their habitat is lost to development (supply). Conservation translocations, by contrast, are motivated by the need to restore the species to an important historical part of their range (demand). An obvious conclusion is that mitigation translocations can become conservation translocations if the relocated animals are released strategically into areas that require reestablishment or supplementation of the species. To maximize the value of mitigation translocations, it is also necessary that practitioners follow best practices established by the science of conservation translocations, including use of hypothesis testing and adaptive management, closer attention to ecological and behavioral needs, and full disclosure of outcomes in reports made available to the public (Germano et al. 2015).

While acknowledging that conservation translocations involve meeting appropriate standards for a number of criteria (Pérez et al. 2012), our review focuses on the scientific and technical aspects of translocation biology and provides a brief overview of some of the plethora of other issues that complicate this strategic conservation tool.

Strategic Planning: Envisioning a Successful Translocation Program

Strategic planning is a key element for the success of biodiversity conservation projects (Nichols and

Armstrong 2012). Inherently conservation interventions, translocations require adequate planning to minimize risks. The IUCN guidelines (IUCN 2013) are the go-to tool for planning a translocation program, covering goal-setting, monitoring programs, feasibility analysis, criteria for success, and recommendations for some best practices in design and implementation, but they remain necessarily general with regard to the latter three topics (which have been better developed in Batson et al. 2015). Structural decision making (SDM) is a powerful planning approach, although it requires extensive information about probabilities of outcomes and degree of uncertainty (Nichols and Armstrong 2012). Another useful planning approach is the Open Standards for the Practice of Conservation (Schwartz et al. 2012). This methodology specifies how the adaptive management cycle should be carried out, and it allows for initiating actions before all knowledge is available. This approach also requires a clear definition of the project's targets, vision, scope, and threats, and provides a roadmap (result chain) to achieve the goals. Both SDM and Open Standards emphasize monitoring and adaptive management and provide the framework necessary to keep the project in focus and create a learning process that will contribute to the science of reintroduction.

The initial planning steps are: assemble a good core team and select the target (species, ecosystem, or ecological process), vision (the end goal or state of affairs to be achieved), and scope (geographic and thematic focus). The core team should be multidisciplinary and will develop, plan, and execute the project. They also identify and select other participants and stakeholders to help with strategies and other procedures. The team selects the appropriate conservation target, the vision, the scope, and the time frame. Projects have a measurable set of goals, enough to achieve the desired conservation end goal within the time frame (Table 11.1). The next set of decisions includes feasibility assessment and cost-benefit analysis with regard to conservation targets (Pérez et al. 2012, Schwartz et al. 2012), which could

lead to project termination. Factors that affect feasibility are lack of good source population, lack of funding, continued threats at the release site, low-quality habitat, negative human perception, or lack of local support. Once the decision is made to go forward with the translocation, a new set of decisions are necessary: select translocation units, develop release tactics, identify and mitigate threats, choose monitoring tools and success indicators, develop a project timeline, and develop an exit strategy. In cases of significant uncertainty, a science-driven project will rely on focused research to make some of these decisions.

The current IUCN guidelines for translocations (IUCN 2013) consider an exit strategy an explicit component of planning a translocation project. An exit strategy helps avoid unacceptable or undesired social or financial outcomes. Whereas criteria for success are relatively well defined, similar criteria for failure have not been established. Because the road to failure may be in stages, a key decision is when to implement the exit strategy. If too early, for example, after initial failures, then conservation goals may be compromised. If too late, then resources are wasted and ethical issues may arise. A temptation will be to commit the "Concorde fallacy," continuing to invest in a program that is failing simply because of the degree of prior investment. Therefore, translocation projects must have clearly defined criteria for success and failure within a specified timeline.

Translocation risks and the associated negative outcomes must be made explicit at the beginning of the project to all stakeholders and participants (Table 11.2). For example, how many changes to management tactics is the group willing to accept before deeming the translocation a failure or unsuccessful? What is the expected lag time for population recovery to show? What are the acceptable losses in terms of survival rates, reproductive rates, descendant survival, or financial costs? Then the exit strategy can be planned accordingly, and must include consideration of negative effects on source populations, funding agencies, the general public,

Table 11.1 Monitoring goals and measures

General approach	Conservation target	Vision or goal	Project scope	Monitoring measure	Monitoring method
Single species	Individual species	Population recovery Population viability	Species range or portion Duration based on life history strategy (generations an appropriate metric)	Population estimate Genetic diversity Dispersal Viability modeling	Observational surveys Mark-recapture Telemetry tracking Molecular genetics Pedigree analysis
Ecological process	Seed dispersal Predator-prey dynamics	Recruitment of target species Seed dispersal Increased biodiversity	Specified geographical area Duration depends on targets	Seedling success Dispersal distance Mortality Population estimate Diversity index	Observational surveys Mark-recapture Telemetry tracking Molecular genetics Observational surveys Vegetation transects and plot surveys
Rewilding	Ecological community	Increased biodiversity index Guild representation	Specified geographical area Duration depends on targets	Diversity index Occupancy Persistence Relative abundance Intervention requirements	Observational surveys Mark-recapture Spatial modeling
Welfare choice	Species or individuals	Well-being improved	Specified geographical area Duration depends on target species	Behavioral / physiological measures of well-being Social interactions Stress	Behavioral observations Physiological measures

colleagues, local stakeholders, and project staff (Kleiman et al. 1991, Christensen 2003).

Monitoring Translocations

Monitoring refers to the collection of necessary and adequate data to evaluate progress and make informed decisions about current and future actions (Fig. 11.1). Without monitoring, valuable learning opportunities are lost, making it difficult to develop an optimal course of action. Adaptive management methodology provides the conceptual and methodological tools for monitoring targets and project success (Salafsky et al. 2001, Nichols and Armstrong 2012). Monitoring should be designed as a question-driven or hypothesis-testing series of (management) activities before and after release of the animals. Several variables could be monitored: survival, re-

production, abundance (relative or absolute), occupancy, presence, index of diversity, persistence over generations, movement or dispersal, and rates or distances, among others. Current technological advances, although potentially costly, in animal marking, telemetry, acoustic monitoring, camera traps, and other forms of detection offer many options for monitoring individuals over time and across large geographical areas (Fig. 11.2).

Monitoring variables must correspond to identified conservation goals and targets (Table 11.1). In a single-species conservation approach, monitoring must include some measure of population size or viability, often defined in terms of demography and genetic diversity. For example, the golden lion tamarin (*Leontopithecus rosalia*) project in Brazil (Fig. 11.3) established a target population size of 2,000 individuals in 25,000 ha of connected and protected forest,

Table 11.2 Human-dimension problems and solutions

Stakeholder	Potential problems & consequences	How to avoid/address problems
Private landowner	Direct conflict with humans (e.g., crop raiding, livestock predation, or attack on humans), economic consequences, or (partial) loss of use of land for specified activities (development). A relationship turned sour with a local landowner can lead to denial of access to private lands, and this can spread to neighboring properties, disrupting monitoring methods or future release plans.	Local landowners can be powerful allies. One promising tool to facilitate community involvement is citizen science (Dickinson et al. 2012). Local community members can be recruited to assist with project implementation, such as monitoring, habitat restoration, or other activities. Participation in such ecological stewardship is perhaps the most powerful tool at our disposal to get lasting support for a conservation projects (Schwartz 2006).
Professional colleagues	Many aspects of translocations remain controversial, such as disease risk, genetic mixing, ecological replacement, assisted migration, welfare impacts, and other first-do-no-harm principles, presenting significant obstacles to what is essentially an interventionist endeavor, and can lead to paralysis affecting achievement of important conservation goals (Meek et al. 2015).	Communication skills are necessary to successfully navigate these situations, and practitioners should be prepared to explain why they are implementing a translocation or choosing certain tactics. Backing up your arguments with preliminary data, case studies, or other scientifically sound evidence supporting your decisions will be important to convince colleagues. It is also important to remain collegial and avoid territorial disputes and other forms of escalation (which serve only to confuse other stakeholders and damage your reputation with them).
Internal stakeholders	Uninformed staff may hold the organization accountable if goals are not met or setbacks occur. Individuals inside an organization may voice opposition or question mission relevance if they are not included early in the planning process.	Communication plans should transparently acknowledge the risks, and emphasize that setbacks should not be considered failures, provided measures are in place to ensure that learning takes place (adaptive management). All internal stakeholders should participate in the strategic planning of the project or be given an opportunity to express their opinion, depending on their role in the project.
Funding agencies	If risks and expected time frame for grant deliverables are not understood, agencies may withdraw funding. Funders are sometimes known to "chase shiny objects," and while early releases may be exciting and attract funders, translocation programs are often a long haul. Projects experiencing donor fatigue may get off to a good start but fail to accomplish all goals if funding dries up.	Many funding relationships rely on only the written word in a grant proposal. It is important to point out that there will be setbacks and delays and to cultivate reasonable expectations (while also generating enthusiasm for a project that can be a winner), but also important to meet funders and discuss the project more openly so they can ask questions and come to a better understanding. In-person meetings also help build trust in the investigator.
Regulatory agencies	Poor understanding of project goals, timeframe, and methods may lead to noncompliance findings, revoked permits, or loss of access to lands.	Clear, detailed, and transparent plans, particularly those using a formal decision-making process, will help cultivate appropriate expectations, avoid surprises, and instill trust that investigators have done due diligence. Diagrams (such as those produced in Open Standards Planning) can be particularly useful for explaining goals, threats, and how decisions were made.
General public	Critical media attention or negative local gossip can hamstring a translocation program. Poor relations with the general public can reflect poorly on the program and its proponents and change opinion or cause problems with other stakeholders with more immediate influence on the program (i.e., the stakeholder categories reviewed above).	A communication plan for engaging the general public is essential, one that conveys risks honestly but also contains positive messaging explaining why the return of a species to the ecosystem is a good thing. Reintroduction of local native species into parks or visited areas may help people enjoy and respect wildlife (Parker 2008). People may respond positively to knowing that wildlife habitats bring a variety of ecosystem services that directly affect health through air and water quality. Chances of long-term success increase with community participation, especially when members of the community are stakeholders and actively participate, are committed to, and rally around goals of the project.

Figure 11.1   Biologists releasing a desert tortoise as part of a San Diego Zoo Global translocation research program, designed to determine the ecological factors at the release site that influence dispersal and survival. Ron Swaisgood

a goal that, when reached, would constitute a viable population capable of persistence with minimal continued intervention. Therefore, indicators of population size, forest area, landscape connectivity, and forest protection status needed to be measured for the project. Within that context, the specific goals of the translocation of golden lion tamarins were to reinforce the existing wild population, to increase the occupancy of forests, and consequently protect the forests (Kierulff et al. 2002).

Ecosystem approaches may monitor other variables. If the goal is to recover ecological processes, such as seed dispersal, then the project must define success in terms of how translocations of other organisms produce changes in seed dispersal and select indicators of those targets. Rewilding projects are often used to restore habitats, including ecological processes. They must monitor biological diversity, specific ecological processes, and resilience, because a goal of rewilding is to remove the need for future human intervention (Corlett 2016). The length of the monitoring is related to species life history and measured in generations (Jones and Merton 2012).

Figure 11.2   This golden lion tamarin, a reproductive male with radio transmitter collar, carries an infant offspring. A studbook managed captive population distributed across more than 100 zoos in Europe and the United States; a research-driven post-release monitoring program is a key element of the release in this successful reintroduction project. Despite some initial losses of animals, approximately 45% of the current free-living global population of tamarins are descendants of the reintroduced groups. Andreia F. Martins

Figure 11.3   The Golden Lion Tamarin Association field team and researchers guide visits of local adults (teachers and landowners) and children to observe descendants of reintroduced golden lion tamarins in a forest on a privately owned property. Involving the local community in the conservation project is necessary to ensure long-term continuity and diversity of conservation activities. Andreia F. Martins

A Multidisciplinary Science

Translocation science is inherently interdisciplinary, incorporating behavior, ecology, genetics, disease, parasitology, veterinary science, nutrition, physiology, welfare science, social science, and many more fields. In this chapter we cannot do justice to all these necessary contributions, but we touch on a few highlights here.

Genetic considerations are of paramount importance in planning and implementing a translocation program (Jamieson and Lacy 2012). When establishing a conservation breeding program or selecting wild individuals for translocation, it is critical to obtain as much genetic representation from the source population(s) as possible. This can be done by sampling animals over a large geographic area or through genetic evaluation from DNA samples. If establishing a breeding program in captivity, a genetic management plan (typically based on pairings with regard to mean kinship) is required to ensure that genetic diversity is not lost through inbreeding, drift, or domestication processes (Woodworth et al. 2002). The fewer generations the species has been in captivity, the less risk there is that important alleles will be lost. Released populations must also contain high levels of the existing genetic diversity to guard against inbreeding depression and to maintain future evolutionary potential and adaptability. This requires release of sufficient numbers of individuals (typically >60; Jamieson and Lacy 2012). Removals from source populations must also consider the genetic diversity remaining so as not to similarly jeopardize the source population's genetic viability.

Good health is a prerequisite for any translocation, and pre-release health screening is necessary to ensure that only animals prepared to thrive are included in the release (Ewen et al. 2012). A standard veterinary examination should include a detailed physical exam, screening for any symptoms of disease, an evaluation of body condition, and in some cases hematological evaluation. As is often the case for rare and endangered species, however, normal baseline parameters are typically not available for comparison. Disease risk is also important for the recipient population. The introduction of novel pathogens could be devastating for any existing resident population or other components of the ecosystem. Thus, disease risk assessments prior to release are critical. The degree of risk will be determined by several factors, including pathogen virulence, the degree to which it is already present at the release site and neighboring areas, the distance animals are moved in the case of wild-to-wild translocation, its probability of spreading, and other epidemiological variables. For example, the combination of novelty and virulence makes chytrid introductions particularly problematic, requiring rigorous biosecurity measures in captive amphibian populations and stringent screening protocols prior to release of organisms potentially carrying this deadly fungus (Pessier and Mendelson 2010). Finally, it is worth noting that stress is prevalent in translocations, and it exacerbates health and disease issues by reducing immunocompetence (Dickens et al. 2010).

Physiology has become increasingly relevant for conservation, and physiological research now figures prominently in translocation (Jachowski et al. 2016a). A number of physiological parameters provide excellent measures for assessing how individual animals are coping with the process of translocation, from holding facilities to the post-release environment. Physiological indicators of stress, such as glucocorticoids, are useful tools for understanding how animals perceive their environment and their process of adapting to the novel environment (Teixeira et al. 2007, Dickens et al. 2010). To the extent that stress is detrimental, these physiological evaluations allow managers to address and mitigate stressors, understand the underlying causal mechanisms, and adapt future release strategies. Other important applications of physiological research include assessment of immunocompetence and monitoring of reproductive function.

Conservation and welfare issues intersect prominently in translocations (Swaisgood 2010, Moehrenschlager and Lloyd 2016). On the surface, welfare and conservation goals may seem to conflict, with individuals suffering the stressors of the translocation process in service of greater conservation goals. There are many opportunities for synergy between conservation and welfare goals, however, and good science applied to translocation can improve both conservation and welfare outcomes.

Behavioral and Ecological Approaches to Translocation

Perhaps no area in conservation biology stands to benefit more from behavioral and ecological research than does translocation. Fortunately, the history of behavioral ecological research contributes to improving conservation outcomes (Kleiman 1989, Swaisgood 2010, Moehrenschlager and Lloyd 2016). The application of behavioral ecology has had a history in devising ex situ breeding programs and is important for identifying the appropriate social and environmental stimuli to support normal behavioral development and prevent erosion of key behaviors necessary for post-release survival (Kleiman 1989). Attention to behavioral details of the species can mean the difference between success and failure in translocation programs. For example, whereas multiple previous attempts to translocate Stephens' kangaroo rats (*Dipodomys stephensi*) failed, the application of behavioral research targeting the social setting of release led to the successful establishment of five new populations (Shier and Swaisgood 2012). How an animal interacts with its environment is intimately tied to translocations because translocations alter the animal-environmental interaction, making this behavioral adjustment a key problem that can be addressed with targeted research. Sociopolitical aspects aside, translocation outcomes will thus be gov-

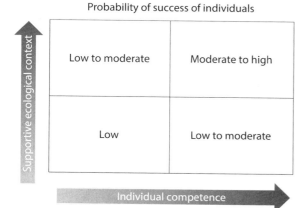

Probability of success of individuals

Low to moderate	Moderate to high
Low	Low to moderate

Supportive ecological context

Individual competence

Figure 11.4 Conceptual diagram depicting the probability of post-release success as a result of the relationship between individual animal capacity to survive and reproduce in the wild (competence) and the ecological context at the release site, such as habitat quality, threats, social environment, and other factors influencing survival and reproduction.

erned largely by a combination of the individual animal's competence and the ecological context (habitat quality, threats, social environment) in which it is released (see Fig. 11.4). Behavioral ecology is a critical issue in pre-release training programs (Shier 2016), on-site holding or acclimation (Moehrenschlager and Lloyd 2016), and post-release settlement and adjustment to the new environment (Swaisgood 2010, Le Gouar et al. 2012). Below we highlight a few of the more important behavioral ecological applications from each of these three stages of translocation.

Antipredator Training

Pre-release training of a wide variety of survival skills is an essential part of translocations relying on source animals bred in captivity, with applications including training antipredator behavior, foraging skills, and more general learning and cognitive skills (Moehrenschlager and Lloyd 2016, Shier 2016). Predation is one of the most important factors limiting translocation success of prey species (Moseby et al.

2011), an unsurprising outcome given that translocated animals often remain unsettled for a period of time as they explore their environment and must learn how to use the habitat to avoid predators, such as locating predator refugia. Captive-bred animals are at a further disadvantage because they lack developmental experience with predators and, as a result, experience predation levels nearly twice as high as wild-caught translocated animals (Harrington et al. 2013). Antipredator behavior may also erode with increasing generations in captivity.

Antipredator training is thus an important tool for training captive-bred animals for release and can be associated with higher survival post-release (Griffin et al. 2000, Shier 2016). Antipredator training takes a variety of forms. Often it uses associative learning, pairing predator stimuli with aversive stimuli to train a fear response. To determine whether an animal has learned something that might be useful following release, it is essential to have a baseline trial presenting an animal with a standardized predator stimulus, a training trial in which stimuli are paired with aversive stimuli, and a post-training trial, identical to the baseline trial, to measure change in behavior that lasts beyond the aversive training trial (learning). Historically aversive stimuli have included loud noises, electric shock, and even being shot with a rubber band, but the best training will likely include biologically relevant stimuli that produce a motivational state similar to that experienced during a real predation event. Good examples include conspecific alarm calls, capture and restraint by humans, chasing by live or robotic predators, or presentation of a "dead" conspecific. It is also important to have target behaviors that constitute an appropriate antipredator response for the species, such as fleeing, hiding, seeking refuge, alarm calling, or mobbing, and measures of what animals should not be doing in a predator context, such as foraging, socializing, or some other activity that detracts from vigilance. Additional contextual modifications to training programs may increase their efficacy. The presence of an experienced conspecific that can serve as a demonstrator can greatly enhance learning and help shape appropriate antipredator responses (observational learning) (Shier 2016).

In part because of lack of proper experimental controls, it is unknown for most programs whether training actually increases post-release survival (Moehrenschlager and Lloyd 2016); thus, it is difficult to recommend best practices. In some cases, antipredator training confers no survival advantage (Moseby et al. 2012), perhaps because antipredator training addresses only a subset of the learning that takes place in avoiding predators. Antipredator training typically focuses on recognizing and responding to predators but rarely teaches animals how to behave appropriately to avoid undetected predators (Moseby et al. 2016). To avoid becoming prey, animals learn to balance vigilance with other activities such as foraging, learn variation in risk and vulnerability associated with different habitats, and exercise greater caution at certain times of day, among other more nuanced forms of antipredator behavior (Cooper and Blumstein 2015).

Acclimation at the Release Site

Acclimation, holding animals in an enclosure at the release site, is one of the most frequently used but relatively poorly tested aspects of translocation protocols. Sometimes referred to as "soft release," allowing animals time to adjust to conditions at the release site before release, has intuitive appeal. Animals may need to adjust to differences in climate or habitat between the source and release sites, and they may learn important aspects about their new environment without direct exposure to risks, such as predation or aggression from resident conspecifics. Holding release animals in acclimation enclosures provides opportunities to promote social familiarity with resident animals in the case of translocations to areas with resident conspecifics. Animals may also become familiar with the surrounding habitat, and thus be more likely to remain near the release site.

The results from studies testing acclimation strategies are mixed (Moehrenschlager and Lloyd 2016). Some cross-species meta-analyses show no effect of acclimation, with translocations using acclimated animals no more successful than those using "hard releases" (i.e., releasing animals directly into an area; Wolf et al. 1998). Individual studies, however, directly comparing hard versus soft releases have demonstrated higher survival rates in some but not all cases (Moehrenschlager and Lloyd 2016). Similarly, acclimation sometimes reduces dispersal distances (Mitchell et al. 2011), but sometimes does not (Richardson et al. 2015). The reasons for these variable effects of acclimation are not always clear, and may involve differences in species life history characteristics and social organization, and differences in situational context. For example, it may be predicted that acclimation will be more efficacious for captive-bred than wild-caught animals because captive animals have less experience with natural environments and therefore have more to gain from learning and adaptation to disparate environmental conditions at the release site.

An important point rarely made is that the effects of acclimation may change in complicated ways for different acclimation periods. Wild-caught animals, for example, may experience elevated stress initially when confined to acclimation enclosures, but this response may gradually diminish as the animal accommodates to its new surroundings. As a result, translocation outcomes may follow a similar pattern, with higher success associated with hard release (no acclimation) and longer acclimation periods, but lower success resulting from short acclimation periods. To the extent that stress compromises post-release outcomes (Teixeira et al. 2007, Dickens et al. 2010), one might predict intermediate durations of acclimation to have more favorable results. By contrast, captive-bred animals may experience less stress when relocated from one captive environment to another, or simply benefit more from the learning opportunities acclimation affords. Here, one might expect a more linear increase of translocation success with longer acclimation periods, with an asymptote at the point in time when sufficient acclimation and learning has taken place. At present, these ideas are largely conjectural, but may help explain the complicated relationship between acclimation and translocation success. Regardless of the mechanisms influencing relationships between acclimation period and translocation outcome, adaptive management approaches wherein acclimation period is experimentally manipulated across a range of durations will be necessary to advance our understanding and application of translocation.

Reducing Post-Release Dispersal

Long-distance dispersal is one of the most important factors limiting translocation success and has been implicated as a problem in at least 20% of translocations (Harrington et al. 2013). Consequently, understanding and dampening post-release dispersal has been nominated as one of the 10 most important questions in translocation biology (Armstrong and Seddon 2008) and as one of the top 50 questions in conservation behavior (Greggor et al. 2016). Why do so many animals apparently reject (often perfectly suitable) habitat at the release site and move rapidly away, sometimes dispersing distances much farther than species-typical natal dispersal (Stamps and Swaisgood 2007, Le Gouar et al. 2012)? More importantly, what tools can managers use to address this problem?

Excessive post-release dispersal is problematic for several reasons. Dispersal distances generally correlate with mortality, often because dispersing animals put themselves at greater cumulative risk (Stamps and Swaisgood 2007, Le Gouar et al. 2012). Dispersing animals may be more vulnerable to predators, be at greater risk of conspecific conflict, have lower food intake rates, and expend more energy. Once settled, animals gain knowledge about their environment that reduces these costs (greater foraging efficiency, use of refuges from predators). Delayed settlement can also delay reproduction, thus influencing reproductive success and population establishment (Shier and Swaisgood 2012), ultimately requiring the release of more individuals to establish a genetically viable population. Dispersing animals may move out of protected areas and into less suitable habitat, which may make post-release support (e.g., supplemental food) and monitoring difficult or create conflict with humans, in the case of dangerous or pest species.

To develop strategies for dampening dispersal, we must first understand why animals leave the release site. Hypotheses for post-release dispersal enjoy varying degrees of empirical support. Animals may reject the release site because they perceive the habitat to be unsuitable. Animals that use the presence of conspecifics as cues for habitat quality may disperse if there are no conspecifics present (Reed and Dobson 1993) or because there are benefits to living in social groups. To address this problem, managers can ensure that release sites contain conspecifics (though not at carrying capacity) or they can experimentally provide conspecific cues (acoustic, visual, olfactory cues) that facilitate settlement by released animals (Swaisgood 2010). Social familiarity, with the release group or between released animals and residents, can also promote settlement. Thus providing opportunities for social interactions between animals before release can reduce dispersal distances, even among asocial species (Shier and Swaisgood 2012).

Animals that develop in one habitat and are released in another may reject that habitat because they have developed a preference for their natal habitat or because they have learned how to more successfully exploit their natal habitat (Stamps and Swaisgood 2007). Although habitat quality is one of the best predictors of successful translocation (Wolf et al. 1998), managers must also realize that it may be important to match habitat at the source and release sites as much as possible. Within areas that generally constitute suitable habitat, there may also be specific resources that govern an animal's decision regarding settlement. Releasing animals into areas with specific or plentiful foraging resources can re-duce post-release movements (Nafus et al. 2016) in the same manner that provision of supplemental food helps anchor released animals (Jones and Merton 2012). Less intuitively, release of animals into areas that provide good background camouflage (to reduce vulnerability to predators) can also reduce dispersal (Nafus et al. 2016).

In addition to manipulation of these key ecological features, translocation managers have used basic soft-release techniques, such as supplemental feeding and on-site acclimation; however, results have been mixed, and these methods have not been shown to reliably reduce post-release movements (Le Gouar et al. 2012). The success of all these strategies for reducing dispersal varies across species and contexts. Thus, expert opinion, captured in structured decision making or a similar planning process (Nichols and Armstrong 2012, Schwartz et al. 2012), can inform possible avenues to prioritize, but controlled experimentation will be necessary to confirm efficacy.

Selecting and Managing Release Habitat

The quality of habitat at the release area is one of the most powerful predictors of translocation success (Wolf et al. 1998), and correspondingly, understanding habitat suitability is one of the most important questions for translocation success (Armstrong and Seddon 2008). In the past, subjective opinion has often substituted for real data on habitat suitability (64% of reported translocations rely on subjective opinion of habitat quality; Wolf et al. 1998), but empirical assessment of habitat quality beyond what can be inferred by geographic information systems is now considered a prerequisite for translocation planning (IUCN 2013). In the Anthropocene, however, habitats are rapidly changing, making it difficult to map occupancy onto suitability (Osborne and Seddon 2012). Habitat suitability models may be misleading if built on data collected from animals living in degraded, less-suitable areas (Battin 2004). Fur-

thermore, present-day suitable habitat may not be suitable in the future, even with protection, because of climate change. Other considerations when selecting release habitat include ensuring that it is large enough to contain sufficient resources to sustain the target population size and that there is connectivity with other populations to ensure genetic and demographic viability (IUCN 2013).

Habitat suitability is often treated as a binary measure: habitat is either suitable or it is not. But habitats containing an abundance of some key resources may be more suitable, allowing animals to thrive. Specifically, habitat is higher quality if individuals exhibit good body condition, survival, and reproductive success that indicate positive demographic parameters and thus reduced likelihood of population extinction. We describe this emphasis on relative habitat quality the "good to great" approach. When selecting release areas, animals must be given every advantage possible, so "suitable" habitat might not be good enough. Translocated animals experience a novel environment and therefore are expected to be less efficient at locating foraging, refuge, and other resources, so higher quality habitat may offset these disadvantages.

Translocations might also be viewed as a "probe" to better understand the habitat factors that support positive outcomes. Data collection on foraging activities, habitat use, and body mass can help evaluate variation in habitat suitability and aid in decision making on the potential value of supplemental feeding (Bright and Morris 1994). For water voles (*Arvicola terrestris*), animals released at more vegetated sites suffered less predation and had higher survival and establishment rates (Moorhouse et al. 2009). For the desert tortoise (*Gopherus agassizii*), several sometimes subtle factors are important for successful translocation outcomes (Nafus et al. 2015, 2016). Releasing tortoises near washes (containing high-quality burrows and forage) leads to higher survival. Less intuitively, the rockiness of the substrate at the release site reduces predation rates because rocky substrate provides opportunities for camouflage that reduces predator detection. Similarly, release of tortoises in areas with high densities of small mammal burrows (where juvenile tortoises also seek refuge) reduces predation rates. Thus, selection of release sites with washes, rocky substrate, and high burrow density can increase success of tortoise translocations. Alternatively, managers can provide artificial shelters to increase access to refuges (Cabezas and Moreno 2007). Careful studies of the distribution or manipulation of resources in the release area can thus enhance ecological knowledge of habitat requirements and improve future translocation outcomes.

Human Dimensions of Conservation Translocations

Any translocation program that does not first consult local human stakeholders, acknowledge the role of human values and perceptions, consider ethical dimensions, or galvanize support in local communities risks failure (Jachowski et al. 2016b). Traditionally overlooked, human dimensions now play a prominent role in conservation, and translocation projects require engagement with a variety of stakeholders (Table 11.2).

When we hear the word "stakeholder" in translocation contexts, most of us will think first of the private landholders owning land at or near the release site. This is especially true if the released species cause conflicts with humans, such as crop raiding and livestock predation, or are large dangerous animals. The reintroduction of golden lion tamarins provides a good case study (Ruiz-Miranda et al. 2010). From its inception, this project has relied almost exclusively on forests within private properties. Early and substantive nongovernmental organization involvement created positive long-term relationships with local landowners through conservation education, home visits, sharing information, listening to concerns, and citizen science.

Professional colleagues are often not considered as part of the stakeholder community, but the opinion of colleagues can influence the success or failure of a translocation program. We have even encountered other scientists who "do not believe" in translocations. Professional critiques and diverging opinions are to be expected and can be healthy for the conceptualization and execution of the project. Buy-in from internal stakeholders is also essential. Animal caretakers are a particularly important stakeholder for captive-release programs, as they have vested emotional interest in the individuals released. Other important stakeholder groups, each with its own set of issues requiring tailored communication strategies, include funding agencies, regulatory agencies, and the general public (Table 11.2).

All stakeholders should be well informed of the risks and potential outcomes of the project, including direct biodiversity conservation gains and benefits to human health and economy.

Summary

Living in the Anthropocene, we must acknowledge that we are also living in an era of conservation interventions (Corlett 2016). Conservation translocations (Seddon et al. 2007) and mitigation translocations (Germano et al. 2015) figure prominently in the current and future conservation agenda. Historically, translocations have been a tool mostly for recovering endangered species in a particular part of their range or an activity conducted by wildlife managers to increase game species, but the scope of this conservation management tool has expanded rapidly; among other uses, it is now employed for relocating nuisance wildlife, rewilding landscapes, and as a climate-adaption strategy (Seddon et al. 2012, IUCN 2013, Corlett 2016). Thus, it is imperative that we develop and use the proper decision-making tools to decide when and where to use translocations, the best tactics and protocols based on biological relevance and good science, and the appropriate communication mechanisms to share lessons learned with practitioners. The science of translocation (social and biological) is the best way to manufacture the tools needed for maximum efficacy in the future. As these sets of tools are honed and improved, an easily accessible repository of evidence-based best practices—such as can be found on the website Conservation Evidence (www.conservationevidence .com)—is needed. Unfortunately, only a few of the growing number of publications on translocation tactics have been included at the time of this writing. We urge the continued development of the science of translocation and increased effort to disseminate findings in a user-friendly forum.

LITERATURE CITED

Armstrong, D.P., and P.J. Seddon. 2008. Directions in reintroduction biology. Trends in Ecology and Evolution 23:20–25.

Batson, W., I. Gordon, D. Fletcher, and A. Manning. 2015. Translocation tactics: A framework to support the IUCN guidelines for wildlife translocations and improve the quality of applied methods. Journal of Applied Ecology 52:1598–1607.

Battin, J. 2004. When good animals love bad habitats: Ecological traps and the conservation of animal populations. Conservation Biology 18:1482–1491.

Bright, P.W., and P.A. Morris. 1994. Animal translocation for conservation: Performance of dormice in relation to release methods, origin and season. Journal of Applied Ecology 31:699–708.

Cabezas, S., and S. Moreno. 2007. An experimental study of translocation success and habitat improvement in wild rabbits. Animal Conservation 10:340–348.

Christensen, J. 2003. Auditing conservation in an age of accountability. Conservation in Practice 4:12–19.

Cooper, W.E.J., and D.T. Blumstein. 2015. Escaping from predators: An integrative view of escape decisions. Cambridge University Press, Cambridge, UK.

Corlett, R.T. 2016. Restoration, reintroduction, and rewilding in a changing world. Trends in Ecology and Evolution 31:453–462.

Dickens, M.J., D.J. Delehanty, and L.M. Romero. 2010. Stress: An inevitable component of animal transloca- tion. Biological Conservation 143:1329–1341.

Dickinson, J.L., J. Shirk, D. Bonter, R. Bonney, R.L. Crain, J. Martin, T. Phillips, and K. Purcell. 2012. The current state of citizen science as a tool for ecological research and public engagement. Frontiers in Ecology and the Environment 10:291–297.

Ewen, J.G., K. Acevedo-Whitehouse, M.R. Alley, C. Carraro, A.W. Sainsbury, K. Swinnerton, and R. Woodroffe. 2012.

Empirical consideration of parasites and health in reintroduction. Pages 290–335 in J. Ewen, D. Arm- strong, K. Parker, and P. Seddon, editors. Reintroduction biology: Integrating science and management. Wiley- Blackwell, Chichester, UK.

Germano, J.M., K.J. Field, R.A. Griffiths, S. Clulow, J. Foster, G. Harding, and R.R. Swaisgood. 2015. Mitigation- driven translocations: Are we moving wildlife in the right direction? Frontiers in Ecology and the Environ- ment 13:100–105.

Greggor, A.L., O. Berger-Tal, D.T. Blumstein, L. Angeloni, C. Bessa-Gomes, B.F. Blackwell, C.C. St. Clair, et al. 2016. Research priorities from animal behaviour for maximising conservation progress. Trends in Ecology and Evolution 31:954–964.

Griffin, A.S., D.T. Blumstein, and C.S. Evans. 2000. Training captive-bred or translocated animals to avoid predators. Conservation Biology 14:1317–1326.

Harrington, L.A., A. Moehrenschlager, M. Gelling, R.P. Atkinson, J. Hughes, and D.W. MacDonald. 2013. Conflicting and complementary ethics of animal welfare considerations in reintroductions. Conservation Biology 27:486–500.

IUCN. 2013. Guidelines for reintroductions and their conservation translocations. IUCN, Gland, Switzerland.

Jachowski, D., S. Bremner-Harrison, D. Steen, and K. Aarestrup. 2016a. Accounting for potential physi- ological, behavioral, and community-level responses to reintroduction. Pages 185–216 in D.S. Jachowski, J.J. Millspaugh, P.L. Angermeir, and R. Slotow, editors. Reintroduction of fish and wildlife populations. University of California Press, Oakland, California, USA.

Jachowski, D.S., R. Slotow, P.L. Angermeir, and J.J. Millspaugh. 2016b. The future of animal reintroduction. Pages 367–380 in D.S. Jachowski, J.J. Millspaugh, P.L. Angermeir, and R. Slotow, editors. Reintroduction of fish and wildlife populations. University of California Press, Oakland, California, USA.

Jamieson, I.G., and R.C. Lacy. 2012. Managing genetic issues in reintroduction biology. Pages 441–475 in J. Ewen, D. Armstrong, K. Parker, and P. Seddon, editors. Reintroduction biology: Integrating science and management. Wiley-Blackwell, Chichester, UK.

Jones, C.G., and D.V. Merton. 2012. A tale of two islands: The rescue and recovery of endemic birds in New Zealand and Mauritius. Pages 33–72 in J. Ewen, D. Armstrong, K. Parker, and P. Seddon, editors. Reintroduction biology: Integrating science and management. Wiley-Blackwell, Chichester, UK.

Kierulff, M.C., P.P. Oliveira, B.B. Beck, and A. Martins. 2002. Reintroduction and translocation as conservation tools for golden lion tamarins. Pages 271–282 in

A.B. Rylands, editor. Lion tamarins: Biology and conservation. AZA, Washington, DC, USA.

Kleiman, D.G. 1989. Reintroduction of captive mammals for conservation. BioScience 39:152–161.

Kleiman, D.G., B.B. Beck, J.M. Dietz, and L.A. Dietz. 1991. Costs of a re-introduction and criteria for success: Accounting and accountability in the golden lion tamarin conservation program. Symposium of the Zoological Society of London 62:125–142.

Le Gouar, P., J.-B. Mihoub, and F. Sarrazin. 2012. Dispersal and habitat selection: Behavioural and spatial constraints for animal translocations. Pages 138–164 in J. Ewen, D. Armstrong, K. Parker, and P. Seddon, editors. Reintroduction biology: Integrating science and management. Wiley-Blackwell, Chichester, UK.

Meek, M.H., C. Wells, K.M. Tomalty, J. Ashander, E.M. Cole, D.A. Gille, B.J. Putman, et al. 2015. Fear of failure in conservation: The problem and potential solutions to aid conservation of extremely small populations. Biological Conservation 184:209–217.

Mitchell, A.M., T.I. Wellicome, D. Brodie, and K.M. Cheng. 2011. Captive-reared burrowing owls show higher site-affinity, survival, and reproductive performance when reintroduced using a soft-release. Biological Conservation 144:1382–1391.

Moehrenschlager, A., and N.A. Lloyd. 2016. Release considerations and techniques to improve conservation translocation success. Pages 245–280 in D.S. Jachowski, J.J. Millspaugh, P.L. Angermeir, and R. Slotow, editors. Reintroduction of fish and wildlife populations. University of California Press, Oakland, California, USA.

Moorhouse, T., M. Gelling, and D. MacDonald. 2009. Effects of habitat quality upon reintroduction success in water voles: Evidence from a replicated experiment. Biological Conservation 142:53–60.

Moseby, K., J. Read, D. Paton, P. Copley, B. Hill, and H. Crisp. 2011. Predation determines the outcome of 10 reintroduction attempts in arid South Australia. Biological Conservation 144:2863–2872.

Moseby, K.E., A. Cameron, and H.A. Crisp. 2012. Can predator avoidance training improve reintroduction outcomes for the greater bilby in arid Australia? Animal Behaviour 83:1011–1021.

Moseby, K.E., D.T. Blumstein, and M. Letnic. 2016. Harnessing natural selection to tackle the problem of prey naïveté. Evolutionary Applications 9:334–343.

Nafus, M.G., T.C. Esque, R.C. Averill-Murray, K.E. Nussear, and R.R. Swaisgood. 2016. Habitat drives dispersal and survival of translocated juvenile desert tortoises. Journal of Applied Ecology 54:430–438.

Nichols, J.D., and D.P. Armstrong. 2012. Monitoring for reintroductions. Pages 223–255 in J. Ewen, D. Arm-

strong, K. Parker, and P. Seddon, editors. Reintroduction biology: Integrating science and management. Wiley-Blackwell, Chichester, UK.

Osborne, P.E., and P.J. Seddon. 2012. Selecting suitable habitats for reintroductions: Variation, change and the role of species distribution modelling. Pages 73–104 in J. Ewen, D. Armstrong, K. Parker, and P. Seddon, editors. Reintroduction biology: Integrating science and management. Wiley-Blackwell, Chichester, UK.

Parker, K.A. 2008. Translocations: Providing outcomes for wildlife, resource managers, scientists, and the human community. Restoration Ecology 16:204–209.

Pérez, I., J.D. Anadón, M. Díaz, G.G. Nicola, J.L. Tella, and A. Giménez. 2012. What is wrong with current translocations? A review and a decision-making proposal. Frontiers in Ecology and the Environment 10:494–501.

Pessier, A., and J. Mendelson. 2010. A manual for control of infectious diseases in amphibian survival assurance colonies and reintroduction programs. IUCN/SSC Conservation Breeding Specialist Group, Apple Valley, California, USA.

Reed, J.M., and A.P. Dobson. 1993. Behavioural constraints and conservation biology: Conspecific attraction and recruitment. Trends in Ecology and Evolution 8:253–255.

Ruiz-Miranda, C., B. Beck, D. Kleiman, A. Martins, J. Dietz, D. Rambaldi, M. Kierulff, P. Oliveira, and A. Baker. 2010. Re-introduction and translocation of golden lion tamarins, Atlantic Coastal Forest, Brazil: The creation of a metapopulation. Global re-introduction perspectives: Additional case-studies from around the globe. IUCN/SSC Re-introduction Specialist Group, Abu Dhabi, UAE.

Salafsky, N., R. Margoluis, and K. Redford. 2001. Adaptive management: A tool for conservation practitioners. Biodiversity Support Program. World Wildlife Fund, Washington, DC, USA.

Schwartz, M.W. 2006. How conservation scientists can help develop social capital for biodiversity. Conservation Biology 20:1550–1552.

Schwartz, M.W., K. Deiner, T. Forrester, P. Grof-Tisza, M.J. Muir, M.J. Santos, L.E. Souza, M.L. Wilkerson, and M. Zylberberg. 2012. Perspectives on the open standards for the practice of conservation. Biological Conservation 155:169–177.

Seddon, P.J., and D.P. Armstrong. 2016. Reintroduction and other conservation translocations: History and future developments. Pages 7–27 in D.S. Jachowski, J.J. Millspaugh, P.L. Angermeir, and R. Slotow, editors. Reintroduction of fish and wildlife populations. University of California Press, Oakland, California, USA.

Seddon, P.J., D.P. Armstrong, and R.F. Maloney. 2007. Developing the science of reintroduction biology. Conservation Biology 21:303–312.

Seddon, P.J., W.M. Strauss, and J. Innes. 2012. Animal translocations: What are they and why do we do them. Pages 1–32 in J. Ewen, D. Armstrong, K. Parker, and P. Seddon, editors. Reintroduction biology: Integrating science and management. Wiley-Blackwell, Chichester, UK.

Shier, D.M. 2016. Manipulating animal behavior to ensure reintroduction success. Pages 275–303 in O. Berger-Tal and D. Saltz, editors. Conservation behavior: Applying behavioral ecology to wildlife conservation and management. Cambridge University Press, Cambridge, UK.

Shier, D.M., and R.R. Swaisgood. 2012. Fitness costs of neighborhood disruption in translocations of a solitary mammal. Conservation Biology 26:116–123.

Stamps, J.A., and R.R. Swaisgood. 2007. Someplace like home: Experience, habitat selection and conservation biology. Applied Animal Behaviour Science 102:392–409.

Swaisgood, R.R. 2010. The conservation-welfare nexus in reintroduction programs: A role for sensory ecology. Animal Welfare 19:125–137.

Teixeira, C.P., C. Schetini de Azevedo, M. Mendl, C.F. Cipreste, and R.J. Young. 2007. Revisiting translocation and reintroduction programmes: The importance of considering stress. Animal Behaviour 73:1–13.

Walters, C.J. 1986. Adaptive management of renewable resources. Macmillan, New York, USA.

Wolf, C.M., T. Garland, and B. Griffith. 1998. Predictors of avian and mammalian translocation success: Reanalysis with phylogenetically independent contrasts. Biological Conservation 86:243–255.

Woodworth, L.M., M.E. Montgomery, D.A. Briscoe, and R. Frankham. 2002. Rapid genetic deterioration in captive populations: Causes and conservation implications. Conservation Genetics 3:277–288.

12

Shekhar K. Niraj
Shreya Sethi
S. P. Goyal
Amar N. Choudhary

Poaching, Illegal Wildlife Trade, and Bushmeat Hunting in India and South Asia

Introduction

Alteration of forest and forest degradation are major causes for the loss of biodiversity across Southeast and South Asia. These habitat alterations and illegal hunting and trade are responsible for a decline in populations of flagship species (tiger, *Panthera tigris*: Karanth et al. 2011; leopard, *Panthera pardus*; pangolins, *Manis* spp.; Table 12.1). Tiger poaching is a recurrent crisis in trade throughout their range, and poaching, illegal hunting, and smuggling have grown into a well-organized syndicate-based international wildlife crime. From 2002 to 2013, 1,755 tiger seizures were made in 810 seizures globally (Stoner et al. 2016); from 2012 to 2016, 66 tiger seizures were recorded in India alone. Local traditions in some states in India encourage people to poach a tiger to take a paw of the right leg, which is used in black magic to gain fortunes from God. Leopards (Fig. 12.1) are substituted for tigers because there are not enough tigers to satiate the market (Niraj et al. 2012). From 2012 to 2016, 650 leopards were poached, and instances of leopard consumption were noticed for the first time in India.

Increasing human population and a growing economy create a greater demand for resources for the populace and an emerging affluent lifestyle.

Poaching and illegal trade in other Asian big cats (Fig. 12.2; snow leopard, *Panthera uncia*: Maheshwari et al. 2016, Maheshwari and Niraj 2016, Taubmann et al. 2016; and clouded leopard, *Neofelis nebulosa*), other large mammals (Asian elephant, *Elephas maximus*; greater one-horned rhinoceros, *Rhinoceros unicornis*), and several lesser-known species (pangolins; monitor lizards, *Varanus* spp.; freshwater turtles; mongooses, Herpestidae; geckos, Gekkonidae) have been recorded in India and surrounding areas. Nearly 450,000 water monitors (*Varanus salvator*) are killed annually for their skins in Indonesia (Koch et al. 2013). Growing illegal demand has been threatening the future existence of the tiger, elephant, leopard, Himalayan black bear (*Ursus thibetanus*), Himalayan brown bear (*Ursus arctos isabellinus*), and one-horned rhino. Illegal trade in noncharismatic and lesser-known species (pangolins, monitor lizards, geckos, turtles, and tortoises; lorises, *Loris tardigradus* and genus *Nycticebus*; birds; corals; and sea cucumbers, *Holothuria scabra, H. spinifera*) has caused these species to also become conservation concerns since 2000. For example, Malayan (*Manis javanica*), Chinese, and Philippine pangolin (*Manis culionensis*) are endangered (IUCN 1996), and overexploitation of all species of pangolins convinced the Convention on International

Table 12.1 Major species harvested for bushmeat from India and the South Asian region, 2016–2017

Name	Estimated population size and trend in India (IUCN)	Major geographical distribution and population	Is the species used in TMS?	International trade: Parts and derivative used	Destination markets	Protection status (CITES/ WPA 1972/ IUCN)	Estimated price (in US$): Parts and whole	Does the species have any special conservation plan?	Is there a conservation concern? Local or global?
Leopard	~ 8,000–10,000 Decreasing	Southern Asia and Sub-Saharan Africa	Yes: Meat by local communities in India, Bangladesh	Yes: skin, claws, bones		Appendix I (trade of skin and product is restricted to 2,560 individuals in 11 countries in Sub-Saharan Africa). Schedule 1 IUCN: VU (2016), NT (2008), LC (2002), T (1988), VU (1986)			Yes (Internationally). Highest protected India (WPA 1972), Nepal, Bangladesh
Asian elephant	41,410–52,345 Decreasing	India 26,390–30,770; Indonesia: 2,400–3,400; Thailand 2,500–3,200; Lao PDR 500–1,000; Malaysia 2,100–3,100; Myanmar 4,000–5,000; Nepal 100–125; Sri Lanka 2,500–4,000; Vietnam 70–150.		Yes: Tusk, skin, teeth, tail hair, meat eating in northeast India	China, Japan, India, Thailand, EU, USA	Appendix 1 Schedule 1 EN (1986–2008)	US$ 2,100 per kg (raw ivory price in China), 2014	Yes (Project Elephant)	Yes
Indian one-horned rhino	2,575 Increasing	India 2,200 Nepal 378	Yes (TCM)	Yes: Horns	Southeast Asia and China	Appendix I (since 1975) Schedule I (Part I) VU (2008) EN(1996)		IUCN SSC Asian Rhino Specialist Group, an Indian Rhino Vision 2020/Nepal Rhino Action Plan	Yes
Tiger	3,890	India 2226; Russia 433; Indonesia 371; Malaysia 250; Nepal 198; Thailand 189; Bangladesh 106; Bhutan 103; China 7; Vietnam 5; Laos 2	Yes	Yes: Skin, bone, claw, canine, fat		Appendix 1 Schedule 1 EN		Yes	Yes
Star tortoise (Geochelone elegans)	Unknown Decreasing	India (Andhra Pradesh, Karnataka, Orissa, Tamil Nadu); Pakistan; Sri Lanka	Yes (TCM)	Yes: Pet	Malaysia, Singapore, Thailand, China, Europe, US, India	Appendix II Schedule IV Vulnerable (2016) Least Concern (2000)	15–50 US$ per animal (in India)		India (WPA1972). Pakistan in Schedule II (Protected Animals) of Sindh Wildlife Protection Ordinance 1972.

Species	Population	Trend	Utilized	Distribution	Parts used	Markets	Legal status (CITES; WPA; IUCN)	Market value	IUCN specialist group
Indian soft shell turtle (*Nilssonia gangetica*)	Unknown	Unknown		Bangladesh, India, Pakistan	Yes: meat, live	China, East Asian markets	Appendix I; Schedule I; VU		
Asian Pangolin	Unknown	Decreasing	Yes	South and Southeast Asia, Pakistan and China	Yes: Scales, skin, bile, nails, and meat	South Asia, Southeast Asia, East Asia mainly in China	Appendix 1; Schedule 1 (Indian pangolin); CE (Chinese and Malayan pangolin); EN (Indian and Palawan pangolin)	US$501/kg raw scales (China)	The IUCN SSC Pangolin Specialist Group
Monitor lizard	Unknown	Decreasing	Yes	south central Asia, Southeast Asia, southeastern Iran	Yes: Skin		Appendix I; Schedule I (Part II); LC		No
Black-naped hare (*Lepus nigricollis*)				Bangladesh, India, Indonesia (Jawa - Present - Origin Uncertain), Nepal, Pakistan, Sri Lanka			Schedule IV; LC (1996)	No	No
Wild boar (*Sus scrofa*)	Unknown	Unknown		India, Nepal, Burma, western Thailand, and Sri Lanka			–; –	No	No
Spotted deer (chital)	Unknown	Unknown		Bangladesh, Bhutan, India (Andaman Is. - introduced), Nepal, Sri Lanka	Yes: Skin, antlers		–; Schedule III; LC	No	No
Sambar deer	Unknown	Decreasing		South, Southeast, and East Asia	Yes: Antlers		–; Schedule III; VU		
Common palm civet (*Paradoxus hermaphroditus*)	Unknown	Decreasing		South and Southeast Asia	Yes: Meat, as pet for civet coffee (especially in Indonesia)	Southeast Asia	Appendix III (India); Schedule II (Part II); LC		

Notes:

CITES: Convention on International Trade in Endangered Species of Wild Fauna and Flora

IUCN: International Union for Conservation of Nature

LC: least concern

NT: near threatened

T: threatened

TCM: traditional Chinese medicine

TMS: traditional medicine system

VU: vulnerable

WPA: Wildlife Protection Act 1972, India

Figure 12.1   Seizures of illegally collected wildlife in India: (A) leopard skins, (B) star tortoises (*Geochelone elegans*), (C) parrots, and (D) elephant tusks. Shekhar Niraj

Figure 12.2   Poaching for bushmeat and sale of pelts is an issue for many large carnivores: (A) tiger that was poached and sectioned in Madhya Pradesh and detected by undercover field investigators for TRAFFIC India; (B) snow leopard pelt for sale at street market in Afghanistan. Shekhar Niraj

Trade in Endangered Species of Wild Fauna and Flora (CITES) policy makers to place them in Appendix 1 (the highest protection) through global campaigns. In addition, more than 450 species of 1,300 Indian bird species have been documented in illegal international and domestic bird trade (Ahmed 2004). Bird trade continues extensively through trade fairs and hidden means (Niraj and Ghosh 2017).

Wildlife trade has also affected reptiles. At least 100 species of freshwater turtles and tortoises native to Asia are actively traded (Bhupathy et al. 2000, Van Dijk 2000, Choudhury 2001), and 33 are listed as threatened (IUCN 1996). Hunting for subsistence is being replaced by a steadily increasing harvest for commercial trade in wildlife to supply markets in China, Thailand, Vietnam, and other countries (Martin and Phipps 1996). Unregulated and often illegal hunting is less easily quantifiable but is a more severe threat to survival of many endangered species (Stewart and Hutchings 1996; Table 12.1). Mammals and birds are at risk of extinction (Fig. 12.3). Hence, out of the 75 mammal, 56 bird, 10 aquatic, and

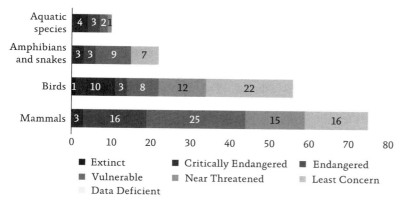

Figure 12.3   Classification of Schedule 1 species of the Wildlife Protection Act of India, 1972, and International Conservation of Nature status 2017

24 amphibian and snake species that are protected under Schedule 1 of the Indian Wildlife Protection Act (WPA) 1972, 20 species are critically endangered, 25 are endangered, 44 are vulnerable, and 27 are near threatened according to the IUCN Red List.

Hunting has a diverse ecological history, and hunters have caused 23% of known species extinctions, while also modifying the landscapes (Sodhi et al. 2009). Every year in India, hundreds of pangolins, lizards, and tortoises are poached; an estimated 700,000 birds are illegally trapped (Ahmed 2004); and approximately 70,000 metric tonnes of sharks are caught (Mundy and Crook 2013). Unfortunately, the levels of exploitation of these species are rarely reported. Market surveys in northeast India estimated thousands of birds and mammals annually at market for trade and consumption (Bhupathy et al. 2013). Large-scale exploitation and minimal information about their population status, poaching, and smuggling trends threatens effective management of these species.

Overharvesting wildlife could be the greatest new challenge to conservation because of improved access to remote forest areas, markets, improved hunting approaches, increased demand for bushmeat in rural and urban areas, and their use in Oriental medicine systems, in traditional Chinese medicine (TCM), and as pets. Bushmeat trade relates hunting to a sharp decline of species, particularly the smaller mammals and reptiles, whose population sizes remain unassessed. Uhm

and Daan (2016) and Warchol (2017) provide an overview of the illegal wildlife trade throughout the world. Our objective in this chapter is to document the illegal trade and use of wildlife using South Asia and India as examples of this worldwide problem.

Bushmeat Hunting

Bushmeat is meat derived from hunting and harvesting wild animals in tropical and subtropical countries for food and for nonfood purposes (medicine; Buck and Hamilton 2011). Limited analysis of wildlife hunting is available in Asia (Harrison et al. 2016), except for a few studies that highlight the significance of unsustainable use of bushmeat on biodiversity conservation. Lee et al. (2014) discuss the use of wildlife in medicine, Pattiselanno and Lubis (2014) describe use of wild meat from marine protected areas (PA), Natusch and Lyons (2012) studied wildlife used for pets, and Shepherd et al. (2007) studied the transport of illegally harvested wildlife.

Studies of the wildlife trade have been distributed in Southeast Asia (61%), South Asia (22%), and East Asia (16%; Lee et al. 2014). Most authors (71%) reported that wildlife trade is a conservation crisis, and most of the bushmeat is used for livelihood and traditional medicine purposes. There is also a growing demand for traditional Chinese medicines in East Asia and other countries. Oriental traditional medicine (OTM) practiced in Korea, Japan, Taiwan, and other countries of East Asia also makes demands for body parts and derivatives of various wildlife species, including the large mammals and lesser-known species from Asia and Africa (Vohora and Khan 1979).

Primary and Secondary Influences for Bushmeat Hunting and Illegal Trade

Primary influences for hunting are those that directly threaten species. Illegal trade in animal products including skins, bones, meat, and tonics constitutes the primary threats to Asian big cats, rhinos, elephants, and pangolins. Other factors (loss of buffer areas around the PAs, retaliatory actions of humans, construction of highways) have secondary-level influences on wildlife and PAs, as they indirectly lead to hunting and consumption of the animal derivatives by promoting conflict of wildlife with humans. Bushmeat hunting is also influenced by low productivity of domestic livestock in the tropical forests and the risks and investment costs associated with animal husbandry. Low purchasing power of human forest dwellers and the proximity to forests makes animal husbandry a nonprofitable option. Thus, wild harvest becomes the most accessible form of protein (Brown et al. 2007). The forest-dwelling communities in tropical countries consider bushmeat staple food. Like other developing countries, secondary influences of bushmeat hunting in India includes opportunistic killing of species because the dependence of villagers on fuelwood brings them in contact with wildlife. An average Indian villager spends < US$105/ month and falls under the lowest 20% of economic levels of Indian households. High densities of humans living in buffer areas of tiger reserves (TR), national parks (NP), and wildlife sanctuaries (WS) in India, and in PAs in Bangladesh and Nepal, create motives for retaliation in negative human-animal interactions, causing illegal hunting. Access is easy to materials used for making snares (cycle wires, telephone cables), and pesticides and fertilizers are readily available for poisoning water to kill wildlife for consumption and trade.

Enforcement Issues

Law enforcement agencies in South Asia have challenges (lack of data, weak scientific monitoring and evaluation protocols) in combating illegal use of bushmeat. The United Nations Office on Drugs and Crime (UNODC 2016) documented the illegal trade around the major hubs in the world. The UNODC report provides analysis of >164,000 seizures related to wildlife crime from 120 countries. Trafficking of wildlife is increasingly recognized as a specialized area of organized crime and a significant threat to plants and animals. The report illustrates diversity of wildlife crime and indicates the presence of 7,000 species in seizures, yet no single species represents >6% of the total, nor does a single country constitute the source of >15% of the seized shipments. Most seizures included mammals (30% of seizures) and reptiles (28%), including their parts and derivatives. To improve understanding of the illegal bushmeat trade from South Asia, there is a need to undertake regional-level analysis, which is absent in most of the literature. We conducted a limited survey of villagers to analyze bushmeat-related take of wild animals at regional levels, using the records of samples received at the Wildlife Institute of India (WII), one of the leading wildlife research institutes in South Asia. The following summarizes submitted records from police departments, enforcement agencies, courts, custom departments, and hospitals.

Most of the poaching cases referred to WII for forensic analyses were from the central Indian region (Fig. 12.4). Samples received for analysis between January 2006 and December 2010 were of hair, bones, meat, claw, canines, shawls, brushes, ivory, and skins (Table 12.1, Fig. 12.4), and are alarming.

Documentation of data on seizures and trade was not evenly distributed across taxa. The majority of seizures was dominated by mammals and their derivatives, constituting approximately 80% ($n = 1,035$) of all seizures. Skins, meat, wool for shawls and furs, mostly of tigers, leopard, Tibetan antelope (*Pantholops hodgsonii*), and deer species, constituted 56%

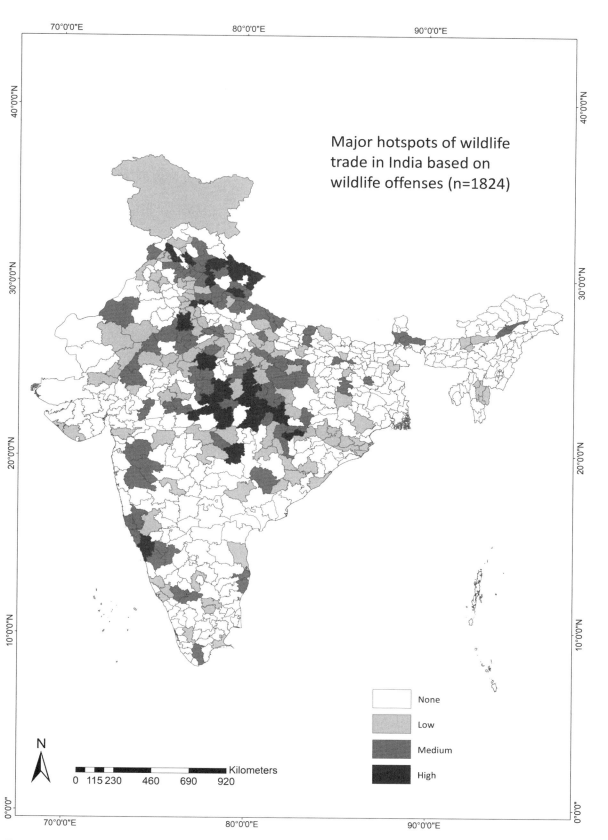

Figure 12.4 Major hotspots for wildlife trade in India based on wildlife cases referred to the Wildlife Institute of India, Dehra Dun, India, 2002–2015.

($n = 1,035$) of mammal product seizures. Seizures of ivory constituted 8% ($n = 1,035$) of all seizures of mammal products. Among the seized specimens, meat (25%, $n = 1,035$) and skin (21%, $n = 1,035$) were seized most, whereas bones, hairs, and blood-soaked parts were also seized. Deer antlers, bear bile, claws, canines, shawls made of shahtoosh, and elephant ivory were commonly reported in the seizures.

Hunting Tools

The country gun, which is unlicensed throughout India and South Asia, is the most preferred hunting tool. Almost all households in all villages have licensed guns (muzzle-loading, 0.303, 0.12, 0.22 caliber) or air guns. Wide availability of guns in villages dates to World War II. During the British-Japan war, modern guns and bullets were left behind by the troops and were picked up by the Nagas, a hunting tribe, in northeast India. Traditional bow and arrows and spears have been replaced by guns as a hunting tool since the 1960s. Traps are also used in nearly all villages, but they are more prevalent in the eastern districts of Nagaland. Sometimes, catapults are used by adult hunters for bird or squirrel hunting. Wildlife populations increase in areas where hunting does not occur. If species' populations (leaf deer, *Muntiacus putaoensis*) are susceptible to the high levels of hunting in a region, the species could be severely at risk. Communities on both sides of India's transborder region hunt a diversity of species, from muntjac or barking deer (*Muntiacus muntjac*) to fulvous fruit bats (*Rousettus leschenaultii*). Most animals hunted are consumed by the hunter, close relatives, and neighbors.

Evaluation of Wildlife Crime

Bangladesh

Birds (67.6%), reptiles (18.4%), and mammals (13.2%) were seized in 2015. Most seizures of bushmeat were deer. In the recorded seizures in Bangla-

desh during 2015, birds had 149 records, plus 77 records of dead budgerigar (*Melopsittacus undulatus*), 500 kg of deer meat, deer skins, bear skins, snake skins, pangolin scales, monitor lizard skins, and articles made of monitor lizard skin and fishing cat (*Prionailurus viverrinus*) skin. Bangladesh has emerged as an important illegal wildlife trade route linking India with Southeast Asia via Myanmar. Smaller cats (fishing cat; jungle cat, *Felis chaus*), and other smaller mammals (jackal, *Canis aureus indicus*) are increasingly taken from the wild for local consumption and in international trade from India.

Sri Lanka

The trade of bushmeat from Sri Lanka during 2013 was extensive. For example, >9,060 kg of wild boar/month (US$10,000) contributed to about 44% of the revenue (Ranjini et al. 2006). This study (Ranjini et al. 2006) is a partial assessment of the level of bushmeat from Sri Lanka but indicates the volume of illegal wildlife trade in South Asia.

Southeast Asia

Among the causes for bushmeat hunting, income is important. In North Myanmar, bushmeat was the highest source of income for 24% of respondents ($n = 84$), following forest-product collection (31%), and farming (45%; Rao et al. 2010). This study also confirmed trade of pangolins, particularly for their scales. The average price of 1.6 kg of scales was US$42.31, implying that hunting serves as a reliable and a profitable backup in times of financial stress (Rao et al. 2010). A study on unsustainable use of bushmeat indicated that, in the rural areas of Southeast Asia, bushmeat consumption remained high, as the price of bushmeat was often less than half the price of domesticated animal meat (Vliet and Mbazza 2011); as forests are reduced and human populations increase, people shift to domestic sources of meat. Rao et al. (2010) concluded that access to market is a key factor for the economic worth of bush-meat, and if prices and profits are high, the local traders would make efforts to transport the meat over longer distances, which provide access to the wealthy sections of society for this luxury.

India and India's Transborder Regions

Hunting is a major challenge facing tropical wildlife, and 143 studies in India across species and geographic locations (Velho et al. 2012) have been conducted. Velho et al. (2012) concluded that 114 mammal species are susceptible to hunting in India. Hunting occurs for cultural reasons (marriage, death, and birth), food, and illegal export. In India, an estimated 25 large mammal species are at risk of extinction (Swamy and Pinedo-Vasquez 2014). The historical and traditional reasons for hunting are strong (Swamy and Pinedo-Vasquez 2014). In Arunachal Pradesh, 20 of the 33 mammals hunted for subsistence and illegal trade are endangered, vulnerable, or near threatened (Aiyadurai et al. 2010). In another study in the Ziro Valley (Arunachal Pradesh), among the 85 households surveyed, about 54% reported hunting for subsistence, 25% for commercial trade, 10% for medicinal use, and 4.7% for pleasure. Most hunted species belonged to the protected lists of WPA 1972. The average age of hunters ranged from 8-year-olds, who hunted birds for consumption, to 55-year-olds who hunted for trade and subsistence (Selvan et al. 2013). The results of the study on the hunting practices of an Indo-Tibetan Buddhist tribe in Arunachal Pradesh reported about 96% ($n = 50$) of respondents preferred wild meat over domestic meat.

Literature on hunting practices in northeast India is mostly concentrated on Arunachal Pradesh (Aiyadurai et al. 2010, Aiyadurai 2012, Selvan et al. 2013, Velho and Lawrence 2013). These studies show that hunting is extensive among major tribes in different regions of Arunachal Pradesh, and it is difficult to quantify all wildlife harvested from these forests.

Major Species Hunted for Bushmeat

Large fruit bats are consumed worldwide, valued for use in traditional medicines, commercial use, and illegal trade. Fruit bats are considered important in forest ecosystems because of their ability for long-distance dispersal of seeds in fragmented landscapes (Corlett 2009). Mickleburgh et al. (2009) reported a serious level of hunting of bats, particularly, fruit bats in the Old World Tropics. Most of the hunted bats belonged to the genus *Pteropus*. Killing fruit bats for trade in north Sulawesi, Indonesia, is a serious conservation issue (Tsang 2015). Because of unsustainable hunting in several Malaysian states, the Bornean subspecies of the large flying fox (*Pteropus vampyrus natuna*) can no longer play a key role in pollination (Struebig et al. 2007), and harvesting other bats could have serious public health implications (Nipah and Hendra viruses emerging as infectious diseases; Harrison et al. 2011).

From 2000 to the 2010s, illegal trade in turtles and tortoises from South Asia was mainly for food and pets. In southern India, eating turtle meat derived from marine green sea turtle (*Chelonia mydas*), leatherback turtle (*Dermochelys coriacea*), hawksbill turtle (*Eretmochelys imbricata*), and olive ridley turtle (*Lepidochelys olivacea*) is mainly under the mythical consideration of their having aphrodisiac properties and providing nutritional strength. The surplus meat is sold at US$9–US$12/kg locally, which could generate large revenue for the poachers and the local traders. Hunting for subsistence is being replaced by a steadily increasing harvest for commercial trade in wildlife.

The global conservation status of the Chinese pangolin is critically endangered, and the Indian pangolin is near threatened, but the status of both species is deteriorating (Duckworth et al. 2008, Molur 2008, Challender 2011). Indian and Chinese pangolin are among the eight pangolin species in the world that are poached for their scales and meat for the international market. Because of growing international demand, price of pangolin scales in local trade in India has risen nearly tenfold between 2010 and 2016 (TRAFFIC India, unpublished data).

Monitor lizards (Table 12.1) are found across Africa and Asia, and in Oceania. Numerous other species are regularly hunted for pet trade. The meat of monitor lizard is eaten by many rural and urban communities in India, Thailand, and West Africa as a supplemental meat source, and in Nepal for medicinal and food purpose (Ghimire et al. 2014).

Monitor lizard meat, particularly the tongue and liver, is eaten in parts of southern India and Malaysia as an alleged aphrodisiac (Ellis 2005). In parts of Pakistan, different parts of monitor lizards are used for a variety of medical purposes (Hashmi and Khan 2013). Large-scale exploitation of monitor lizards is undertaken for their skins, which are extensively used in the international leather industry.

The decision taken by CITES, in March 2013, to list Oceanic whitetip (*Carcharhinus longimanus*), scalloped hammerhead (*Sphyrna lewini*), great hammerhead (*S. mokarran*), smooth hammerhead (*S. zygaena*), and porbeagle (*Lamna nasus*) sharks, and two species of manta rays (*Manta* spp.) under Appendix II of CITES was based on surveys conducted by several CITES member countries. Only the Appendix II species, and Appendix I species with defined quota, are permitted for international trade. Shark meat consumption within South Asia is not large; however, sharks are caught for fins, which are very popular in East Asian countries as an expensive food delicacy. Recent campaigns have led to several countries banning shark fin soup and several commercial airlines refusing to carry shark fin cargo. Until 2016, India was the second-largest shark-catching nation in the world next to Indonesia, catching an average of 70,000 metric tons/year (Dent and Clarke 2015). Elasmobranchs are traded mainly for three purposes in India: fins, manta ray and devil ray gill plates, and meat. Approximately 65 species of sharks are reported in Indian waters, and 18 of them are fished intensively.

About 12 edible species of the animals commonly known as sea cucumbers are from Indian waters and are exploited to cater to food delicacy markets in China, Japan, Taiwan, and Korea. The entire genus of the holothurians was brought into Schedule 1 of the WPA in 2001 to give the genus protection from illegal trade and overexploitation. The species has a low fecundity (James 2004) and is difficult to breed in captivity.

The South Asian subcontinent has the largest variety of deer species in the world. Deer are abundant in several TRs, NPs, and WSs in India, Bhutan, Nepal, and Sri Lanka. They also have high densities in non-PA forest areas. Chital (*Axis axis*), sambar deer (*Cervus unicolor*), swamp deer (*Rucervus duvaucelii*), four-horned antelope (*Tetracerus quadricornis*), barking deer, mouse deer (*Moschiola meminna*), leaf deer, Himalayan serow (*Capricornis thar*), and Himalayan goral (*Naemorhedus goral*) are poached for bushmeat.

Deer meat is the biggest contributor to bushmeat consumption in India and South Asia. Sambar is the second-most sought after species, after chital, for bushmeat (Madhusudan and Karanth 2002). Musk deer (*Moschus moschiferus*) is found in the high altitudes of the Himalayas and is hunted for the musk, part of the musk gland, which is considered the world's costliest wildlife commodity (Ellis 2005).

Since the 1970s rhinos have also been subjected to extensive poaching for their horns. Hunting and degradation of habitats reduced the geographical limits of the one-horned rhinoceros to Assam and North Bengal, India, in the 1970s, and to Chitwan and Bardia NPs and Shuklaphanta National Park in Nepal (DNPWC 2009). As part of conservation efforts in 1984, a rhino translocation program was undertaken in Dudhwa National Park, abundant with grasslands in Uttar Pradesh, India, bordering Nepal. Rhinos were also brought from Chitwan NP in Nepal to India.

Like the other rhino species, the greater one-horned rhinoceros has been hunted by Asians and Europeans for centuries, mainly to meet the demand for traditional Chinese medicines. The horn of the greater one-horned rhino may grow up to 0.56 meters in length and can weigh around 800–900 grams in an adult (Ellis 2005). The skin of this rhino is used for making a variety of articles (e.g., spice containers, flowerpots), and rhino penis is considered a strong aphrodisiac among the Nepalese communities. Many Nepalese people also believe drinking rhino urine can cure asthma, attacks, congestion, stomach disorders, and other ailments (Ellis 2005). Nepalese, however, use most of the rhino parts, unlike other communities in the world (Martin 1985). In contrast, black rhinos are hunted in Namibia and other parts of southern and western Africa also as a source of protein.

Subsistence versus Commercial Demand Levels

China and India are the world's fastest-growing markets for bushmeat (Virmani 1999, 2004). An effect of strong economic growth in many emerging economies is the growing and prospering middle class. Free trade, retail credit, foreign direct investments, and privatization have helped the middle class upgrade their lifestyles and improve quality of life. But the communities living along the fringe of a PA or a forest habitat have not attained proportional growth in living standards. Public services and social facilities have not been improving commensurately. The resulting situation has created microeconomic pressure on the urban and rural poor that has influenced them toward adopting quicker and illegal means for making money and mobilizing other resources to provide a sense of equality with the urban rich. As a result, a subsistence-commercial continuum has emerged. Because poaching is demand-driven, supply rests on demand, and makes species more vulnerable as poaching becomes targeted to maximize profit and to continue the demand-driven supply chain. The poaching-trade dynamic incurs fewer operational costs and is low in investment. Poaching and illegal wildlife trade practices are also opportunistic.

Summary

The Veblen idea of conspicuous consumption (Friedman 2005) hold that demand is strengthened with a rise in price or weakened with a fall in price. Consumption of natural resources and extremely high per capita demand of resources continue to cause loss of biodiversity (Gossling 1999). Market forces affect wildlife transactions (Niraj 2009, Niraj et al. 2010). If the growth rate of the economy (as measured by the gross domestic product) exceeds the growth rate in population, then per capita income grows steadily (Virmani 2002). Many developmental economists predict that in all quarters of the twenty-first century, Asian countries, led by India and China, will dominate economic growth in the world (Bloom and Williamson 1998, Swaminathan 2001, World Bank 2005). Asia's rapid economic development has expanded the number of people who are able to afford traditional medicines and other wildlife products (Hillstrom and Hillstrom 2003). Bison (*Bos bison*), passenger pigeon (*Ectopistes migratorius*), and whale population declines and extinctions were the result of market forces (Bolen and Robinson 1999). The free market is an incredibly powerful economic force (Moulton and Sanderson 1997) and directly affects wildlife (Niraj 2009). Illegal hunting has a deleterious effect on species richness (Bodmer et al. 1997), and with the adaptation of more hunting techniques by man, local extinctions from overhunting are more common (Robinson and Redford 1991). Unsustainable hunting has also been triggered by market demand (Bodmer et al. 1994, Jenkins et al. 2011), which also affects the species richness. Regardless of the decline of species richness, incentives to the hunter prevail. The more animals are hunted, the greater a hunter's bushmeat take, which results in a high proportion of meat being sold (Kumpel et al. 2010), providing an important alternative source of income for most hunters.

Overharvesting is one of the major threats to the tropical vertebrates worldwide (Bennett and Robinson 2000, Milner-Gulland et al. 2003). Although being targeted for bushmeat is a direct challenge and immediate effect of hunting on a species, there are other cascading, indirect effects. In most cases, the hunted animal is a frugivore that disperses tree and shrub seeds (Redford 1992, Chapman and Chapman 1995, Stoner et al. 2007a, b). A study conducted in Northern Thailand reported that, as mammalian frugivores are reduced or removed from tropical forests, the success of trees that depend on their seed dispersal for population and persistence is hindered (Brodie et al. 2009). Further, a decline in dispersal of seeds implies a reduction in large woody plants that act as carbon stores; thus, loss of seed dispersal indirectly affects the carbon balance (Brodie and Gibbs 2009, Jansen et al. 2010).

Rampant hunting for meat and game has pushed large numbers of animal species to the brink of extinction (Madhusudan and Karanth 2002). Hunting by indigenous people is no longer sustainable in many regions, especially in tropical countries where humans have lived and hunted for thousands of years (Pangua-Adam and Noske 2010, Tidemann and Gosler 2010). In the past, cultural factors that are now lost, mostly involving religious taboos, had a major influence on hunting practices of indigenous peoples (Aiyadurai et al. 2010). Traditional societies employ food taboos on animal species for several reasons, and many threatened populations, including endemic and keystone species, benefit from such taboos (Colding 1998). Mahawar and Jaroli (2008) reported 109 animal species with 270 uses in traditional medicine in different parts of India.

Family wealth also influences hunting wild animals. The most important factor influencing bushmeat consumption was household income: households with higher income participated more in purchasing and consuming bushmeat. (Wilkie and Godoy 2001, Apaza et al. 2002). Hunting could result in extinction of many species, with the recent extinction of Formosan clouded leopard (*Neofelis nebulosa brachyura*) in Taiwan as an example (https://blogs.scientificamerican.com/extinction -countdown/clouded-leopards-confirmed-extinct -taiwan/?redirect=1).

Literature on hunting practices in South Asia is sparse, and within India, only the hunting practices in Arunachal Pradesh have been consistently studied (Aiyadurai et al. 2010, Aiyadurai 2012, Selvan et al. 2013, Velho and Lawrence 2013). These studies report that hunting is extensive among major tribes in different regions of Arunachal Pradesh, and it is difficult to quantify sustainable harvesting from these forests. Hunting depends on the socioeconomic background of the hunters and has linkages to tradition. For northeastern tribes, various wild animal parts are used in traditional attire and in customary social ceremonies such as those for marriage, death, and birth. Linkage to the market economy, however, means that the take from the wild largely exceeds the sustainable level, and grows as the human population grows. It has been demonstrated worldwide that market hunting is not sustainable for wildlife.

With rapidly growing human population in the world and huge demands for wildlife for consumption, traditional medicines, health supplements, clothing, and traditional adornments globally, survival of many species is questionable. India is the largest country on the South Asian continent and hosts a rich diversity of wild animals with several iconic species. Poaching induced by increased illegal trade and growth in consumption today threatens these iconic species and several of the lesser-known species. Bushmeat hunting, a traditional practice since historical time, no longer remains at subsistence level, as it is increasingly linked to market dynamics. Demand far exceeds supply. Even as our understanding of bushmeat hunting remains incomplete, with a serious dearth of studies in South Asia, some studies point to overharvest and overexploitation of several wildlife species, a situation prevalent in several other parts of the world. Overhunting could also lead to reduction of habitat quality. South Asia's wildlife could be safer if the international demand in each species is limited by a systematic demand-reduction campaign, along with multilateral policy initiatives. An effective strategy to conserve biodiversity cannot overlook imminent threats to other lesser-known species to feed illegal trade and the growing bushmeat trade in India and other South Asian countries.

LITERATURE CITED

Ahmed, A. 1997. Live bird trade in northern India. WWF/TRAFFIC-India, New Delhi, India.

Ahmed, A. 2004. Illegal bird trade in India. Pages 66–70 in M. Z. Islam, and A. R. Rahmani, editors. Important Bird Areas in conservation: Priority sites for conservation. Indian Birds Conservation Network, Bombay Natural History Society, and Bird Life International. Bombay Natural History Society, Mumbai, India.

Aiyadurai, A. 2012. Bird hunting in Mishmi Hills of Arunachal Pradesh, Northeast India. Indian Birds 7:134–137.

Aiyadurai, A., N.J. Singh, and E.J. Milner-Gulland. 2010. Wildlife hunting by indigenous tribes: A case study from Arunachal Pradesh, Northeast India. Oryx 44:564–572.

Apaza, L., D. Wilkie, E. Byron, T. Huanca, W. Leonard, E. Pérez, V. Reyes-García, V. Vadez, and R. Godoy. 2002. Meat prices influence the consumption of wildlife by the Tsimane'Amerindians of Bolivia. Oryx 36:382–388.

Bennett, E.L., and J. G. Robinson. 2000. Hunting of wildlife in tropical forests: Implications for biodiversity and forest peoples. AGRIS, World Bank, Washington, DC, USA.

Bhupathy, S., B.C. Choudhry, F. Hanfee, S.M. Kaylar, M.H. Khan, S.G. Platt, and S.M.A. Rashid. 2000. Turtle trade in south Asia: Regional summary (Bangladesh, India, and Myanmar). Pages 101–105 in P.P. Van Dijk, B.L. Stuart, and A.G.J. Rhodin, editors. Asian turtle trade: Proceedings of a workshop on conservation and trade of freshwater turtles and tortoises in Asia. Chelonian Research Foundation, Lunenberg, Massachusetts, USA.

Bhupathy, S., S.R. Kumar, P. Thirumalainathan, J. Paramanandham, and C. Lemba. 2013. Wildlife exploitation: A market survey in Nagaland, Northeastern India. Tropical Conservation Science 6:241–253.

Bloom, D.E., and J.G. Williamson. 1998. Demographic transitions and economic miracles in emerging Asia. World Bank Economic Review 12:419–455.

Bodmer, R.E., T.G. Fang, L. Moya, and R. Gill. 1994. Managing wildlife to conserve Amazonian forests: Population biology and economic considerations of game hunting. Biological Conservation 67:29–35.

Bodmer, R.E., J.F. Eisenberg, and K.H. Redford. 1997. Hunting and the likelihood of extinction of Amazonian mammals. Conservation Biology 11:460–466.

Bolen, E.G., and W.L. Robinson. 1999. Wildlife ecology and management. Fourth edition. Prentice-Hall, Upper Saddle River, New Jersey, USA.

Brodie, J.F., and H.K. Gibbs. 2009. Bushmeat hunting as climate threat. Science 326:364–365.

Brodie, J.F., O.E. Helmy, W.Y. Brockelman, and J.L. Maron. 2009. Bushmeat poaching reduces the seed dispersal and population growth rate of a mammal-dispersed tree. Ecological Applications 19:854–863.

Brown, D., J.E. Fa, L. Gordon, and ODI-Durrel. 2007. Assessment of recent bushmeat research and recommendations to Her Majesty's Government. http://static.zsl.org/files/odi-assessment-of-bushmeat-research-2007-719.pdf.

Buck, M., and C. Hamilton. 2011. The Nagoya Protocol on access to genetic resources and the fair and equitable sharing of benefits arising from their utilization to the Convention on Biological Diversity. Review of European Community and International Environmental Law 20:47–61.

Challender, D.W.S. 2011. Asian pangolins: Increasing affluence driving hunting pressure. Traffic Bulletin 23:92–93.

Chapman, C.A., and L.J. Chapman. 1995. Survival without dispersers: Seedling recruitment under parents. Conservation Biology 9:675–678.

Choudhury, A. 2001. Primates in northeast India: An overview of their distribution and conservation status. Environmental Information System Bulletin: Wildlife and Protected Areas 1:92–101.

Colding, J. 1998. Analysis of hunting options by the use of general food taboos. Ecological Modelling 110:5–17.

Convention on Biological Diversity. 2011. Outcomes of the Joint Meeting of the CBD liaison group on bushmeat and the CITES Central Africa Bushmeat Working Group. Nairobi. https://www.cbd.int/meetings/LGBUSHMEAT-02.

Corlett, R.T. 2009. Seed dispersal distances and plant migration potential in tropical East Asia. Biotropica 41:592–598.

Dent, F., and S. Clarke. 2015. State of the global market for shark products. Fisheries and Aquaculture Technical Paper 590. Food and Agricultural Organization, Rome, Italy.

DNPWC (Department of National Parks and Wildlife Conservation). 2009. Koshi Tappu Wildlife Reserve and buffer zone management plan (2009–2013). Government of Nepal, Ministry of Forests and Soil Conservation, Department of National Parks and Wildlife

Conservation Koshi Tappu Wildlife Reserve, Kathmandu, Nepal.

Duckworth, J.W., R. Steinmitz, A. Pattanavibool, Z. Than, D. Tuoc, and P. Newton. 2008. *Manis pentadactyla.* IUCN Red List of Threatened Species. http://www .iucnredlist.org/.

Ellis, R. 2005. Tiger bone and rhino horn: The destruction of wildlife for traditional Chinese medicines. Island Press, Washington, DC, USA.

Friedman, M. 2005. The optimum quantity of money. Transaction Publishers, New Brunswick, New Jersey, USA.

Gössling, S. 1999. Ecotourism: A means to safeguard biodiversity and ecosystem functions? Ecological Economics 29:303–320.

Ghimire, H.R., S. Phuyal, and K.B. Shah. 2014. Protected species outside the protected areas: People's attitude, threats and conservation of the yellow monitor (*Varanus flavescens*) in the far-western lowlands of Nepal. Journal for Nature Conservation 22: 497–503.

Harrison, R.D. 2011. Emptying the forest: Hunting and the extirpation of wildlife from tropical nature reserves. BioScience 61:919–924.

Harrison, R.D., R. Sreekar, J.F. Brodie, S. Brook, M. Luskin, H. O'Kelly, M. Rao, B. Scheffers, and N. Velho. 2016. Impacts of hunting on tropical forests in Southeast Asia. Conservation Biology 30:972–981.

Hashmi, M.U.A., and M.Z. Khan. 2013. Studies of basking activity in monitor lizard (*Varanus bengalensis*) from Thatta of Sindh. International Journal of Fauna and Biological Studies 1:32–34.

Hillstrom, K., and L.C. Hillstrom. 2003. Asia: A continental overview of environmental issues. ABC-CLIO, Santa Barbara, California, USA.

IUCN (International Union for Conservation of Nature). 1996. IUCN Red List of threatened species. IUCN, Gland, Switzerland. http://www.iucnredlist.org.

James, D.B. 2004. Captive breeding of the sea cucumber, *Holothuria scabra,* from India. Fisheries Technical Paper 463:385–395. Food and Agriculture Organization, Rome, Italy.

Jansen, P.A., H.C. Muller-Landau, and S.J. Wright. 2010. Bushmeat hunting and climate: An indirect link. Science 327:30.

Jenkins, R.K., A. Keane, A.R. Rakotoarivelo, V. Rakotombo-avonjy, F.H. Randrianandrianina, H.J. Razafimanahaka, S.R. Ralaiarimalala, and J.P. Jones. 2011. Analysis of patterns of bushmeat consumption reveals extensive exploitation of protected species in eastern Madagascar. PLOS One 6:27570.

Karanth, K.U., A.M. Gopalaswamy, S.N. Kumar, S. Vaidyanathan, J.D. Nichols, and D.I. MacKenzie. 2011. Monitoring carnivore populations at the landscape scale: Occupancy modelling of tigers from sign surveys. Journal of Applied Ecology 48:1048–1056.

Koch, A., T. Ziegler, W. Boehme, E. Arida, and M. Auliya. 2013. Pressing problems: Distribution, threats, and conservation status of the monitor lizards (Varanidae: *Varanus* spp.) of Southeast Asia and the Indo-Australian archipelago. Herpetological Conservation and Biology 8:1–62.

Kümpel, N.F., E.J. Milner-Gulland, G. Cowlishaw, and J.M. Rowcliffe. 2010. Incentives for hunting: The role of bushmeat in the household economy in rural Equatorial Guinea. Human Ecology 38:251–264.

Lee, T.M., A. Sigouin, M. Pinedo-Vasquez, and R. Nasi. 2014. The harvest of wildlife for bushmeat and traditional medicine in East, South and Southeast Asia: Current knowledge base, challenges, opportunities and areas for future research. Centre for International Forestry Research, Bogor, Indonesia.

Madhusudan, M.D., and K.U. Karanth. 2002. Local hunting and the conservation of large mammals in India. AMBIO: A Journal of the Human Environment 31:49–54.

Mahawar, M.M., and D.P. Jaroli. 2008. Traditional zootherapeutic studies in India: A review. Journal of Ethnobiology and Ethnomedicine 4:17.

Maheshwari, A., and S.K. Niraj. 2016. Conservation and adaptation in Asia's high mountain landscapes and communities. Melting the snow: Monitoring illegal trade in snow leopards. TRAFFIC-India, World Wildlife Foundation, India Secretariat, New Delhi, India.

Maheshwari, A., S.K. Niraj, S. Sathyakumar, M. Thakur, and L. Sharma. 2016. Snow leopard illegal trade in Afghani-stan: A rapid survey. CAT News 64:22–23.

Martin, E.B. 1985. Religion, royalty and rhino conservation in Nepal. Oryx 19:11–16.

Martin, E.B., and M. Phipps. 1996. A review of the wild animal trade in Cambodia. TRAFFIC Bulletin—Wildlife Trade Monitoring Unit 16:45–60.

Mickleburgh, S., K. Waylen, and P. Racey. 2009. Bats as bushmeat: A global review. Oryx 43:217–234.

Milner-Gulland, E.J., and E.L. Bennett. 2003. Wild meat: The bigger picture. Trends in Ecology and Evolution 18:351–357.

Molur, S. 2008. South Asian amphibians: Taxonomy, diversity and conservation status. International Zoo Yearbook 42:143–157.

Moulton, M.P., and J. Sanderson. 1997. Wildlife issues in a changing world. St. Lucie Press, Delray Beach, Florida, USA.

Mundy-Taylor, V., and V. Crook. 2013. Into the deep: Implementing CITES measures for commercially-valuable sharks and manta rays. Report prepared for the

European Commission, TRAFFIC International, Cambridge, UK.

Natusch, D.J., and J.A. Lyons. 2012. Exploited for pets: The harvest and trade of amphibians and reptiles from Indonesian and New Guinea. Biodiversity and Conservation 2:2899–2911.

Niraj, S.K. 2009. Sustainable development, poaching, and illegal wildlife trade in India. Dissertation, University of Arizona, Tucson, Arizona, USA.

Niraj, S.K., and S. Ghosh. 2017. Illegal wildlife trade amidst the biggest animal fair of Asia: The sonepur animal mela in Bihar. http://www.conservationindia.org/articles /sonepur.

Niraj, S.K, V. Dayal, and P.R. Krausman. 2010. Applying methodological pluralism to wildlife and the economy. Ecological Economics 69:1610–1616.

Niraj, S. K, P.R. Krausman, and V. Dayal. 2012. Temporal and spatial analysis of wildlife poaching in India from 1992–2006. International Journal of Ecological Economics and Statistics 24:194–219.

Pangau-Adam, M., and R.A. Noske. 2010. Wildlife hunting and bird trade in north-east Papua (Irian Jaya), Indonesia. Pages 73–86 in S. Tidemann, A. Gosler, and R. Gosford, editors. Ethno-Ornithology: Birds, indigenous peoples, culture and society. Earthscan, London, UK.

Pattiselanno, F., and M.I. Lubis. 2014. Hunting at the Abun regional marine protected areas: A link between wildmeat and food security. Hayati Journal of Biosciences 21:180–186.

Ranjini, M., T. Baheerathi, S. Wijeyamohan, and C. Santiapillai. 2006. An assessment of the bushmeat trade in Northern Sri Lanka. Tiger Paper 33:17–20.

Rao, M., S. Htun, T. Zaw, and T. Myint. 2010. Hunting, livelihoods and declining wildlife in the Hponkanrazi Wildlife Sanctuary, North Myanmar. Environmental Management 46:143–153.

Redford, K.H. 1992. The empty forest. BioScience 42:412–422.

Robinson, J.G., and K.H. Redford. 1991. Sustainable harvest of Neotropical forest mammals. Neotropical Wildlife Use and Conservation 1991:415–429.

Selvan, K.M., G.G. Veeraswami, B. Habib, and S. Lyngdoh. 2013. Losing threatened and rare wildlife to hunting in Ziro Valley, Arunachal Pradesh, India. Current Science 104:1492–1495.

Shepherd, C.R., J. Compton, and S. Warne. 2007. Transport infrastructure and wildlife trade conduits in the GMS: Regulating illegal and unsustainable wildlife trade. Pages 1–8 in Biodiversity conservation corridors initiative. International Symposium Proceedings, Bangkok, China.

Sodhi, N.S., B.W. Brook, and C.J. Bradshaw. 2009. Causes and consequences of species extinctions. Princeton Guide to Ecology 1:514–520.

Stewart, A.J., and M.J. Hutchings. 1996. Conservation of populations. Conservation Biology 1996:122–140.

Stoner, C., T.M. Caro, S. Mduma, C. Mlingwa, G. Sabuni, and M. Borner. 2007a. Assessment of effectiveness of protection strategies in Tanzania based on a decade of survey data for large herbivores. Conservation Biology 21:635–646.

Stoner, K.E., K. Vulinec, S.J. Wright, and C.A. Peres. 2007b. Hunting and plant community dynamics in tropical forests: A synthesis and future directions. Biotropica 39:385–392.

Stoner, S., K. Krishnasamy, T. Wittmann, S. Delean, and P. Cassy. 2016. Reduced to skin and bones re-examined. Full analysis: An analysis of tiger seizures from 13 range countries from 2000–2015. TRAFFIC, Southeast Asia Regional Office, Petaling Jaya, Selangor, Malaysia.

Struebig, M.J., M.E. Harrison, S.M. Cheyne, and S.H. Limin. 2007. Intensive hunting of large flying foxes *Pteropus vampyrus natunae* in Central Kalimantan, Indonesian, Borneo. Oryx 41:390–393.

Swaminathan, M.S. 2001. From Rio de Janeiro to Johannesburg: Action today and not just promises for tomorrow. East West Books, Madras, India.

Swamy, V., and M. Pinedo-Vasquez. 2014. Bushmeat harvest in tropical forests: Knowledge base, gaps and research priorities. Occasional Paper. http:// www.cifor.org/publications/pdf_files/OccPapers /OP-114.pdf.

Taubmann, J., K. Sharma, K.Z. Uulu, J.E. Hines, and C. Mishra. 2016. Status assessment of the endangered snow leopard *Panthera uncia* and other large mammals in the Kyrgyz Alay, using community knowledge corrected for imperfect detection. Oryx 50:220–230.

Tidemann, S., and A. Gosler. 2010. Ethno-Ornithology: Birds, indigenous peoples, culture and society. Earthscan, London, UK.

Tsang, S.M. 2015. Quantifying the bat bushmeat trade in North Sulawesi, Indonesia, with suggestions for conservation action. Global Ecology and Conservation 3:324–330.

Uhm, V., and P. Daan. 2016. The illegal wildlife trade: Inside the world of poachers, smugglers and traders. Springer, Switzerland.

UNODC (United Nations Office on Drugs and Crime). 2016. World wildlife crime report: Trafficking in protected species, United Nations. https://www.unodc.org/documents/data-and-analysis

/wildlife/World_Wildlife_Crime_Report_2016_final
.pdf.

Van Dijk, P.P. 2000. The status of turtles in Asia. Chelonian Research Monographs 2:15–23.

Velho, N., and W.F. Laurance. 2013. Hunting practices of an Indo-Tibetan Buddhist tribe in Arunachal Pradesh, north-east India. Oryx 47:389–392.

Velho, N., K.K. Karanth, and W.F. Laurance. 2012. Hunting: A serious and understudied threat in India, a globally significant conservation region. Biological Conservation 148:210–215.

Virmani, A. 1999. Star performers of the 20th (21st) century: Asian tigers, dragons or elephants. Indian Council for Research on International Economic Relations (ICRIER) Occasional paper. India Habitat Centre, New Delhi, India.

Virmani, A. 2004. Economic performance, power, potential and global governance: Towards a new international order. Working paper 150. December 2005. Indian Council for Research on International Economic Relations, Delhi, India.

Vliet, N.V., and P. Mbazza. 2011. Recognizing the multiple reasons for bushmeat consumption in urban areas: A necessary step toward the sustainable use of wildlife for food in central Africa. Journal of Human Dimensions 16: 45–54.

Vohora, S.B., and S.Y. Khan. 1979. Animal origin drugs used in Unani medicine. Vikas Publishing House, New Delhi, India.

Wilkie, D.S., and R.A. Godoy. 2001. Income and price elasticities of bushmeat demand in Lowland Amerindian societies. Conservation Biology 15: 761–769.

World Bank. 2005. China: Integration of national product and factor markets. Economic benefits and policy recommendations. Poverty Reduction and Economic Management Unit East Asia and Pacific Region, World Bank, 31973-CHA.

— *13* — Management of Migratory Wildlife and Others Influenced by Borderlands

ANDREA SANTANGELI
SHAMBHU PAUDEL

Species Distributions across Political Borders—The Case of Birds Worldwide

To effectively tackle the current biodiversity crisis, it is important to identify the main subjects (typically countries) to which conservation responsibilities are assigned. This helps inform the allocation of often limited resources and better track progress toward specific objectives. Administrative borders, however, typically cross ecological biomes. As a result, such biomes, and their associated species, often fall within multiple countries. Conservation decisions are generally made at the country level; consequently there is a risk that, when species occur across multiple countries, the responsibility for the conservation of those species is not clearly assigned, hindering progress in ameliorating their status. This may be particularly the case for widely distributed and highly vagile taxa, such as birds. For example, by counting the number of countries intersected by the overall distribution range of each bird species (BirdLife International and NatureServe 2015), it is possible to quantify the extent to which each bird species is reliant on multiple countries. Across all the bird species ($n = 10{,}174$), approximately 73% occur in more than one country, and 21% in 10 or more countries (Fig. 13.1). Among the threatened and near threat-

ened birds ($n = 2{,}292$ species) on the International Union for Conservation of Nature (IUCN) Red List of 2014 (with categories from near threatened to critically endangered), 44% occur in more than one country. Among the migratory bird species (Fig. 13.1) of the world ($n = 1{,}864$; classified as migratory by BirdLife International), however, 92% of species cross at least two countries, and 42% cross at least 10 countries, as expected by their ecology. One could repeat the same exercise by considering discrete geographic regions ($n = 8$; Antarctica, East Asia and Pacific, Europe and Central Asia, Latin America and Caribbean, Middle East and North Africa, North America, South Asia, Sub-Saharan Africa) instead of national borders: about 25% of all bird species are distributed across at least two of the above eight regions (Fig. 13.2). Among the IUCN threatened and near threatened species, 12% distribute across more than one region. Conversely, 59% of all the migratory species occur in more than one region (Fig. 13.2). These statistics strongly highlight the importance of cross-national cooperation, not only between individual countries in the same region, but also among countries from different regions, to ultimately achieve effective conservation of wide-ranging species. This is particularly relevant for migratory species, for which conservation actions

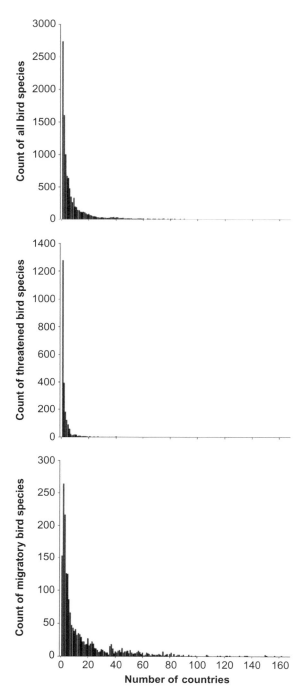

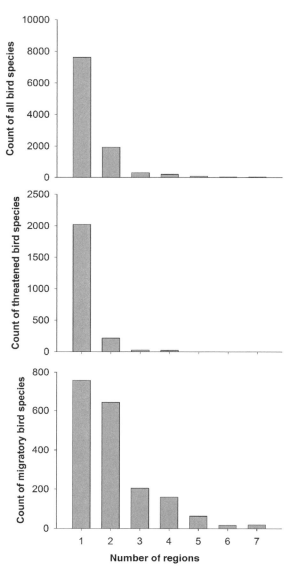

Figure 13.1 Count of bird species worldwide by the number of countries they occur in, from 1 (the count of species occurring in only one country) to 167 (species occurring in 167 countries), considering (top) all the world bird species together; (middle) threatened species only (IUCN categories from near threatened to critically endangered); and (bottom) migratory bird species.

Figure 13.2 Count of bird species worldwide by the number of regions in which they occur (e.g., Middle East and North Africa), from 1 (the count of species occurring in only one region) to 167 (species occurring in 167 regions), considering (top) all the world bird species together; (middle) threatened species only (IUCN categories from near threatened to critically endangered); and (bottom) migratory bird species.

targeted across their entire flyway have recently received much attention in the conservation community (Runge et al. 2014).

Threats Acting within and across National Boundaries

The threats affecting migratory species are qualitatively the same as those affecting worldwide biodiversity. The magnitude of the effect of threats may often be higher, however, for migratory than for nonmigratory species because migration entails higher exposure and vulnerability to threats (Newton 2008). Migratory species face a multitude of anthropogenic pressures along their passage routes that may lower their chances for survival. During this critical phase of the life cycle, migrating species often cross regional and administrative borders, facing threats of various intensity and scope. Migratory birds of Europe, for example, have been reported to be illegally killed in huge numbers along their Mediterranean passage areas, which they use to transit between their breeding and nonbreeding grounds (Brochet et al. 2016). Additionally, wind energy infrastructure development is expanding, largely along areas of high and regular winds that also represent migration bottlenecks for many large-bodied birds, and smaller ones, and migrating bats (Pearce-Higgins and Green 2014, Thaxter et al. 2017, Santangeli et al. 2018). As a result, large numbers of birds also fall victim to collision mortality with wind turbines (Northrup and Wittemyer 2013, Thaxter et al. 2017). Although this threat may be affecting mostly migratory species, it may also represent a high threat to resident nonmigrant species (Carrete et al. 2009, Dahl et al. 2012, Santangeli et al. 2018). Moreover, while residential and commercial development cause habitat loss and fragmentation, such as the loss and deterioration of wetlands used as passage points for many waterbirds (Amano et al. 2017, Xia et al. 2017), human infrastructure development also causes direct mortality of many bird species that migrate at night through light pollution, which attracts birds to infrastructures, causing collision mortality (Cabrera-Cruz et al. 2018). One of the most commonly identified or presumed important threats to migratory species, however, is habitat loss or deterioration on their nonbreeding grounds (Vickery et al. 2014). Several influences underlie this process, such as expansion and intensification of agriculture, particularly in sensitive areas under high vulnerability to changing climate (Atkinson et al. 2014, Finch et al. 2014). A typical example is the deterioration of the Sahel region in Africa caused by the increasing pressure posed on this fragile environment by a growing human population (Wilson and Cresswell 2006). This region represents a crucial wintering area for many European breeding birds, the decline of which has been attributed to changes in the habitat occurring in Africa (Atkinson et al. 2014, Vickery et al. 2014). Land-use changes in these areas may result in depleted food resources for wildlife, a critical factor underlying survival during the nonbreeding season (Schlaich et al. 2016). More generally, the movements of species in highly affected landscapes may often be constrained, impairing migration and other movements, such as dispersal or foraging activities. A recent study reported that mammal movements across the world are significantly reduced in landscapes where the human footprint is high, compared with more wild landscapes (Tucker et al. 2018), a result attributed largely to behavioral changes in animals and to the constraints exerted on long-distance movements of species by highly degraded landscapes (Tucker et al. 2018). Ultimately, what unifies the threats to migratory species is the spatial scope of their movements, which typically spans administrative borders. An example of international commitment to protect migratory species across borders is represented by the recently adopted Multi-species Action Plan to Conserve African-Eurasian Vultures (Botha et al. 2017). These highly endangered and wide-ranging species face a number of widespread threats that are driving their populations to extinction (Buechley and Şekercioğlu 2016). The action plan promotes concerted and collaborative international actions for the conservation of the 15 species of vultures across the 128 relevant countries in Africa and Eurasia. Designing and implementing wide-scale cross-national concerted actions is thus the only way to effectively address threats and avert further population declines of migratory species (Runge et al. 2014, Lopez-Hoffman et al. 2017, Runge et al. 2017, Anderson et al. 2018b).

The Tragedy of Parochialism: How International Threats Hinder National Conservation Efforts

It is clear by now that a large share of biodiversity and ecosystem processes and their associated threats act across large areas that typically cross sociopolitical boundaries (Dallimer and Strange 2015). Such boundaries may range from single land parcel ownership within a small region to subnational, national, and continental boundaries. The common link among these varied scale boundaries is that they artificially divide the landscape into discrete entities that are typically managed independently based on uncoordinated decision making, causing a steep divide in ecosystem properties on each side of the boundary (for an extreme example of the remarkable impact that different land management policies between neighboring countries can have on wildlife, see Arrondo et al. 2018). The spatial scale at which boundaries are relevant for the conservation of ecosystems and biodiversity depends on the individual properties of the species or ecosystem considered. For example, for the conservation of a localized and endemic carabid beetle species (coral pink sand dunes tiger beetle, *Cicindela albissima*), land units separating different private or local ownerships may be most relevant (Knisley and Hill 2001). Conversely, for the conservation of a long-distance migratory species, national and continental boundaries may be most salient (see more examples in the case studies below). Failure to recognize and minimize the mismatch between the spatial scale relevant for biodiversity and ecosystem conservation and artifi-

cial sociopolitical boundaries may hinder conservation efforts and result in inefficient use of already scarce resources (Kark et al. 2009, Montesino Pouzols et al. 2014, Kukkala et al. 2016a). We term this mismatch and the resulting consequences "the tragedy of parochialism." This term has some parallels with another recognized issue in nature conservation and sustainable resource use: the tragedy of the commons (Hardin 1968). In the latter, commonly shared resources are depleted by individuals behaving according to self-interest instead of in the interest of the community. In the tragedy of parochialism, sociopolitical boundaries define the individual actors (e.g., countries), which have the responsibility and make decisions on the use and management of the natural resources within their boundaries. Such boundaries thus define the individual actors of the tragedy of parochialism, and the preservation of wide-ranging resources partly depends on decisions taken beyond those boundaries. Therefore, if on one side the existence of boundaries may allow for the identification and assignment of conservation responsibilities, it may, on the other side, hinder implementation of conservation measures at the ecologically relevant scale. Examples of such mismatch in the scale and extent of conservation actions hindering success can be found in the conservation of long-distance migratory Palearctic birds (Atkinson et al. 2014, Vickery et al. 2014). These species breed in Western Europe but migrate across the Sahara Desert to spend the nonbreeding period in Africa. Many conservation efforts have been and currently are under way to protect these species on their breeding grounds in Europe (Sanderson et al. 2016). Recent evidence from multiple studies strongly suggests that populations of these trans-Saharan migrant species are steeply declining; the cause of the negative trend is attributed to threats acting on the nonbreeding areas for the species (Sanderson et al. 2006, Vickery et al. 2014). Failure to address threats beyond the breeding areas of these species may undermine conservation efforts implemented in Europe.

Practical Wildlife Conservation across Sociopolitical Borders

To make evidence-based decisions for species conservation, it is fundamental that their basic ecology and life history be well understood. Particularly for the conservation of highly mobile and migratory species, elucidating their distribution and movements across critical life stages is paramount for defining where, when, and how to protect the species. Gathering information on the occurrence and movements of highly mobile species is challenging because of their vast geographical range. During the recent decade, however, the unprecedented advent and advancement of technology has opened new and promising avenues for rapidly filling many of the knowledge gaps for key species (Pimm et al. 2015). For example, it is now possible to remotely count penguin colonies or track threats (such as forest loss) using satellite images or follow the journey of individuals of highly mobile species through their life cycle across continents via tracking technology. It is also possible to survey highly elusive species, such as many nocturnal mammals, with camera-trap technology. At a larger scale, citizen science–based programs harness technology to quickly collect and make available large amounts of observational data on species occurrence (Pimm et al. 2015). All these technologies are allowing a rapid accumulation of high-resolution and large-scale information that can then be used to infer species occurrence and their movements across countries, ultimately leading to science-informed conservation. Among the most striking examples of how technology can inform science was the discovery of a flock of over 3,000 individuals of the critically endangered sociable lapwing (*Vanellus gregarius*) in southeastern Turkey. This discovery was made possible by fitting one bird on its breeding ground in Kazakhstan with a satellite tracking unit and following its migration journey to Turkey. There, ornithologists searched the area where the bird was recorded staging and found the large and unknown flock. The area has since become the target of on-the-ground conservation efforts. Similarly, satellite telemetry information is also used to uncover the exposure of migratory species to threats acting on their nonbreeding grounds, and to assess how well these sites are covered by the current network of protected areas (Limiñana et al. 2012, 2014). These types of additional satellite telemetry data were also reported to greatly improve the accuracy of spatial priority areas for conservation of species, as demonstrated by the case of sea turtle conservation in the Mediterranean (Mazor et al. 2016).

Ultimately, all newly acquired geospatially referenced data can be fed into cross-country conservation policies and legal instruments scaled at the appropriate level and aimed to protect target species and their habitats. The challenges for implementing coordinated conservation actions across different and often contrasting geopolitical units (countries) are many and important (Dallimer and Strange 2015), but some progress has been made. At the global level, agreements for biodiversity conservation, such as the Convention on International Trade in Endangered Species of Wild Fauna and Flora (CITES), the Convention on Wetlands of International Importance (Ramsar Convention), and the Convention on Biological Diversity (CBD) provide examples. CITES is an international agreement among signatory governments aimed at regulating the international trade in wild animal and plant specimens (www.cites.org). Entered into force in 1975, CITES now counts 183 signatory countries and accords protection to more than 35,000 species of animals and plants. This convention is among the most wide-scale examples of conservation agreements, providing a legally binding framework for regulating trade of wild species to prevent their overexploitation and ensure their persistence across the multiple countries in which they occur and into the future. The Ramsar Convention, entered into force in 1975, has been ratified by most members of the United Nations. It provides a framework for international collaboration and national actions for the conservation and sustainable use of wetlands and their associated resources, including birds. International cooperation is among the three pillars of the Ramsar Convention, placing a strong emphasis on transboundary wetland systems that promote shared and coordinated management across countries. Such systems can be particularly important for the conservation of migratory species such as shorebirds (Szabo et al. 2016).

The CBD is arguably the broadest in scope and extent of all conventions targeting the conservation of species and their habitats. The CBD came into force in 1993; in 2010, the 193 parties of the convention agreed to the adoption of strategic goals and targets, to be met by 2020, with the aim of averting the ongoing biodiversity decline (www.cbd.int/sp/targets). Among these, Strategic Goal A aims to address the causes of the loss of biodiversity and mainstream biodiversity conservation across governments and society; Aichi Target 11 dictates the level to which the current protected area network should be expanded. The latter is particularly relevant for the conservation of wildlife across national borders. Although this target provides an important opportunity to improve the protection status for threatened biodiversity, recent studies have shown that it could be best achieved if the expansion of the current protected area network is coordinated internationally rather than each country making conservation decisions independently (Montesino Pouzols et al. 2014). Excellent examples of the relevance and benefits of internationally designated protected areas are the transfrontier parks, such as the one between Kenya and Tanzania that covers large tracts of the Mara-Serengeti ecosystem of East Africa, and the Kgalagadi Transfrontier Park between South Africa and Botswana, among others.

Agreements made at the regional level, whereby cooperation is promoted among the relevant countries of the region, can also yield important benefits for the conservation of wildlife. An example of such regional agreements is represented by the Natura 2000 network of protected areas across the European Union. The primary aim of the network is to protect biodiversity irrespective of national or political boundaries, thereby promoting the designation of sites for protection based on criteria and priorities that are relevant at the European Union level. Although the effectiveness of this network could be further improved, the coordination among interested countries has played an important role in aiding the coverage of regionally important species and improving their status (Kukkala et al. 2016a,b, Sanderson et al. 2016).

Another instrument for making a positive contribution to the conservation of highly mobile and migratory species is represented by multilateral agreements, such as the Convention on the Conservation of Migratory Species of Wild Animals (CMS). The CMS is the only global convention focused on the conservation of migratory species, their habitats, and their migration routes. It provides a legal framework for the international coordination of actions for the conservation of migratory species throughout their ranges. The convention has so far been signed by 124 countries, with a good representation of nations from the Southern Hemisphere and Europe (http://www.cms.int/en/parties-range-states). Signatory parties to the CMS are required to enforce strict protection measures for a list of target migratory species and to develop multilateral agreements for the conservation of migratory species that could capitalize on internationally coordinated efforts. Among these, the Agreement on the Conservation of African-Eurasian Migratory Waterbirds (AEWA) is a relevant example. The AEWA targets important areas for waterbird conservation across their entire flyway, providing a platform for coordinating actions across countries along the flyway, thereby taking a holistic approach that is essential for the conservation of migratory species. This reduces the risk of falling into the tragedy of parochialism, a situation whereby, for example, a country may enforce strict protection measures for a target species that is overharvested in other parts of the flyway. The AEWA has achieved some success, at least for some target species groups, such as wild geese populations in Europe (Fox and Madsen 2017).

Case Study 1: Migratory Montagu's Harrier (*Circus pygargus*) Conservation across the Flyway

Among all bird groups, migratory species are of high conservation concern because of the wide and diverse areas they use across their life cycles and the various threats they face within those areas (Runge et al. 2014). Over the past decade, one group identified of high conservation concern is Afro-Palearctic migrant birds (Sanderson et al. 2006). These long-distance migrant species breed in Europe but typically migrate across the Sahara to spend the nonbreeding season in Sub-Saharan Africa

(Fig. 13.3). Their populations have been reportedly declining in many areas of Europe, with the cause of the downward trend largely or partly attributed to processes and threats acting beyond the breeding grounds (Sanderson et al. 2006, Laaksonen and Lehikoinen 2013). A species that has received much conservation attention, particularly within Western Europe, is the Montagu's harrier. This ground-nesting raptor breeds across large open areas (farmland and steppe habitats) of Eurasia. In Western Europe, the species is strongly affected by farming intensification, which, through widespread mechanical crop harvesting practices, causes nest destruction and impairs breeding success

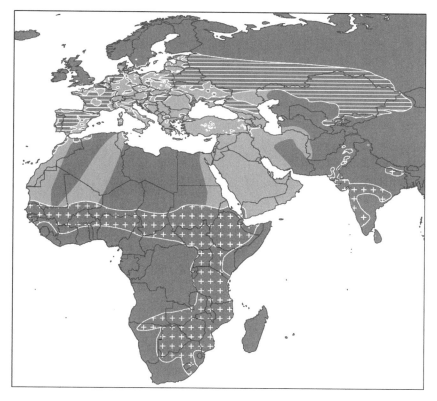

Figure 13.3   The range of the Montagu's harrier, spanning three continents and several countries. The figure shows the breeding range of the species (areas filled with white horizontal continuous line), the nonbreeding range at southern latitudes (areas filled with white crosses), and the passage areas (light gray colour). Dark gray continuous lines define the boundaries of each country. Range map of the Montagu's harrier was obtained from BirdLife International and NatureServe (2015).

(Arroyo et al. 2002, Santangeli et al. 2014, 2015). Large-scale intensive management efforts, however, are under way to protect nests of this bird on cultivated land from being destroyed by farming practices (Arroyo et al. 2002, 2004). Consequently, the breeding success of Montagu's harriers, especially those breeding in Western Europe, is strongly supported by expensive and intensive management efforts (Arroyo et al. 2002, Santangeli et al. 2014, 2015, Torres-Orozco et al. 2016). Although intensive expensive management may be important in the short term (Santangeli et al. 2014, Torres-Orozco et al. 2016), in the long run an ecosystem based on an internationally coordinated approach should be initiated to protect this species across its breeding range. Long-term sustainability of breeding populations could be achieved by means of European Union conservation tools, such as the Natura 2000 network of protected areas (Kukkala et al. 2016b) and agri-environment schemes (Batáry et al. 2011). These schemes, if properly designed, could not only cover the nesting habitat, but could also allow for retaining landscape features that are important in determining the foraging habitat of the species (Guixé and Arroyo 2011).

All the above-mentioned conservation efforts could ultimately be ineffective if threats on the nonbreeding ground are not addressed properly. The nonbreeding period is particularly crucial for long-lived migratory species, as it affects the survival of juveniles and adults. In the case of the Montagu's harrier, as in most other long-lived species, adult survival is a key demographic factor strongly affecting population trajectories (Santangeli et al. 2014). This further underscores the importance of addressing threats acting outside the breeding grounds. Montagu's harrier populations breeding in Western Europe spend their nonbreeding period largely in the African Sahel after crossing the Mediterranean Sea and the North African regions (Santangeli et al. 2014, Schlaich et al. 2016). Across these areas, the harriers are likely ex-

posed to different threats than those faced on their breeding grounds. Along the migratory route, the biggest threat is arguably represented by the illegal killing and taking of birds across the Mediterranean and North Africa (Brochet et al. 2016). Like most other Afro-Palearctic species spending the nonbreeding season in the Sahel, Montagu's harriers there are affected by anthropogenic threats. Such threats include illegal killing and taking but also, to a larger and wider extent, the rapid and progressive landscape transformation taking place in the Sahel region, where increasing pressure from a growing human population on natural and agricultural ecosystems, coupled with the progressive reduction in rainfall under current climate change, may cause large-scale degradation and loss of important habitat for wintering harriers (Schlaich et al. 2016). The challenge for protecting this species' habitat is further exacerbated by the large within-year movements of individual harriers through the Sahel, making the target area for conservation extremely wide (see Fig. 13.3; Limiñana et al. 2014, Schlaich et al. 2016). Moreover, this area is also reported to be largely uncovered by the network of protected areas (Limiñana et al. 2012). Ultimately, the massive extent of the focal region and the scope of the threats call for a strong holistic conservation approach acting at different scales, from local to regional and global. This can be achieved only if international cooperation is promoted, for example, through the enforcement of multilateral agreements such as the CMS, of which signatory parties cover large tracts of the Montagu's harrier flyway (http://www.cms.int/en/legalinstrument/cms). Ultimately, to achieve effective long-term conservation of Montagu's harrier populations, international collaboration will be paramount. The species is heavily dependent on intensive management on its breeding grounds, which may soon be in vain if increasing pressures on its wintering grounds in Africa are not rapidly addressed.

Case Study 2: Marine Mammal Management in International Waters

Migratory marine and freshwater mammal species, particularly the pelagic ones, often range across wide areas; therefore different phases of their life cycles occur in various countries with different jurisdictions and levels of protection (Hyrenbach et al. 2000). Consequently, international collaboration is crucial for their conservation. As an example, the International Whaling Commission (IWC) gathers its signatory member states in annual scientific meetings to update, promote, and coordinate marine mammal conservation efforts in international waters. Regional collaborative efforts among a restricted number of involved parties have also taken place with regard to marine vertebrate conservation. An example is represented by the formation of the East African Marine Ecoregion, a cooperative frame-

work among Tanzania, Mozambique, and South Africa aimed at establishing transboundary conservation mechanisms for marine mammals in the region (Guerreiro et al. 2010, Paudel et al. 2015a).

An interesting example of the challenges faced in protecting freshwater mammals across political borders is represented by the conservation of the Ganges river dolphin (*Platanista gangetica gangetica*) in Asia. The species is under severe threat due to infrastructure development, namely large hydropower projects blocking the main rivers where extant dolphin populations live (Smith et al. 1994, Paudel et al. 2015a,b). These development projects have been taking place under different political agreements at or near the border between the countries of Nepal, India, and Bangladesh (Fig. 13.4). The dams blocked the movement of the dolphins (Fig. 13.5), fragmenting populations and driving their decline in numbers and distribution (Paudel et al. 2015a). Poor

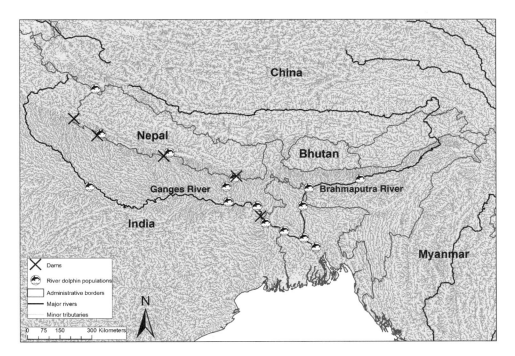

Figure 13.4 The location of large dams that fragment and isolate populations of the Ganges river dolphin (*Platanista gangetica gangetica*), once fully connected along the Ganges and Brahmaputra Rivers and their tributaries.

Figure 13.5 Sapta Koshi River barrage at the border between Nepal and India. Such barrages block movement of migratory animals like river dolphins and increase the risk of extinction by increasing chances of inbreeding in the small local populations created after barrage contruction. Shambhu Paudel

coordination between Nepal and India with regard to mitigation and conservation measures for the dolphin populations under pressure because of dam construction jeopardized conservation efforts to preserve the last remaining populations of this species in the region. Unfortunately, this is only one of the many cases of habitat fragmentation of freshwater species caused by hydropower dams along major waterways (Anderson et al. 2018a). Progress in the right direction, however, is also been made. In the case of river dolphins, recently the South Asia River Dolphin Task Force was formed during the meeting of the scientific committee (SC67A, May 2017) of the IWC. This task force, comprising members from relevant countries, aims to design and consolidate joint cross-national efforts for the conservation of migratory river dolphins in the South Asia river systems. Although India declared the river dolphin its national aquatic animal (Sinha and Kannan 2014), the country modifies each year the flow regime of the Sapta Koshi River at the border between Nepal and India. This activity, performed using dredging, affects the dolphins and their physical habitat directly, jeopardizing conservation efforts for the species practiced elsewhere.

Summary

Large gaps remain in our knowledge of the threats to biodiversity, and they need to be filled (Joppa et al. 2016). Similarly, more evidence is needed on the effectiveness of actions to make sure that efforts result in positive and measurable effects (Sutherland et al. 2004). During a time of such pervasive paucity of information on threats and actions, knowledge of the ecology of and threats to migratory wildlife is growing rapidly, thanks to the use of new tracking technology and quantitative methods (Pimm et al. 2015). Migratory species of fish, birds, bats, ungulates, and marine mammals experience directly the challenges of multinational, cross-border existence (Hidinger 2009, Hays et al. 2016, Trouwborst et al. 2016). Border fences, walls, and other semipermeable or impermeable barriers can potentially restrict movements of terrestrial species (Flesch et al. 2010, Linnell et al. 2016). As a result, new approaches for the conservation of migratory species are being embraced and can provide important insight for international collaboration on management and conservation efforts. A very positive example among these is the concept of species conservation at the flyway level, which allows for holistic protection of species across the areas used during their life cycle. Although such approaches sound extremely promising in theory, their implementation presents several challenges posed by the multitude and diversity of the relevant stakeholders, from regional authorities and national and subnational administrations down to communities and single private landowners (Runge et al. 2014, 2017). These challenges have been recognized and acknowledged by the conservation community. Now the momentum must be harnessed to build or consolidate an effective and efficient infrastructure to promote the type of international cooperation shown by the examples in this chapter. This is the best way forward to secure the long-lasting preservation of wildlife across administrative boundaries.

LITERATURE CITED

Amano, T., T. Székely, B. Sandel, S. Nagy, T. Mundkur, T. Langendoen, D. Blanco, C.U. Soykan, and W.J. Sutherland. 2017. Successful conservation of global waterbird populations depends on effective governance. Nature 553:199.

Anderson, E.P., C.N. Jenkins, S. Heilpern, J.A. Maldonado-Ocampo, F.M. Carvajal-Vallejos, A.C. Encalada, J.F. Rivadeneira, et al. 2018a. Fragmentation of Andes-to-Amazon connectivity by hydropower dams. Science Advances 4:eaao1642.

Anderson, M.G., R.T. Alisauskas, B.D. Batt, R.J. Blohm, K.F. Higgins, M.C. Perry, J.K. Ringelman, et al. 2018b. The Migratory Bird Treaty and a century of waterfowl conservation. Journal of Wildlife Management 82:247–259.

Arrondo, E., M. Moleón, A. Cortés-Avizanda, J. Jiménez, P. Beja, J.A. Sánchez-Zapata, and J. Donázar. 2018. Invisible barriers: Differential sanitary regulations constrain vulture movements across country borders. Biological Conservation 219:46–52.

Arroyo, B., J.T. García, and V. Bretagnolle. 2002. Conservation of the Montagu's harrier (*Circus pygargus*) in agricultural areas. Animal Conservation 5:283–290.

Arroyo, B., J.T. García, and V. Bretagnolle. 2004. Montagu's harrier. Birds of the Western Palearctic Update 6:41–55.

Atkinson, P.W., W.M. Adams, J. Brouwer, G. Buchanan, R.A. Cheke, W. Cresswell, C.M. Hewson, et al. 2014. Defining the key wintering habitats in the Sahel for declining African-Eurasian migrants using expert assessment. Bird Conservation International 24:477–491.

Batáry, P., A. Báldi, D. Kleijn, and T. Tscharntke. 2011. Landscape-moderated biodiversity effects of agri-environmental management: A meta-analysis. Proceedings of the Royal Society B: Biological Sciences 278:1894–1902.

BirdLife International and NatureServe. 2015. Bird species distribution maps of the world. Version 4.0, BirdLife International, Cambridge, UK, and NatureServe, Arlington, Virginia, USA.

Botha, A., J. Andevski, C.G.R. Bowden, M. Gudka, R.J. Safford, J. Tavares, and N.P. Williams. 2017. Multi-species action plan to conserve African-Eurasian vultures. CMS Raptors MOU Technical Publication No. 5. Coordinating Unit of the CMS Raptors MOU, Abu Dhabi, UAE.

Brochet, A.-L., W. Van Den Bossche, S. Jbour, P.K. Ndang'ang'a, V.R. Jones, W.A. Abdou, A.R. Al-Hmoud, et al. 2016. Preliminary assessment of the scope and scale of illegal killing and taking of birds in the Mediterranean. Bird Conservation International 26:1–28.

Buechley, E.R., and C.H. Şekercioğlu.2016. The avian scavenger crisis: Looming extinctions, trophic cascades, and loss of critical ecosystem functions. Biological Conservation 198:220–228.

Cabrera-Cruz, S.A., J.A. Smolinsky, and J.J. Buler. 2018. Light pollution is greatest within migration passage areas for nocturnally-migrating birds around the world. Scientific Reports 8:3261.

Carrete, M., J.A. Sánchez-Zapata, J.R. Benítez, M. Lobón, and J.A. Donázar. 2009. Large scale risk-assessment of wind-farms on population viability of a globally endangered long-lived raptor. Biological Conservation 142:2954–2961.

Dahl, E.L., K. Bevanger, T. Nygård, E. Røskaft, and B.G. Stokke. 2012. Reduced breeding success in white-tailed eagles at Smøla windfarm, western Norway, is caused by mortality and displacement. Biological Conservation 145:79–85.

Dallimer, M., and N. Strange. 2015. Why socio-political borders and boundaries matter in conservation. Trends in Ecology and Evolution 30:132–139.

Finch, T., J.W. Pearce-Higgins, D.I. Leech, and K.L. Evans. 2014. Carry-over effects from passage regions are more important than breeding climate in determining the breeding phenology and performance of three avian migrants of conservation concern. Biodiversity and Conservation 23:2427–2444.

Flesch, A.D., C.W. Epps, J.W. Cain III, M. Clark, P.R. Krausman, and J.R. Morgart, 2010. Potential effects of the United States–Mexico border fence on wildlife. Conservation Biology 24:171–181.

Fox, A.D., and J. Madsen. 2017. Threatened species to super-abundance: The unexpected international implications of successful goose conservation. Ambio 46:179–187.

Guerreiro, J., A. Chircop, C. Grilo, A. Viras, R. Ribeiro, and R. Van Der Elst. 2010. Establishing a transboundary network of marine protected areas: Diplomatic and management options for the east African context. Marine Policy 34:896–910.

Guixé, D., and B. Arroyo. 2011. Appropriateness of Special Protection Areas for wide-ranging species: The importance of scale and protecting foraging, not just nesting habitats. Animal Conservation 14:391–399.

Hardin, G. 1968. The tragedy of the commons. Science 162:1243.

Hays, G.C., L.C. Ferreira, A.M. Sequeira, M.G. Meekan, C.M. Duarte, H. Bailey, F. Bailleul, et al. 2016. Key questions in marine megafauna movement ecology. Trends in Ecology and Evolution 31:463–475.

Hidinger, L. 2009. To fence or not to fence. Frontiers in Ecology and the Environment 7:350–351.

Hyrenbach, K.D., K.A. Forney, and P.K. Dayton. 2000. Marine protected areas and ocean basin management. Aquatic Conservation: Marine and Freshwater Ecosystems 10:437–458.

Joppa, L.N., B. O'Connor, P. Visconti, C. Smith, J. Geldmann, M. Hoffmann, J.E. Watson, et al. 2016. Filling in biodiversity threat gaps. Science 352:416–418.

Kark, S., N. Levin, H.S. Grantham, and H.P. Possingham. 2009. Between-country collaboration and consideration of costs increase conservation planning efficiency in the Mediterranean Basin. Proceedings of the National Academy of Sciences of the United States of America 106:15368–15373.

Knisley, C.B., and J.M. Hill. 2001. Biology and conservation of the coral pink sand dunes tiger beetle, Cicindela limbata albissima Reumpp. Western North American Naturalist 61:381–394.

Kukkala, A.S., A. Arponen, L. Maiorano, A. Moilanen, W. Thuiller, T. Toivonen, L. Zupan, L. Brotons, and M. Cabeza. 2016a. Matches and mismatches between national and EU-wide priorities: Examining the Natura

2000 network in vertebrate species conservation. Biological Conservation 198:193–201.

Kukkala, A.S., A. Santangeli, S.H.M. Butchart, L. Maiorano, I. Ramirez, I.J. Burfield, and A. Moilanen. 2016b. Coverage of vertebrate species distributions by Important Bird and Biodiversity Areas and Special Protection Areas in the European Union. Biological Conservation 202:1–9.

Laaksonen, T., and A. Lehikoinen. 2013. Population trends in boreal birds: Continuing declines in agricultural, northern, and long-distance migrant species. Biological Conservation 168:99–107.

Limiñana, R., A. Soutullo, B. Arroyo, and V. Urios. 2012. Protected areas do not fulfil the wintering habitat needs of the trans-Saharan migratory Montagu's harrier. Biological Conservation 145:62–69.

Limiñana, R., B. Arroyo, J. Terraube, M. Mcgrady, and F. Mougeot. 2014. Using satellite telemetry and environmental niche modelling to inform conservation targets for a long-distance migratory raptor in its wintering grounds. Oryx 49:329–337.

Linnell, J.D., A. Trouwborst, L. Boitani, P. Kaczensky, D. Huber, S. Reljic, J. Kusak, et al. 2016. Border security fencing and wildlife: The end of the transboundary paradigm in Eurasia? PLOS Biology 14: p.e1002483.

Lopez-Hoffman, L., C.C. Chester, and R. Merideth. 2017. Conserving transborder migratory bats, preserving nature's benefits to humans: The lesson from North America's bird conservation treaties. Bioscience 67:320–321.

Mazor, T., M. Beger, J. McGowan, H.P. Possingham, and S. Kark. 2016. The value of migration information for conservation prioritization of sea turtles in the Mediterranean. Global Ecology and Biogeography 25:540–552.

Montesino Pouzols, F., T. Toivonen, E. Di Minin, A.S. Kukkala, P. Kullberg, J. Kuusterä, J. Lehtomäki, H. Tenkanen, P.H. Verburg, and A. Moilanen. 2014. Global protected area expansion is compromised by projected land-use and parochialism. Nature 516:383–386.

Newton, I. 2008. Migration ecology of birds. Academic Press, London, UK.

Northrup, J.M., and G. Wittemyer. 2013. Characterising the impacts of emerging energy development on wildlife, with an eye towards mitigation. Ecology Letters 16:112–125.

Paudel, S., P. Pal, M.V. Cove, S.R. Jnawali, G. Abel, J.L. Koprowski, and R. Ranabhat. 2015a. The endangered Ganges river dolphin Platanista gangetica gangetica in Nepal: Abundance, habitat and conservation threats. Endangered Species Research 29:59–68.

Paudel, S., Y.P. Timilsina, J. Lewis, T. Ingersoll, and S.R. Jnawali. 2015b. Population status and habitat occupancy of endangered river dolphins in the Karnali River system of Nepal during low water season. Marine Mammal Science 31:707–719.

Pearce-Higgins, J.W., and R.E. Green. 2014. Birds and climate change: Impacts and conservation responses. Cambridge University Press, Cambridge, UK.

Pimm, S.L., S. Alibhai, R. Bergl, A. Dehgan, C. Giri, Z. Jewell, L. Joppa, R. Kays, and S. Loarie. 2015. Emerging technologies to conserve biodiversity. Trends in Ecology and Evolution 30:685–696.

Runge, C.A., T.G. Martini, H.P. Possingham, S.G. Willis, and R.A. Fuller. 2014. Conserving mobile species. Frontiers in Ecology and the Environment 12:395–402.

Runge, C.A., E. Gallo-Cajiao, M.J. Carey, S.T. Garnett, R.A. Fuller, and P.C. McCormack. 2017. Coordinating domestic legislation and international agreements to conserve migratory species: A case study from Australia. Conservation Letters 10:765–772.

Sanderson, F.J., P.F. Donald, D.J. Pain, I.J. Burfield, and F.B.J. Van Bommel. 2006. Long-term population declines in Afro-Palearctic migrant birds. Biological Conservation 131:93–105.

Sanderson, F.J., R.G. Pople, C. Ieronymidou, I.J. Burfield, R.D. Gregory, S.G. Willis, C. Howard, P.A. Stephens, A.E. Beresford, and P.F. Donald. 2016. Assessing the performance of EU nature legislation in protecting target bird species in an era of climate change. Conservation Letters 9:172–180.

Santangeli, A., E. Di Minin, and B. Arroyo. 2014. Bridging the research implementation gap: Identifying cost-effective protection measures for Montagu's harrier nests in Spanish farmlands. Biological Conservation 177:126–133.

Santangeli, A., B. Arroyo, A. Millon, and V. Bretagnolle. 2015. Identifying effective actions to guide volunteer-based and nationwide conservation efforts for a ground-nesting farmland bird. Journal of Applied Ecology 52:1082–1091.

Santangeli, A., S.H.M. Butchart, M. Pogson, A. Hastings, P. Smith, M. Girardello, and A. Moilanen. 2018. Mapping the global potential exposure of soaring birds to terrestrial wind energy expansion. Ornis Fennica 95:1–14.

Schlaich, A.E., R.H.G. Klaassen, W. Bouten, V. Bretagnolle, B.J. Koks, A. Villers, and C. Both. 2016. How individual Montagu's harriers cope with Moreau's paradox during the Sahelian winter. Journal of Animal Ecology 85:1491–1501.

Sinha, R.K., and K. Kannan. 2014. Ganges river dolphin: An overview of biology, ecology, and conservation status in India. Ambio 43:1029–1046.

Smith, B.D., R.K. Sinha, and U. Regmi. 1994. Status of Ganges river dolphins (Platanista gangetica) in the Karnali, Mahakali, Narayani and Sapta Kosi Rivers of Nepal and India in 1993. Marine Mammal Science 10:368–375.

Sutherland, W.J., A.S. Pullin, P.M. Dolman, and T.M. Knight. 2004. The need for evidence-based conservation. Trends in Ecology and Evolution 19:305–308.

Szabo, J.K., C.Y. Choi, R.S. Clemens, and B. Hansen. 2016. Conservation without borders: Solutions to declines of migratory shorebirds in the East Asian-Australasian Flyway. Emu 116:215–221.

Thaxter, C.B., G.M. Buchanan, J. Carr, S.H. Butchart, T. Newbold, R.E. Green, J.A. Tobias, et al. 2017. Bird and bat species' global vulnerability to collision mortality at wind farms revealed through a trait-based assessment. Proceedings of the Royal Society B: Biological Sciences 284:20170829.

Torres-Orozco, D., B. Arroyo, M. Pomarol, and A. Santangeli. 2016. From a conservation trap to a conservation solution: Lessons from an intensively managed Montagu's harrier population. Animal Conservation 19:436–443.

Trierweiler, C., W.C. Mullié, R.H. Drent, K.M. Exo, J. Komdeur, F. Bairlein, A. Harouna, M. de Bakker, and B.J. Koks. 2013. A Palaearctic migratory raptor species tracks shifting prey availability within its wintering range in the Sahel. Journal of Animal Ecology 82:107–120.

Trouwborst, A., F. Fleurke, and J. Dubrulle. 2016. Border fences and their impacts on large carnivores, large herbivores and biodiversity: An international wildlife law perspective. Review of European, Comparative and International Environmental Law 25:291–306.

Tucker, M.A., K. Böhning-Gaese, W.F. Fagan, J.M. Fryxell, B. Van Moorter, S.C. Alberts, A.H. Ali, et al. 2018. Moving in the Anthropocene: Global reductions in terrestrial mammalian movements. Science 359:466–469.

Vickery, J.A., S.R. Ewing, K.W. Smith, D.J. Pain, F. Bairlein, J. Skorpilova, and R.D. Gregory. 2014. The decline of Afro-Palaearctic migrants and an assessment of potential causes. Ibis 156:1–22.

Wilson, J.M., and W. Cresswell. 2006. How robust are Palearctic migrants to habitat loss and degradation in the Sahel? Ibis 148:789–800.

Xia, S., X. Yu, S. Millington, Y. Jia, L. Wang, X. Hou, and L. Jiang. 2017. Identifying priority sites and gaps for the conservation of migratory waterbirds in China's coastal wetlands. Biological Conservation 210:72–82.

14

JOHN F. ORGAN
GONZALO MEDINA-VOGEL
TSUYOSHI YOSHIDA

International Organizations and Programs for Wildlife Conservation

Introduction

Historically, private nongovernmental entities in the developed world have taken interest in wildlife in developing nations. Nineteenth- and early-twentieth-century explorers of Africa, Asia, and South America reported fascinating encounters with magnificent species that had been previously cryptic or unknown (Robinson et al. 2017). Organizations such as the Royal Geographical Society in the United Kingdom (a sponsor of Charles Darwin's expedition), the Boone and Crockett Club in North America, and numerous museums in the United States and Europe sponsored chronicled expeditions to remote places under the goal of discovery and collection.

Concurrent with the quest for discovery of wild places and species new to science was the drive to find and secure timber, mineral, energy, and other resources to fuel the growing demands of development brought on by the Industrial Revolution (Lucas 2009, Stearns 2013). This led to exploitation, and overexploitation, which led to growing concerns over the loss of species, including species not yet known to Western science. During the latter half of the twentieth century, in response to these concerns, private nongovernmental organizations formed to promote conservation through science, advocacy, and

technical assistance. The focus of these organizations varies, from working independently to forming collaborations across the public and private sphere. Our objective in this chapter is to outline some of the major conventions, major organizations, and programs that promote engagement in wildlife conservation and management at the international level.

International Wildlife Conventions

International conventions are treaties or agreements among nation states, the primary actors in international law. In the United States, the US Constitution, under the Supremacy Clause (Article VI, Clause 2), elevates international treaty supremacy over any other law (Bean 1983). International conventions have significant effects on wildlife conservation globally. The major international conventions are outlined below.

Convention on International Trade in Endangered Species of Wild Fauna and Flora (CITES)

The Convention on International Trade in Endangered Species of Wild Fauna and Flora is a treaty designed to protect species from becoming endangered by international trade. Originally drafted in 1963 at a meeting of the International Union for the Conservation of Nature (IUCN), it was signed in 1973 and went into effect in 1975 (Bean 1983). Currently, 183 nations are party to CITES (https://www.cites.org).

CITES requires certain controls and restrictions to be placed on international trade in wildlife. More than 30,000 species are categorized according to threat within three appendices to the convention. Appendix I lists species threatened with extinction and allows trade under only the strictest of circumstances. Appendix II controls trade in species that, although not threatened with extinction, could have their status threatened by trade. Appendix II also includes species such as the bobcat (*Lynx rufus*) and Nearctic river otter (*Lontra canadensis*), whose status is secure but are similar in appearance to species listed in the appendices. Trade in such species is not necessarily restricted, but official tags indicating origin must be affixed to individuals or parts to be legally traded. Appendix III lists species that, although not threatened globally with extinction, are protected when one party asks other parties for assistance in controlling trade (Energy and Biodiversity Initiative 2003).

The CITES Conference of the Parties (CoP) is held every three years, whereas CITES committees (e.g., Animals) meet annually. Additions and deletions of species, movement among appendices, and amendments to the convention can be made at the CoP.

Migratory Bird Treaty (MBT)

The Convention between the United States and Great Britain (for Canada) for the Protection of Migratory Birds, also known as the Migratory Bird Treaty, was signed in 1916 (Hawkins et al. 1984). Similar agreements were subsequently signed with Mexico (1936), Japan (1972), and Russia (1976). The Migratory Bird Treaty provides the basis for nations with bird species in common that migrate across their borders to agree on protective measures and regulation of take (Energy and Biodiversity Initiative 2003).

The Ramsar Convention

The Ramsar Convention, officially the Ramsar Convention on Wetlands of International Importance especially as Waterfowl Habitat, provides a framework for international cooperation in the conservation of wetlands (Matthews 1993). The idea for the Ramsar Convention originated in 1962 at a waterfowl conservation conference, and was adopted in 1971 in Ramsar, Iran. It is the only global convention that addresses a specific habitat. Upon joining the Ramsar Convention, a party must designate at least one wetland site within their territory for inclusion in the List of Wetlands of International Importance. Sites on this list attain national and international status and the government's commitment to maintain their ecological character. The Ramsar Convention provides criteria for identifying wetlands of international importance, guidance on management of sites, and measures to address threats. As of 2017, there were 169 parties to the convention, with 2,288 designated sites totaling 220,922,033 ha (Energy and Biodiversity Initiative 2003).

The Bonn Convention

The Convention on the Conservation of Migratory Species of Wild Animals, also known as the Bonn Convention, aims to conserve terrestrial, marine, and avian species throughout their range. The treaty was framed in 1983 under the auspices of the United Nations Environment Programme and includes more than 120 parties from Africa, Asia, Central and South America, Europe, and Oceania (Energy and Biodiversity Initiative 2003).

Agreements (legally binding treaties), Memoranda of Understanding specific to individual species or groups of species (e.g., Eurobats), and Special Species Initiatives are the primary mechanisms under the Bonn Convention. Appendix I of the Bonn Convention lists migratory species threatened with extinction. Migratory species that have need for or would benefit from international cooperation are listed in Appendix II (United Nations Environment Programme 1994).

Convention on Biological Diversity (CBD)

The Convention on Biological Diversity was signed in Rio de Janeiro, Brazil, in 1992. As of 2017, it has been ratified by 196 parties, 168 signatories, and 30 nations. The three main goals of the Convention on Biological Diversity are the conservation of biological diversity, the sustainable use of biological resources, and the fair and equitable sharing of benefits derived from genetic resources. The Convention on Biological Diversity was an outcome of the 1992 United Nations Conference on Environment and Development (Earth Summit) in Rio de Janeiro in 1992 (Energy and Biodiversity Initiative 2003).

The CBD is a benchmark because it recognizes biodiversity at the ecosystem, species, and genetic levels. It requires nations to develop national biodiversity strategies and incorporate its components into policy in sectors that could potentially have effect. These strategies and action plans are the primary tools for implementing the convention (Cropper 1993).

United Nations Convention on the Law of the Sea

The United Nations Convention on the Law of the Sea, or the Law of the Sea Convention, adopted in 1982, establishes the rights and responsibilities of all nations that use oceans. Regulations under the convention are designed to protect ocean resources from overuse. The convention also codifies national and international water boundaries and exclusive economic zones for all countries that have ocean coastlines.

The Fish Stocks Agreement (UNFSA) is an elaboration of the Law of the Sea Convention focused on high-seas fishing for highly migratory species such as tuna (tribe Thunnini). The UNFSA obligates nations to minimize pollution and wastes and requires they minimize effects to nontarget species.

United Nations Framework Convention on Climate Change (UNFCCC)

The United Nations Framework Convention on Climate Change, adopted at the Rio Earth Summit in 1992, is the main international agreement on climate action. Within the purview of the UNFCCC are the Kyoto Protocol and the Paris Agreement. The Kyoto Protocol, agreed to in 1997, introduced legally binding emission reduction targets. The Paris Agreement of 12 December 2015 represented a new global agreement on climate change, with an action plan to limit global warming to less than 2°C.

World Heritage Convention

The 1972 World Heritage Convention links the concepts of nature conservation and the preservation of cultural properties. The convention defines the kinds of sites that can be included on the World Heritage List (http://whc.unesco.org/en/list/). As of January 2018, 1,073 properties in 167 nations had been listed; of these, 206 sites were listed purely as natural sites.

International Nongovernmental Wildlife Organizations

Nongovernmental organizations focused on international wildlife conservation are abundant and too great in number to entirely chronicle here. They vary from advocacy organizations with no field presence to rigorous science-based enterprises. Below are some of the more notable organizations, listed from global to more regional in scope, who have demonstrated an ability to successfully engage and partner with governments and other organizations.

International Union for Conservation of Nature and Natural Resources (IUCN)

The International Union for Conservation of Nature and Natural Resources, formerly known as the World Conservation Union, was formed in 1948 in Fontainebleau, France. Currently, it is headquartered in Gland, Switzerland, with offices in more than 50 countries. The mission of the IUCN is to influence, encourage, and assist societies throughout the world to conserve the integrity and diversity of nature and to ensure that any use of natural resources is equitable and ecologically sustainable. The IUCN is a member-based organization and includes more than 1,400 governmental and nongovernmental entities and 16,000 scientists and experts who participate as volunteers in its core work (IUCN 1980).

The IUCN comprises its member organizations, its governing body or secretariat, and six commissions designed to assess the state of the world's natural resources and provide advice and recommendations on conservation issues. The largest commission is the Species Survival Commission, with more than 10,000 volunteer experts participating in more than 140 species specialist groups. These groups serve as a clearinghouse for current scientific information and for illuminating knowledge gaps, and provide insights on the significance of species to society and livelihoods. Information gathered by the specialist groups is fed into the IUCN Red List of Threatened Species, a major database that provides status assessment for species globally. The IUCN Red List provides countries with a framework from which to develop their own red lists; many countries adopt the Red List designations (Peru).

The World Conservation Congress is convened every 4 years by IUCN and alternately hosted by one of the member nations. Several thousand members convene in a variety of forums to share knowledge, develop policy, and provide direction on key global conservation issues.

Wildlife Conservation Society (WCS)

The Wildlife Conservation Society was founded in 1895 as the New York Zoological Society. The WCS employs approximately 200 scientists throughout the world, engaged in more than 500 field conservation projects, and works closely with governments to fill gaps in scientific knowledge that can lead to progressive conservation policies.

The WCS has competitive grant programs. Most of these grant programs are designed to build capacity and develop the next generation of conservation leaders in developing countries (https://www.wcs .org/).

World Wildlife Fund (WWF)

The World Wide Fund for Nature, also known as the World Wildlife Fund, was founded in 1961. With more than five million supporters, it is considered the world's largest conservation organization, active in more than 100 countries and supporting more than 1,300 projects. Its mission is to stop the degradation of the planet's natural environment and build a future where humans live in harmony with nature.

The WWF receives funding support from a mixture of private donations, governments, and corporations. It has headquarter offices in Gland, Switzerland, and Washington, DC, USA. WWF activities focus largely on working to influence policy at the highest levels of government for eliminating wildlife trade, conserving flagship species and environments, and promoting climate action (https://www .worldwildlife.org/).

Safari Club International Foundation (SCIF)

The Safari Club International Foundation is a nonprofit organization dedicated to wildlife conservation, outdoor education, and humanitarian services. It provides funding support to wildlife conservation projects across the world, most notably in Africa, Asia, and North America. Many of the projects funded by SCIF explore predator-prey dynamics, study sustainable uses, develop science tools for wildlife management, support anti-poaching efforts, and build local capacity for conservation.

The SCIF has offices in Tucson, Arizona, USA, and Washington, DC. Its funding is derived through grants from Safari Club International (SCI), donor gifts, events, and other sources. The SCIF is the nonprofit branch of SCI, and is a separate legal entity (http://safariclubfoundation.org/).

Frankfurt Zoological Society (FZS)

The Frankfurt Zoological Society is an international nonprofit organization committed to preserving wildlands and biological diversity. The FZS is active in portions of central and Eastern Europe, East Africa, central South America, and Southeast Asia, with close to 30 conservation programs in 18 countries. The FZS has a significant presence in Tanzania and Peru, where it focuses on conserving iconic species and landscapes.

The FZS has approximately 300 employees, with a small staff at its headquarters in Frankfurt, Germany, and the remainder in the different project countries. It was formed in 1858 and became active in global conservation during the 1950s (https://fzs.org/en/).

Zoological Society of London (ZSL)

The Zoological Society of London is a nonprofit organization founded in 1826 with a long commitment to conservation and science around the world (https://www.zsl.org/). Within the United Kingdom, ZSL manages the London and Whipsnade Zoos. In addition, the scientific journals *Journal of Zoology* and *Animal Conservation* and the associated Institute of Zoology support its mission of international conservation. Research grants are provided through the Darwin Initiative for international conservation and training (Fig. 14.1)

Conservation International (CI)

Conservation International is a nonprofit organization whose goal is to protect nature as a source of food, fresh water, livelihoods, and a stable climate. Conservation International partners with numerous agencies, corporations, and private entities to initiate conservation projects, inform conservation and development policies, and empower local people.

Conservation International is headquartered in Arlington, Virginia, USA, and employs more than 1,000 people, with partners in 30 countries. It supports the establishment and management of protected areas in terrestrial and marine environments and administers a Global Conservation Fund used to support regional programs and partners (https://www.conservation.org/Pages/default.aspx).

Oceana

Oceana, established in 2001, is the largest international ocean conservation and advocacy organization in the world. Oceana's focus is ocean protection and restoration through targeted policy campaigns. Oceana, headquartered in Washington, DC, USA, with several satellite offices around the world, was established by several charitable trusts who collectively were concerned that a very small proportion of environmental advocacy was directed toward ocean conservation.

Ducks Unlimited (DU)

Ducks Unlimited is a volunteer-based nonprofit organization founded in 1937 to conserve waterfowl and wetland habitat. Focused in Canada, Mexico, and the United States, DU has developed diverse public and private partnerships to address a range of factors related to conservation of waterfowl and wetland habitat. In addition to an extremely large volunteer base that sponsors more than 4,000 fund-raising events annually, DU has a professional scientific staff headquartered in Memphis, Tennessee, USA. Ducks

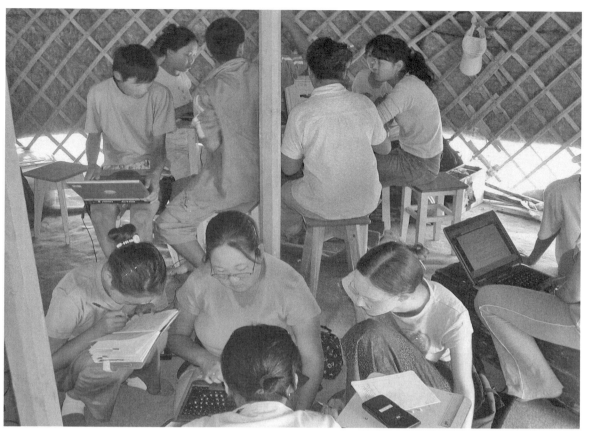

Figure 14.1   Training programs to increase in-country capacity are important. The Zoological Society of London's Steppe Forward Programme provided training to university students from Mongolia through immersion into major ecosystems where lecture material and field techniques were taught. Here, students work in groups within a traditional *ger* (dwelling) to analyze data collected during the day's fieldwork. John Koprowski

Unlimited has sister organizations in Canada (Ducks Unlimited Canada) and Mexico (Ducks Unlimited de Mexico, DUMAC), and, collectively, they have conserved millions of hectares of habitat in those countries.

International Association for Bear Research and Management (IBA)

The International Association for Bear Research and Management is a nonprofit organization open to professional biologists, wildlife managers, and others dedicated to the conservation of all bear species. The IBA has more than 550 members from more than 50 countries. It supports the scientific management of bears through research and information sharing, and sponsors an international conference, publishes a journal (*Ursus*), and awards grants for bear research. The IBA is closely affiliated with the IUCN Bear Specialist Group (http://www.bearbiology.com/).

Bat Conservation International (BCI)

Bat Conservation International is an international private organization dedicated to conserving bat species worldwide through conservation, education, and research. Founded in 1982, BCI is currently headquartered in Austin, Texas, USA, with members in 60 countries; it supports a staff of 30 biologists, educators, and administrators.

Rainforest Alliance

The Rainforest Alliance was founded in 1987 and works to conserve biodiversity and ensure sustainable livelihoods through promoting land-use practices, business practices, and consumer behavior. Based in New York City, USA, with offices in North and South America, Asia, Africa, and Europe, its work in relation to wildlife involves primarily the development and promotion of sustainable forestry and agricultural practices to reduce deforestation and conserve important habitats (https://www.rainforest -alliance.org/).

African Wildlife Foundation (AWF)

The African Wildlife Foundation was established in 1961 with the initial goal of training African wildlife conservation professionals. In recent years, its focus has been on protecting habitat by supporting existing protected areas, creating private land trusts, and working with local communities. Priority is given to landscape-level projects. In 2016, 60 species research and conservation projects were supported by AWF (http://www.awf.org/).

Amazon Conservation Association

The Amazon Conservation Association's purpose is to conserve the biodiversity of the Amazon Basin through supporting new scientific understanding, sustainable resource management, and rational land-use policy. The organization strives to develop field research sites from high-elevation cloud forests to lowland Amazon forests. The association maintains the Wayqecha Cloud Forest Biological Station, Peru's only permanent research station focused on cloud forest ecology and management, and has awarded 56 grants for research work at Wayqecha (http://www .amazonconservation.org/).

International Council for Game and Wildlife Conservation (CIC)

The International Council for Game and Wildlife Conservation, based in Hungary, is a nonpolitical advisory body aiming to preserve wildlife through promotion of sustainable-use principles. The CIC unites 26 member nations, represented by their ministries responsible for wildlife management and conservation, with a wide range of nongovernmental organizations (http://www.cic-wildlife.org/).

European Wildlife

European Wildlife is a nonprofit organization dedicated to nature preservation and environmental protection, based in the Czech Republic. Its principal activities involve partnering to create reserves and corridors to connect biodiverse areas, with its primary effort focused on creating a European center of biodiversity, a large nature reserve in central Europe. Many flagship wildlife species are featured in its programs, but the main focus of its efforts are on habitat protection (http://www.eurowildlife.org/).

Wildlife Asia

Wildlife Asia is a nonprofit organization based in Australia whose mission is the protection and enhancement of the natural habitat of orangutans (*Pongo* spp.), gibbons (Hylobatidae), Asian rhinos (*Rhinoceros* spp.), bears, elephants (*Elephas maximus*), tigers (*Panthera tigris*), and other Asian wildlife. Wildlife Asia programs are focused on rainforest habitat protection, conservation of priority wildlife species, and sustainable development for local communities. It is active primarily in Southeast Asia.

Typical wildlife projects for Wildlife Asia involve monitoring and assessing wildlife through biodiversity surveys, training and deploying wildlife protection personnel, and capacity-building for fieldwork and administration in local areas (http://www .wildlifeasia.org.au/).

Other Organizations

Other smaller funding programs with worldwide coverage are also supporting conservation efforts. For example, the Rufford Foundation (https://www.rufford.org/), a UK-registered charity that funds nature conservation projects across the developing world, has to date awarded grants to more than 4,000 projects in 156 countries. Other small grant sources for wildlife research include initiatives like IDEA WILD (http://www.ideawild.org/), which seeks to minimize the loss of biodiversity by awarding small equipment grants to conservation professionals around the world. Each year, IDEA WILD distributes more than US$300,000 worth of equipment—including digital cameras, spotting scopes, mist nets, tents, binoculars, and global positioning system units—to more than 400 conservation projects worldwide. IDEA WILD has developed a unique and effective model for empowering conservation and has successfully implemented it in areas of the world where it will have an effect on wildlife. Since its inception in 1991, IDEA WILD has successfully supported more than 4,500 conservation projects in more than 100 countries. Because recipients pass on their equipment to at least two other conservation efforts, it is estimated that IDEA WILD has effectively served more than 13,500 conservation projects around the globe.

Cooperation among Nations and Organizations

Cooperation across borders among different conservation actors is necessary to maximize the effect of limited resources in addressing contemporary conservation challenges (Guo 2018). The IUCN, described above, is a major catalyst for cooperative efforts among nations and nongovernmental organizations. The United States Agency for International Development (USAID) is also a major catalyst for conservation efforts; it assists developing nations in improving livelihoods. Under its Environment and Global Climate Change program, USAID supports projects that promote responsible natural resource management and biodiversity conservation (https://www.usaid.gov/).

The Peru Forest Sector Initiative (PFSI) is an example of USAID support for cooperative efforts to ensure sustainable management of Peru's biodiversity. The PFSI was initiated as a result of efforts to develop a free-trade agreement between Peru and the United States. Concerns were raised over unregulated trade in forest and wildlife products, and the Peruvian government passed legislation to regulate forestry and wildlife activities. The PFSI was funded by USAID and supported by the US Forest Service (USAID 2014). Staffed by Peruvian conservation specialists, PFSI collaborated with national and regional governments, with support from US foresters and wildlife biologists, to develop regulations to implement the law and develop training programs for regional government forest and wildlife specialists to conduct forest and wildlife inventories (https://www.fs.fed.us/sites/default/files/peru-program-overview.docx).

The Kenya Wildlife Conservation Project is a partnership to support the Kenya Wildlife Service's management of Kenya's national parks and reserves and to promote community-based wildlife management. The multifaceted project, supported by the African Wildlife Foundation and USAID, involves habitat and species monitoring, mapping, and implementation of national conservation strategies for black rhinos (*Diceros bicornis*), cheetahs (*Acinonyx jubatus*), wild dogs (*Lycaon pictus*), spotted hyena (*Crocuta crocuta*), sea turtles (Chelonioidea), and other species. The project has also implemented the wildlife management information system, a digital monitoring system for reporting and analyzing human-wildlife conflict, poaching incidents, and wildlife movements (https://www.usaid.gov/sites/default/files/documents/1860/Kenya%20Wildlife%20Conservation%20FACT%20SHEET_May 2013.pdf).

The Asian Species Action Partnership (ASAP) is an interagency coalition designed to address the extinction risk of the most threatened terrestrial and freshwater vertebrates in Southeast Asia (IUCN 2014). An analysis of the IUCN's Red List of Threatened Species showed that Southeast Asia has the highest concentration of species nearing extinction of any region in the world. The ASAP, with more than 60 organizational partners, aims to catalyze immediate action to reduce threats, strengthen ongoing action by facilitating partnerships, convene and support dialogue among stakeholders, improve conservation actions through development of best management practices, and safeguard existing populations through captive breeding programs (https://www.iucn.org/theme/species/our-work/asian-species-action-partnership-asap).

The Collaborative Partnership on Sustainable Wildlife Management (CPW) is a voluntary partnership of 14 international organizations to promote the sustainable use and conservation of wildlife resources. The mission of CPW, originated as a joint project of the CIC and the Food and Agriculture Organization of the United Nations (FAO), is to promote conservation through sustainable management of terrestrial vertebrate wildlife in all biomes and geographic areas. The partnership has four main thematic priorities: wildlife, food security, and livelihoods; human-wildlife conflict; illegal-unsustainable hunting; and partnership coordination and outreach. The CPW has developed many resources to aid in development of sustainable wildlife management, including white papers and guidance on topics ranging from bushmeat to poaching to human-wildlife conflict (http://www.fao.org/forestry/wildlife-partnership/en/).

Research Networks

Kessler et al. (1998) reviewed international trends in university wildlife education, noting substantive differences across the globe. These differences did not merely reflect what one might expect within the context of postindustrial versus developing nations, where differences would be influenced by infrastructure and financial dichotomies. Differences arose also from the varied sociocultural contexts influenced by the importance of wildlife to peoples' lives and livelihoods. Bonner (1993) discusses how Western approaches to conservation, management, and education may run counter to values and cultural norms of people in other parts of the world, and the importance of recognizing and being sensitive to these differences. Acknowledging that one nation's approach may not fit another's needs is essential, as it can open avenues for learning and innovation.

Kessler et al. (1998) asserted that traditional wildlife education programs needed to become global in context and expand beyond purely science and technology training to incorporate exposure to political, economic, and cultural realities that influence wildlife conservation issues around the world. They conclude by stressing the need for partnerships among universities, agencies, nongovernmental organizations, industries, and affected communities to achieve real gains in global wildlife conservation. The examples of cooperation among nations and organizations provided above illustrate the value of such partnerships.

In North America, it would likely be difficult to find a major university graduate program in wildlife conservation that did not have ongoing graduate research projects in other countries. Formal agreements between universities in different nations can provide a basis for collaborative research, although without dedicated funding, they can be difficult to implement. Many nations have government-sponsored scholarships and grants to provide opportunities for students to study abroad. In Peru, for example, the National Fund for Scientific and Technological Development (FONDECYT), administered by the National Commission for Scientific and Technological Research (CONICYT), provides significant funds for Peruvian students doing research at foreign universities. Funds can be used to cover living expenses and research materials, but recipients must commit to returning to Peru to work for a time period equivalent to the duration of their research support.

In Chile, like Peru, CONICYT provides scholarships for Chilean students to go abroad, to go to Chilean universities, and for overseas students to come to Chile for postgraduate studies. The CONICYT funds PhD research dissertations, postdoctoral research initiatives, and senior research initiatives, which in many cases involve wildlife ecology, disease, or health. In Central and South America, Brazil is the country with by far the most funding, research initiatives, and scientific productivity on biology, ecology, and management and conservation research in wildlife. In Brazil, the primary funding resources are the National Board for Science and Technology Development (CNPq—Conselho Nacional de Desenvolvimento Científico e Tecnológico), the Boticario Fund for Nature Protection (Fundação Grupo Boticário de Proteção à Natureza), and the Brazilian Fund for Biodiversity (Funbio—Fundo Brasileiro para a Biodiversidade).

Some universities have taken a novel approach to international research. In the PhD program in conservation medicine at Andrés Bello University in Santiago, Chile, roughly one-fifth of the faculty are visiting professors from institutions outside of South America who teach regular courses on compressed schedules and supervise graduate student research. This model brings perspectives and expertise from other nations into the graduate program, fosters collaborative international research opportunities, and brings South American advances in conservation medicine back to the visiting professors' institutions. In a sense, it is a form of academic and scientific introgressive hybridization.

Organizations such as the IUCN, through its network of species specialist groups, cultivate relationships among scientists globally that can often lead to collaborative and comparative research endeavors. Other organizations, such as The Wildlife Society, Society for Conservation Biology, the Bombay Natural History Society (the oldest conservation organization in the world), International Union for Game Biologists, Australasian Wildlife Society, and the Mammal Society of Japan, host meetings (Fig. 14.2)

and publish journals that are international in scope, even though the host organization may have a more regional focus. Annual meetings of these organizations provide excellent opportunities for networking with researchers from other nations and becoming familiar with related research.

In the cumulative experience of the authors, and our anecdotal observations from colleagues, we observe several key actions that contribute to successful international research engagements.

- Build a relationship with a foreign researcher or wildlife manager.
- Identify common ground on science questions or conservation issues (invasive species, human-wildlife conflict, sustainable use).
- Seek to contribute through intellectual or material resources that will be value added to the relationship.
- For graduate student research, sponsor a foreign student at your institution, with collaboration from scientists or wildlife managers from the student's host country.
- Invest in gaining at least conversational fluency in the appropriate foreign language.
- Be willing to invest time and energy in securing outside grant funding for student and research support.
- Focus on how your engagement will advance conservation in the area or nation you are working in, rather than how the enterprise will help your career. If you succeed at the former, the latter will be achieved.

Considerations for International Academic Engagement

Foreign students may encounter barriers to graduate research in another nation. Many universities in the United States require prospective foreign graduate students to take the Test of English as a Foreign Language (TOEFL) or International English Language Testing System exam (IELTS) and achieve an

Figure 14.2   Early career Asian mammalogists assembled as part of a symposium designed specifically for networking at the Fifth International Wildlife Management Congress held in Sapporo, Japan, in 2015. Collaboration among professional societies—in this case, the Mammal Society of Japan and The Wildlife Society—can be instrumental in providing international opportunities. Masaharu Motokawa

acceptable score. This can be a challenge for many, and even those who pass can be at a disadvantage keeping up with other students in class. Advisers need to be aware prior to accepting a foreign student that significant additional time may be required on their part, editing grant proposals, theses, and manuscripts. In our experience, we are amazed at how well foreign students do despite language and cultural barriers, and believe the extra effort required in mentoring these students is more than offset by the knowledge and perspectives they bring to the graduate program, and the benefits other students gain through association. A diverse graduate student body provides benefits to all and, ultimately, should advance conservation globally.

Summary

There is a subset of conservation organizations that is particularly influential in advancing science and policy: those for which a key characteristic is their ability to engage and partner with governments and other organizations. The growth of the IUCN and its subsidiary bodies has been a key catalyst in developing networks and partnerships. Incentives for multilateral wildlife collaborations have come from developed nations as the recognition of the economic importance of global wildlife issues has heightened (Goldman 2005). Sustainability of wildlife populations worldwide is an ever-growing concern, and global collaboration will be essential (Prugh et al. 2000, Karesh et al. 2005, Bennett et al. 2006). The private sector has contributed significantly, and its role will likely be magnified in the future as public interest in international wildlife conservation grows and global trade in wildlife increases.

LITERATURE CITED

Bean, M.J. 1983. The evolution of national wildlife law. Praeger Publishers, New York, USA.

Bennett, E.L., E. Blencowe, K. Brandon, D. Brown, R.W. Burn, G. Cowlishaw, G. Davies, et al. 2006. Hunting for consensus: Reconciling bushmeat harvest, conservation, and development policy in west and central Africa. Conservation Biology 21:884–887.

Bonner, R. 1993. At the hand of man: Peril and hope for Africa's wildlife. Alfred A. Knopf, New York, USA.

Cropper, A. 1993. Convention on biological diversity. Environmental Conservation 20:364–365.

Energy and Biodiversity Initiative. 2003. International conventions. Conservation International, Washington, DC, USA. http://www.theebi.org/pdfs/conventions.pdf.

Goldman, M. 2005. Imperial nature: The World Bank and struggles for social justice in the age of globalization. Yale University Press, New Haven, Connecticut, USA.

Guo, R. 2018. Cross-border resource management. Third edition. Elsevier, Cambridge, Massachusetts, USA.

Hawkins, A.S., R.C. Hanson, H.K. Nelson, and H.M. Reeves, editors. 1984. Flyways. US Department of the Interior, US Fish and Wildlife Service, Washington, DC, USA.

IUCN (International Union for Conservation of Nature). 1980. World conservation strategy: Living resource conservation for sustainable development. International Union for Conservation of Nature, Gland, Switzerland.

IUCN (International Union for Conservation of Nature). 2014. Averting the imminent extinction of south-east Asian vertebrate species: Asian species action partnership (ASAP). Traffic 26:15–17.

Karesh, W.B., R.A. Cook, E.L. Bennett, and J. Newcomb. 2005. Wildlife trade and global disease emergence. Emerging Infectious Diseases 11:1000–1002.

Kessler, W.B., S. Csanyi, and R. Field. 1998. International trends in university education for wildlife conservation and management. Wildlife Society Bulletin 26:927–936.

Lucas, R.E. Jr. 2009. Trade and the diffusion of the Industrial Revolution. American Economic Journal 1:1–25.

Matthews, G.V.T. 1993. The Ramsar Convention on wetlands: Its history and development. Ramsar Convention Bureau, Gland, Switzerland.

Prugh, T., R. Constanza, and H. Daly. 2000. The local politics of global sustainability. Island Press, Washington, DC, USA.

Robinson, P.T., G.L. Flacke, and K.M. Hentschel. 2017. The pygmy hippo story: West Africa's enigma of the rainforest. Oxford University Press, New York, USA.

Stearns, P.N. 2013. The Industrial Revolution in world history. Fourth edition. Westview, Boulder, Colorado, USA.

United Nations Environment Programme. 1994. Convention on the conservation of migratory species of wild animals (Bonn Convention or CMS). Third revision. United Nations Environment Program, New York, USA.

USAID (United States Aid for International Development). 2014. Enhancing forestry governance in the Peruvian Amazon: Mid-term evaluation of the Peru Forest Sector Initiative. US Agency for International Development, AID-527-T-11-00001, Washington, DC, USA.

— 15 — Local Approaches and Community-Based Conservation

JOHN L. KOPROWSKI
JOSÉ F. GONZÁLEZ-MAYA
DIEGO A. ZÁRRATE-CHARRY
UDAY R. SHARMA
CRAIG SPENCER

International wildlife conservation often requires considerable collaboration, from local to national to regional to international (Chapter 14; Western and Wright 1994, Berkes 2007). Proponents exist for efforts at each of these levels, and effective conservation strategies usually require integration of conservation activities and support across levels. Although Yellowstone National Park is generally credited as the world's first "national park" set aside for conservation, many nations, in particular Western nations, have a history of designating public lands for nonconsumptive uses (Nash 1970). Cultural and legal perspectives that influence land and wildlife ownership have considerable influence on the effect of this centralized approach (Chapter 2; Berkes 2007). Such "top-down" centralized approaches to conservation are also known as government-based conservation and have been the predominant approach to management and conservation for the previous century and beyond (Western and Wright 1994a, Berkes 2007). More recently, large multinational reserves have resulted from NGO and international conventions that promote the centralized approaches at scales beyond a single nation (Berkes 2004). Government-based approaches have yielded success in the preservation of important wildlands through periods of considerable infrastructure development and anthropogenic pressure (Bruner et al. 2001, but see Muhumuza and Balkwill 2013), but the inflexible nature of their locations creates vulnerability to challenges such as climate change and necessitates planning and adaptation (Jantarasami et al. 2010). Community-based conservation is a bottom-up decentralized approach that includes protection of natural resources and biodiversity by, for, and with the local community, the primary goal of which is the coexistence of people and nature, as distinct from a protectionist philosophy that segregates people and nature (Western and Wright 1994b). Government-based and community-based conservation are not mutually exclusive strategies (Berkes 2007), although commitment to one approach to the neglect of the other has been referred to as people-free versus people-centered conservation (Wilshusen et al. 2002).

Community-based conservation has been used by humans who have a vested interest in the resources available from the land (Western and Wright 1994). Community-based conservation is not a panacea for management and conservation efforts (Berkes 2007); however, the approach needs to be in the toolbox of professionals, for it can provide an important perspective that welcomes collaboration (Gavin et al.

2015). Although not always successful, community-based conservation can have transformative effects when employed under the proper conditions (Muhumuza and Balkwill 2013, Sharma 2013). Training students and early career professionals to develop the skill set of communication, team building, and leadership should also be a goal of those skilled in international wildlife research (Chapter 1).

Herein, we review three distinctly different community-based conservation programs on four continents as case studies of successful approaches. First, we examine the adoption of a community-based conservation approach for buffer zones and inter-reserve areas that was created in the 1990s within Nepal and has become the dominant paradigm for the nation. Next, we assess the international marketing of coffee grown in Latin America using predator-friendly farming techniques in Costa Rica and Colombia. Finally, we examine the first all-female community-based anti-poaching team in South Africa. These three examples highlight a geographically diverse assemblage of projects and demonstrate a breadth of approaches that engage communities in conservation with a multitude of benefits.

Case Study 1: Adoption of Community-Based Conservation throughout Nepal

Nepal has created an impressive array of protected areas to include viable samples of the biodiversity found in the Himalayas. Thirty-three protected areas (national parks, reserves, conservation areas, and buffer zones) cover about 23.4% of the country's land (Department of National Parks and Wildlife Conservation Nepal, www.dnpwc.gov.np), with dramatic altitudinal variation from fewer than than 100 m above sea level in the subtropical Terai of the south to the world's highest mountain, Mount Sagarmatha (Mount Everest: 8,848 m). Management of protected areas has progressed through several major phases that include reconciling the needs and aspirations of local people, harnessing the economic opportunities offered by tourism, and addressing the

associated threats (Wells and Sharma 1998). Nepal, over three decades, has used several conservation paradigms (Sharma 2013). The current mode is community-based conservation through two distinct models: the integrated conservation and development (ICD) concept and the impact zone (IZ) concept. The ICD concept as applied in montane conservation areas, such as Annapurna, has demonstrated that conservation programs can be complementary to local development efforts. The IZ concept, developed from research at the University of Arizona in 1991 (Sharma and Shaw 1995), calls for strict control of forests in the reserve, together with intensified agriculture and forestry on public properties outside the park with the intent to increase production of natural resources that are in local demand. These models are distinct in that the ICD approach is focused primarily on improving local resident livelihoods around conservation, whereas the IZ approach emphasizes ecological function and conservation as the primary goal (Robinson and Redford 2004, Sharma 2013).

Nepal has incorporated the IZ concept in all national parks and reserves by creating special legal provisions for this zone. The fourth amendment of the Nepal National Parks and Wildlife Conservation Act in 1993 provided the legal support for the shift from the traditional approach of government-based conservation and management to a fully participatory approach. Surrounding landscapes and villages of a park or reserve, up to 5 km, are declared to function as the buffer zone. The communities form institutions, such as user committees and user groups, to mobilize people for intensive, sustainable management of available natural resources and undertake local development within the buffer zone. The park provides necessary assistance and shares 30%–50% of park-generated revenue with the surrounding communities to be used for these activities. Nepal has showcased how conservation with the active involvement of local communities can persist even during periods of national and social unrest and conflicts.

Figure 15.1   Tharu of Chitwan, an indigenous community that practices community-based conservation in Nepal. Uday R. Sharma

Nepal has remained successful in the protection of biodiversity, while also creating a positive collaboration between the park authorities and local communities. The experience of Chitwan National Park demonstrates that communities such as the Tharu people (Fig. 15.1) are not simply dependent on the park's returned revenues, but also generate community capital through savings and credit programs (Sharma 2013). The success of the IZ management model of national parks such as Chitwan is reflected in several periods of 365 days with zero poaching incidents on rhinoceros since 2011. Advanced youth group patrols are being considered to curb poaching in the buffer zone. Such increased community involvement could provide flexibility to the park management in enhancing relationships with park neighbors by providing the opportunity for park law enforcement and local communities to collaborate for effective control of illegal activities.

Case Study 2: Jaguar (*Panthera onca*) and Coffee: Community-Based Jaguar Conservation through a Comprehensive Jaguar-Friendly Eco-Label Scheme

Biodiversity conservation in Latin America is a serious challenge, considering its enormous biological richness but also the vast social and economic limitations of most Neotropical countries and the need for adequate and substantial development for local communities (González-Maya et al. 2011, Balvanera et al. 2012). As development in general, and specifically agriculture, increases and expands in the Neotropics, the remaining natural habitats are transformed (Phalan et al. 2014). Such transformation increases threats to species and ecosystems, especially in critical biodiversity hotspots in the Neotropics (Betts et al. 2017). Even with these socioeconomic constraints, Neotropical countries that harbor the highest level of biodiversity also represent unique opportunities for new approaches that include conservation and achieving economic growth. Currently, considerable challenges emerge for conservation practitioners proposing and promoting alternatives for coupling development with conservation to secure biodiversity-rich areas for the long term (Pirard 2012, Tayleur et al. 2017).

One of the most charismatic and representative species of Neotropical biodiversity is the jaguar, considered a keystone predator in most ecosystems across its distribution (Ceballos et al. 2015). Despite its importance, both ecologically and culturally, in most Latin American countries (Ceballos et al. 2015), the species is imperiled (Caso et al. 2008). Of the 34 subpopulations across its range, 33 are considered to be critically endangered or endangered, and the range has declined nearly 55% in the last century (de la Torre et al. 2018). Efficient and effective conservation measures are needed to retain the species in Latin America.

Three main threats exist for jaguars: habitat loss and fragmentation, prey hunting (de la Torre et al. 2018), and direct conflict derived from the depredation by jaguars on domestic animals, which usually results in the retaliatory killing of the predator (Inskip and Zimmermann 2009). These threats are all related to human activities and are present in the extensive areas of overlap of jaguars and humans, especially in landscapes under current land-cover change. Sadly, the trend does not show signs of reduction in threat, or easy short-term solutions, especially given the urgent socioeconomic challenges faced by most Latin American communities, including income, extensive habitat transformations by industrial monocultures, and local-scale poaching for subsistence. Any proposed conservation solution in Latin America requires consideration of local livelihoods and conservation needs within agricultural production areas, ensuring secure safe habitats, either as corridors or core habitats (Zárrate-Charry et al. 2016).

To achieve effective solutions, strategies need to incorporate habitat areas and connectivity corridors into conservation in addition to reducing conflict, while maintaining productive units and improving economic gains for farm owners. Coffee plantations are the dominant commodity in the focal regions in Costa Rica and Colombia and have provided the most important and stable economic activity in the last 40 years (Philpott and Dietsch 2003). The canopy structure of coffee plantations is of ecological importance and harbors significant biodiversity (Numa et al. 2005, De Beenhouwer et al. 2013). Certified shade-grown coffee is the production system with the most significant potential for jaguar conservation, coupled with continued economic development. Economic (costs, revenues, costs of opportunity) characterization of coffee farms in the context of adopting a wildlife-friendly certification scheme suggested the feasibility of participation in zones with jaguars (D.A. Zárrate-Charry, Oregon State University, personal communication). Both Colombian and Costa Rican coffee are known globally for high quality, and coffee represented the best opportunity to produce an internationally recognized commodity with a strong conservation concept. A market-based eco-label for shade-grown coffee was produced that would benefit jaguars and people in these ecoregions (González-Maya et al. 2016b, Zárrate-Charry et al. 2016), as determined by surveys of local communities to assess potential incentive schemes to benefit jaguars and people and generate broad support (Gómez-Junco 2015).

The Jaguar Friendly certification, supported and endorsed by the Wildlife Friendly Enterprise Network (WFEN) and the Phoenix Zoo (Fig. 15.2), is a green certificate that secures jaguar core and connectivity habitats in priority areas while also providing economic and management incentives for producers that can change antagonistic behaviors toward jaguar to jaguar-friendly perceptions. The certificate incorporates four main elements: reduce deforestation and fragmentation by respecting and improving riparian habitats and other natural-cover areas within the farm; improve cover, in terms of structure and composition, through improvement of the shade

Figure 15.2   The Jaguar Friendly eco-label for jaguar conservation products and a package of Jaguar Friendly coffee ready for marketing. José González-Maya

Figure 15.3   (A) Local producers harvesting Jaguar Friendly Coffee; (B) Melanistic jaguar recorded on one of the Jaguar Friendly farms. José González-Maya

canopy within crop areas; reduce hunting and poaching from owners and neighbors to zero; and improve domestic animal management for conflict reduction and mitigation. Once these requirements are fulfilled, the farms are granted certification and provided access to international green markets via direct trading, where profits are directly transferred to producers, given their contribution to jaguar conservation.

The first batch of Jaguar Friendly coffee was exported to the United States late in 2017 and is for sale in fair-trade markets in Arizona. Because coffee is a single-harvest-per-year crop in most regions, we can secure coffee batches only once a year, which also provides the necessary time to standardize farms for certificate requirements and ensure productive practices to guarantee quality. Furthermore, the protocol requires continued monitoring of the certified farms through camera-trapping and certification audits to ensure that requirements are maintained and that jaguars regularly use these properties (Fig. 15.3).

The results to date are promising. Local producers in Costa Rica and Colombia continue to support certification, and buyers in external markets embrace the product. Success of this initiative resides on the sustained maintenance of markets willing to pay extra for green gourmet products, which will continue to recruit producers to participate (Sun et al. 2017). Habitats with increased tolerance of landowners remain available for jaguar and will undoubtedly improve if local producers find this species an asset that helps maintain and enhance their livelihoods in human-jaguar dominated landscapes.

Case Study 3: All-Female Community-Based Anti-Poaching Unit

Rhinoceros (Rhinocerotidae) poaching in South Africa in 2016 was at a historically high rate of three animals killed/day and had been on a rapid increase since 2007 because of the perceived medicinal value of rhino horns that results in its price of US$56,000/kg (Reuter and Bisschop 2016). Rhino poachers targeted the Balule Nature Reserve in 2013 to illegally harvest reintroduced black rhinoceros (*Diceros bicornis*) and resident white rhinoceros (*Ceratotherium simum*). Balule is a protected area of 56,000 ha and is open to the Kruger National Park, with 136 km of the park's western boundary under its direct management. The Balule landscape comprises five public access gates, the Olifants River corridor that cannot be fenced, active mines, railway lines, and power cables accessed day and night by contractors. Furthermore, the almost 500 historical subsistence farms located there still had families living on them, with little or no income—not an easy landscape to manage for wildlife security.

By 2013, 24 rhino had been poached on a small section of the Balule landscape. The rate of poacher effort was increasing exponentially, and the anti-poaching resource was unable to keep pace. A unique approach to dealing with the emergence of rhino poaching on Balule was sought, as poaching is not a new threat to wildlife, and it most certainly will

not be ended with guns and bullets. What was needed was a relationship with the local communities, where values can be exchanged and fostered. The 136 km of boundary fence does not separate people from the park. Tourists race to Kruger National Park and the associated unfenced border reserves to see the incredible biodiversity of large African mammals; however, local people do not value the wildlife in the same nonconsumptive way (Carruthers 1995).

The poaching threat has three basic levels: bushmeat poachers, who set snares; rhinoceros poachers, who use guns; and informants from inside the park, who sell intelligence to the poachers. All levels of participants are exposed to the same risks if captured or identified. A solution was required that would eliminate wildlife poaching in the short term and address the challenges in the communities that result from a false economy (poacher-obtained wildlife products) and can lead to a collapse in values toward the park and local wildlife. Balule Nature Reserve created the Black Mambas, Africa's first all-female anti-poaching unit (Fig. 15.4). These women are young mothers with families. The Mambas live in the same communities that harbor the poachers. Their children go to the same schools. The goal is to nurture the idea that mothers tucking their children into bed at night will tell wondrous stories of the elephants and antelope that they saw at work and share the beauty of the African wilderness. Women traditionally have a crucial role in the raising of children and caring for the sick and elderly in local communities. By employing young women, three generations are targeted. The offer of prestigious employment within the wildlife conservation and management industry provides a consistent salary; leadership experience and considerable training is desirable; and the team of more than 30 Mambas is maintained. Standard training includes law enforcement instruction in hand-to-hand combat, search and seizure, arrest and court procedures, creation of statements and dockets, and handling and chain of custody of evidence, in addition to conservation-related protocols for snake handling, invasive vegetation

Figure 15.4 Two members of the Black Mambas check remote camera traps in Balule, South Africa. Craig Spencer

removal, snare detection and removal, weapons handling, and general field safety.

Black Mambas work on a 21-day rotating schedule; they receive room, board, and apparel in staff compounds on the reserve; and they spend 10 days of leave in their home communities. Black Mambas are armed not with guns but with pepper spray. They are not deployed in ambushes or other tactical situations that could result in human casualties. Mambas are tracked in real time and operate their own command-and-control center to alleviate speculation of corruption, enable rapid response to a crisis, and capture real-time data on their patrols to inform strategic planning and assessment of the patrol regime. The Mambas patrol on foot during the day and by vehicle after sunset. Routine anti-poaching duties include boundary patrols to seek evidence of poacher activities; roadblocks to search people and vehicles entering and leaving the landscape; systematically covering the landscape to detect suspicious behavior and destroy snares; visiting homes of people living on inholdings and building sites to inspect the premises and question locals; and conducting listening posts and vehicle patrols at night at locations where poachers often traverse. Additionally, Mambas visit schools each day and attend tribal functions in their official capacity and give speeches. Mambas have

a rank structure and a division of labor, with bonuses obtained based on success.

Success is apparent in decreased and displaced poacher activity, by significant reduction in the window of opportunity with 24-hour surveillance, which has resulted in movement from Balule to neighboring reserves with reduced levels of patrols. The catch-per-unit-effort modeling of required poaching effort documents significant inefficiency in success by poachers in the presence of the Black Mambas. Snaring and bushmeat collection efforts have been nearly eliminated, as have incidents of minor theft and damage.

Social benefits that suggest the potential for long-term change in attitudes toward poaching and wildlife conservation are the result of the Mambas' community outreach efforts. The Mambas are frequently requested to perform their stylized parade for tribal functions and traditional ceremonies. The Black Mambas have performed remarkably well in the public arena and are solicited for a multitude of popular media shows and articles that have led to their status as icons for women in South Africa and abroad. Their massive social media following is primarily among women from around the globe. Finally, the Mambas reach out to young children in 11 adopted primary schools, where they conduct lessons and parades such that more than 900 children are engaged every week. Each school has a Mamba classroom that has been decorated and improved to create a stimulating learning environment. The Black Mambas were recognized with the United Nations Champions of the Earth award in 2015.

Although the Black Mamba concept cannot replace the traditional militarized anti-poaching unit approach, community-based conservation monitoring is an essential component of any protected area that has a traditional community living within its midst. The successes of community-based units should be assessed on parameters designed to evaluate their objectives, and not those of a militarized unit. The success of the Black Mambas is evident not only in decreased conservation crimes in the present but also in effecting cultural change through the prestigious nature of the positions within local communities and contact with schoolchildren. The model works well for community-based conservation in Balule under this set of sociopolitical pressures but could be tailored for any protected area to meet local community objectives.

Community-Based Conservation as a Tool

Community-based conservation has been practiced by cultures around the world for millennia as humans lived as stewards or consumers of products from landscapes (Western and Wright 1994). Community-based conservation approaches typically began as a means of reducing threats to conservation by communities based around a protected area but tend to evolve to provide direct incentives for conservation of local communities (Salafsky and Wollenberg 2000). Community-based conservation is probably no more effective as a panacea than is exclusively government-based conservation, because both approaches fail to explicitly incorporate the multilevel nature of linkages and partners required for success (Berkes 2007). The debate on the appropriate combination of government-based, integrated conserva-

tion and development and community-based conservation is needed and, if these approaches are compatible in any combination (Robinson and Redford 2004), will continue through demonstration projects and innovative approaches. Berkes concludes that conservation "cannot be conceived and implemented only at one level, because community institutions are only one layer in a multilevel world" (2007:15192). Successful programs must find ways to continue to integrate increasing segments of society—especially the youth, when possible—so that all experience the benefits and have a vested interest in the success of the approach (Pyakurel et al. 2017). Furthermore, flexibility must be incorporated in the planning phase and integrated early in implementation of the approach to respond to emerging challenges and opportunities that were unanticipated at conception (Adhikari et al. 2016). More recently, a biocultural approach to conservation has been developed that logically considers conservation actions made in the service of sustaining the biophysical and sociocultural components of dynamic, interacting, and interdependent social-ecological systems through the integration of co-management (negotiated shared management) and integrated conservation and development (specific linked conservation and human community livelihood objectives) in addition to community-based conservation (Gavin et al. 2015). Biocultural approaches to conservation add additional layers of complexity but show promise in addressing the multiplicity of needs of landscapes, human populations, and ecosystems.

Summary

Humans have long recognized the need for collaboration to effectively conserve and preserve wildlife and their habitats. Local approaches to conservation that incorporate local communities into decision making, self-governance, and revenue sharing show promise as viable strategies for conservation. Community-based conservation is a bottom-up decentralized approach that includes protection of natural resources and biodiversity by, for, and with the local community, the primary goal of which is the coexistence of people and nature, as distinct from a protectionist government-based philosophy that segregates people and nature. Community-based conservation is probably no more effective as a panacea than is exclusively government-based conservation, because both approaches fail to explicitly incorporate the multilevel nature of linkages and partners required for success. Case studies of jaguar-friendly coffee in Costa Rica, natural area planning in Nepal, and community-based anti-poaching efforts in South Africa demonstrate the potential for success of local community approaches. Future efforts must continue to expand linkages across levels and among partners to maximize impact and sustainability of wildlife conservation through community-based efforts.

LITERATURE CITED

Adhikari, J., H. Ojha, and B. Bhattarai. 2016. Edible forest? Rethinking Nepal's forest governance in the era of food insecurity. International Forestry Review 18:265–279.

Balvanera, P., M. Uriarte, L. Almeida-Leñero, A. Altesor, F. DeClerck, T. Gardner, and M. Vallejos. 2012. Ecosystem services research in Latin America: The state of the art. Ecosystem Services 2:56–70.

Berkes, F. 2004. Rethinking community-based conservation. Conservation Biology 18:621–630.

Berkes, F. 2007. Community-based conservation in a globalized world. Proceedings of the National Academy of Sciences 104:15188–15193.

Betts, M. G., C. Wolf, W.J. Ripple, B. Phalan, K.A. Millers, A. Duarte, and T. Levi. 2017. Global forest loss disproportionately erodes biodiversity in intact landscapes. Nature 547:441–444.

Bruner, A.G., R.E. Gullison, R.E. Rice, and G.A. Da Fonseca. 2001. Effectiveness of parks in protecting tropical biodiversity. Science 291:125–128.

Carruthers, J. 1995. Kruger National Park: A social and political history. University of KwaZulu-Natal Press, Durban, South Africa.

Caso, A., C. Lopez-Gonzalez, E. Payan, E. Eizirik, T. de Oliveira, R. Leite-Pitman, M. Kelly, and C. Valderrama. 2008. *Panthera onca*. The IUCN Red List of Threatened Species 2008: e.T15953A5327466. http://dx.doi.org/10.2305/IUCN.UK.2008.RLTS.T15953A5327466.en.

Ceballos, G., H. Zarza, J.F. González-Maya. 2015. El potencial del jaguar como especie sustituta en la conservación de ecosistemas tropicales. Pages 503–520 in C. González Zuarth, A. Vallarino, J. Pérez, and A. Low, editors. Bioindicadores: Guardianes de nuestro futuro ambiental. El Colegio de la Frontera Sur, Instituto Nacional de Ecología y Cambio Climático, San Cristobal de las Casas, Chiapas, México.

de la Torre, J.A., J.F. González-Maya, H. Zarza, G. Ceballos, and R.A. Medellín. 2018. The jaguar's spots are darker than they appear: Assessing the global conservation status of the jaguar Panthera onca. Oryx 52:300–315.

Gavin, M.C., J. McCarter, A. Mead, F. Berkes, J.R. Stepp, D. Peterson, and R. Tang. 2015. Defining biocultural approaches to conservation. Trends in Ecology and Evolution 30:140–145.

Gómez-Junco, G.P. 2015. Herramientas de gestión territorial y su potencialidad para promover medios de vida sostenibles como estrategia de conservación y desarrollo local: El caso de las fincas del Distrito Pittier, Cantón de Coto Brus, Costa Rica. MS thesis. Centro Agronómico Tropical de Investigación y Enseñanza—CATIE, Turrialba, Costa Rica.

González-Maya, J.F., O. Chassot, A. Espinel, and A.A. Cepeda. 2011. Sobre la necesidad y pertinencia de la gestión integral de paisajes en Latinoamérica. Revista Latinoamericana de Conservación 2:1–6.

González-Maya, J.F., D.A. Gómez-Hoyos, G.P. Gómez-Junco, J. Schipper, and D.A. Zárrate-Charry. 2016. Connecting felid populations between La Amistad and Osa priority conservation areas in Costa Rica. Wild Felid Monitor 9:21.

Inskip, C., and A. Zimmermann. 2009. Human-felid conflict: A review of patterns and priorities worldwide. Oryx 43:18–34.

Jantarasami, L.C., J.J. Lawler, and C.W. Thomas. 2010. Institutional barriers to climate change adaptation in US national parks and forests. Ecology and Society 15:33.

Jha, S., C.M. Bacon, S.M. Philpott, V. Ernesto Mendez, P. Laderach, and R.A. Rice. 2014. Shade coffee: Update on a disappearing refuge for biodiversity. BioScience 64:416–428.

Muhumuza, M., and K. Balkwill. 2013. Factors affecting the success of conserving biodiversity in national parks: A review of case studies from Africa. International Journal of Biodiversity 2013:798101.

Nash, R. 1970. The American invention of national parks. American Quarterly 22:726–735.

Numa, C., J.R. Verdú, and P. Sánchez-Palomino. 2005. Phyllostomid bat diversity in a variegated coffee landscape. Biological Conservation 122:151–158.

Phalan, B., R. Green, and A. Balmford. 2014. Closing yield gaps: Perils and possibilities for biodiversity conservation. Philosophical Transactions of the Royal Society B: Biological Sciences 369:20120285.

Philpott, S.M., and T. Dietsch. 2003. Coffee and conservation: A global context and the value of farmer involvement. Conservation Biology 17:1844–1846.

Pirard, R. 2012. Market-based instruments for biodiversity and ecosystem services: A lexicon. Environmental Science and Policy 19–20:59–68.

Pyakurel, U., B.P. Bhattarai, S. Adhikari, and S. Shrestha. 2017. Youth involvement in community based forest management in Nepal. Nepali Journal of Contemporary Studies 17:59–87.

Reuter, E., and L. Bisschop. 2016. Keeping the horn on the rhino. Pages 149–186 in G. Potter and A. Nurse, editors. Geography of environmental crime. Palgrave Macmillan, London, UK.

Robinson, J.G., and K.H. Redford. 2004. Jack of all trades, master of none: Inherent contradictions among ICD approaches. Pages 10–34 in T.O. McShane and M.P. Wells, editors. Getting biodiversity projects to work: Towards more effective conservation and development. Columbia University Press, New York, USA.

Salafsky, N., and E. Wollenberg. 2000. Linking livelihoods and conservation: A conceptual framework and scale for assessing the integration of human needs and biodiversity. World Development 28:1421–1438.

Sharma, U.R. 2013. Biodiversity conservation: Changing paradigm. Pages 146–153 in P.K. Jha, F.P. Neupane, M.L. Shrestha, and I.P. Khanal, editors. Biological diversity and conservation. Nepalpedia Series No. 2. Nepal Academy of Science and Technology, Khumaltar, Nepal.

Sharma, U.R., and W. W. Shaw. 1995. The "impact zone" concept: A regional approach for managing Royal Chitwan National Park, Nepal. Pages 246–249 in J.A. Bissonette and P.R. Krausman, editors. Integrating people and wildlife for a sustainable future. Proceedings of the First International Wildlife Management Congress. The Wildlife Society, Bethesda, Maryland, USA.

Sun, C.H.J., F.S. Chiang, M. Owens, and D. Squires. 2017. Will American consumers pay more for eco-friendly labeled canned tuna? Estimating US consumer demand for canned tuna varieties using scanner data. Marine Policy 79:62–69.

Tayleur, C., A. Balmford, G.M. Buchanan, S.H.M. Butchart, H. Ducharme, R.E. Green, and B. Phalan. 2017. Global coverage of agricultural sustainability standards, and their role in conserving biodiversity. Conservation Letters 10:610–618.

Wells, M.P., and U.R. Sharma. 1998. Socio-economic and political aspects of biodiversity conservation in Nepal. International Journal of Social Economics 25:226–243.

Western, D., and R.M. Wright. 1994a. Natural connections: Perspectives in community-based conservation. Island Press, Washington, DC, USA.

Western, D., and R.M. Wright. 1994b. Background to community-based conservation. Pages 1–12 in D. Western and R.M. Wright, editors. Natural connections: Perspectives in community-based conservation. Island Press, Washington, DC, USA.

Zárrate-Charry, D.A., I.G. Ochoa, J.S. Jiménez-Alvarado, A. Massey, M. Calderon, A. Hurtado-Moreno, J. Prieto, I. Aconcha-Abril, I.M. Vela-Vargas, and J.F. González-Maya. 2016. Strategies for human-jaguar conflict resolution in agricultural areas of Sierra Nevada de Santa Marta, Colombia. Wild Felid Monitor 9:20–27.

16 — Getting Involved
Advice for Students and Wildlife Professionals

ROBERT A. MCCLEERY
JULIE T. SHAPIRO
KAREN BAILEY
THOMAS K. FRAZER

Introduction

There is no shortage of ways for students and professionals to begin their engagement in international wildlife and conservation issues. However, there are probably more formalized opportunities for students and young professionals. The sooner in your career that you get started, the easier it will be for you to begin having rewarding international experiences. In this chapter, we detail ways for students and professionals (young or established) to start getting involved internationally. From our experiences, we also provide some advice and expectations to help everyone understand the challenges and issues that they may face as they start to work overseas. We present a number of ways to get involved in wildlife issues outside of North America, but we particularly focus on engagement in the developing world, because these are the areas with the greatest need for wildlife professionals and where wildlife professionals can likely have the greatest effect.

Challenges and Rewards

It is a big beautiful world, with incredible opportunities to grow, learn, and contribute to the understanding, management, and conservation of wildlife around the planet (Fig. 16.1). These opportunities come at a price, however, and you will want to consider the trade-offs and sacrifices before you start your international engagement. It is nothing new for wildlife biologists to spend long periods of time in remote places, but doing this internationally can pose additional strains. Simply traveling to a remote location on the other side of the planet can be a monumental challenge, and arriving with all your gear, food, equipment, and water can sometime seem nearly impossible. Communicating across languages and cultures with your collaborators can also be frustrating and challenging. These same challenges occur regardless of whether you are working in an urban environment or at a remote field camp. You are likely to struggle finding the foods and comforts that you are used to back home. More seriously, working overseas comes with some notorious health challenges. Some health issues, like jetlag and travelers' diarrhea, can cause discomfort and annoyance but are not a substantive problem. Alternatively, other issues such as malaria, schistosomiasis, and other tropical disease can be more severe. Finally, traveling away from friends, family, and loved ones can sometimes strain relationships and be emotionally challenging. We think it is important that people recognize that international work comes with these

Figure 16.1 Students and faculty from Sewatini and the United States enjoy the camaraderie of international collaboration in fieldwork. Samantha Wisely

and other sacrifices. We also think that the experiences, relationships, and professional and personal growth that comes from international work greatly outweigh the challenges. Nonetheless, we think it is important that everyone consider what they are willing to sacrifice for this incredibly rewarding work.

If you have decided that you would like to pursue opportunities that take you abroad, we have some general advice to help get you started. One of the most important things you can do is start by getting experience at home. International work in wildlife, especially paid work, is competitive, but building a strong resume with relevant work experience at home can help give you an edge on job openings. If you have the resources, we encourage you to travel and engage in different cultures outside of work. Traveling around the developing world can be done with minimal resources if you can afford a plane ticket to get there. When you are traveling, do not go in large groups or on tours. Spend your time engaging with people who are different from you. If you are adventurous, you might even consider traveling by yourself. This is not for everyone, but it is an excellent way to immerse yourself in a different culture. Traveling can be a rewarding and fun experi-

ence, but learning how to navigate in different countries and cultures is critical to being successful in an international wildlife career and is excellent preparation.

Language and Learning

No matter what your goals for working or traveling abroad, learning the local language will enrich your experience and increase opportunities. Even if most people speak English, speaking the local language will change the nature of your interactions. People appreciate a genuine effort to speak the local language, and it will allow you to more easily integrate with groups of colleagues and be a part of group conversations, when people are more likely to speak their first language. Finally, speaking other languages will make you a more competitive candidate for jobs, grants, and other opportunities, many of which prefer applicants who are proficient in the language of the country they will work in.

Today, many resources are available for language learning. Most universities have multiple language departments, and students can take classes for credit at a variety of levels. Majoring or minoring in foreign languages or regional studies is an option, especially for those with an interest in studying wildlife in a particular country or region of the world. Such a course of study can provide historical and cultural context, which is essential when working abroad. Additionally, many study abroad programs offer language classes. For those who already have some level of proficiency, directly enrolling in a foreign university as a part of study abroad or exchange can help enhance language skills in speaking, reading, and writing. For those no longer in school, local community colleges and adult learning centers often offer classes in foreign languages. In some cases, there are also in-country options for language learning, such as local language schools, which often can provide the possibility of learning a new language at a reasonable cost. Some language schools can also arrange local homestays, which increase exposure to language and culture. Finally, online courses and computer programs and software, such as Rosetta Stone and Duolingo, can aid or supplement other methods of language learning.

Building Your International Network

If you have the time and can afford it, professional meetings and conferences can be an ideal way to make new connections, connect faces to names, and follow up with contacts. Many large conferences often have working groups and symposium dedicated to aspects of international research. More importantly, they often provide ample opportunity for informal interactions among professionals and students that can facilitate the development of your international network. Individuals looking to recruit international research collaborators and students are more likely to bring someone on board once they have met and chatted face to face.

If conferences are not an option, almost everyone who is working in the world today can be reached digitally. If you know of or find people with your passions, doing the things you would like to do,

reach out to them. Ask them about their work and how you might be able to get involved in the future. These types of conversations can provide you with contacts, useful information, and even opportunities to work overseas. It might be difficult to contact strangers, but this is no time to be shy. When you are reaching out to people, sending in applications, or just waiting to hear back from an international collaborator, it is important to be patient. Most of the world does not work on a North American time schedule, and this is important to remember when communicating with foreign contacts and collaborators. We commonly find that our emails are answered a week or two after they are sent. Nonetheless, it is also important to be persistent and to send follow-up communications. Being persistent can help show that you are serious about your international engagement.

By getting relevant work experience, traveling, learning a language, and reaching out to people, you will only increase your chances of getting an opportunity for international engagement. When these opportunities come and you need to decide whether you are going to take them, it is important to remember that many times this is a leap of faith. There will likely be little information on most of the details for each opportunity, even opportunities from large aid organizations like the US Peace Corps. This is not like researching what restaurant you want to go to, where you can examine hundreds of reviews before making a decision. Additionally, it is almost impossible to tell how you will fit in to any particular culture and situation until you have been there. So when you are laboring over a decision to engage abroad, we encourage you not to dwell on the specifics, because they are unknown and likely to be different for each individual. This can be hard to do, coming from a culture where information is ubiquitous, but if the important issues fit your requirement (time commitment, region of world, general area of expertise), we encourage you not to pass on such opportunities.

Advice for Students and Young Professionals: Start Early

You will never have better opportunities to start international engagement than you do as a student. We detail many of these opportunities below, but realize that they are ephemeral and, as you progress in your career, you will have to make your own opportunities. The decisions and choices that you make during your schooling and at the beginning of your career can do more to help you become engaged internationally than at any other time during your career. The decisions that we made at that stage in our lives are still shaping our work and careers 20 and 30 years later. You should consider shaping your curriculum to help you achieve your international aspirations. As mentioned above, universities often teach courses on culture, language, and history that can be taken as electives or audited. One of our colleagues used their natural resource education to learn Mandarin. She then used these skills to work in China and become an expert on panda (*Ailuropoda melanoleuca*) conservation. Finding the time and money for these courses becomes increasing difficult after you enter the work force.

Conservation Is a Two-Way Knowledge Exchange

Sometimes when students think about the challenges of international wildlife conservation, it is hard for them to picture how they can contribute. Students should understand that working overseas may, at first, be more about personal and professional growth than about shaping policies and practices on the ground. However, through their education and experience, most students likely have obtained knowledge and skills that can contribute to international wildlife projects. In many parts of the developing world, North American students have received more formal classroom training in wildlife ecology and conservation than the people addressing these issues on the ground.

Local practitioners are likely to have an unmatched breadth and depth of knowledge on local natural history, but that does not mean that students may not be able to contribute some of the technical or conceptual skills they learned in the classroom. Additionally, many of the planet's wildlife and conservation problems do not need highly technical solutions; they need people who are passionate, can solve problems, and can communicate effectively. Students should bring their passion, energy, interpersonal skills, and willingness to their international wildlife experiences. Furthermore, it is becoming clear that diverse groups of people solve problems better than homogeneous ones (Cox and Blake 1991, Jehn et al. 1999, Nielsen and Nielsen 2013, Cheruvelil et al. 2014). A more diverse group of voices will aid us as we face our ongoing conservation challenges. Just as international perspectives can help us address our wildlife conservation issues in North America, North American university student have the potential to contribute their voices and passions to international conservation issues.

When we travel to a new and unfamiliar place, we face unique opportunities to learn and grow as a person and a professional. Working and learning from local practitioners and scientists is probably the best way for students to learn field skills, natural history, effective communication, and problem solving in a real-world context. However, a common mistake made by many North American students is that they do not take the time to absorb the ecological and cultural knowledge around them. Instead, some students deal with new (often challenging and confusing) experiences by talking about home and verbally comparing the world around them to what they know. Do not fall into this trap; take the time to listen to, smell, and observe the world around you. The key to getting the most out of international experiences, and novel situations more broadly, is to be patient and open-minded. The key is to seek first to understand, and then to be understood. Be prepared to listen to people who understand the local ecosystem and have different beliefs than you, beliefs and ideas that may conflict with your ideals. Listen and learn and then share your experiences. In this manner, you are much more likely to reap the rewards and enable effective conservation. There are many examples of failed conservation efforts that ignored seemingly obvious aspects of ecology, culture, history, economics, or values. They ignored the need for local participation, discounted traditional ecological knowledge, or proposed solutions in direct conflict with local culture (Gibson and Marks 1995, Muhumuza and Balkwill 2013, Nilsson et al. 2016). These could have been more successful had they employed conscientious listening and learning.

Ways for Students and Young Professionals to Get Started: Social Sciences and Human Dimensions

In addition to a solid background in wildlife sciences and languages, students that are eager to work abroad should consider some training or coursework in social sciences and the human dimension of wildlife conservation. These courses can provide context and an understanding of people and cultures around the world and how they interact with nature. Courses such as environmental anthropology and rural sociology can help students better understand how to work with people on conservation projects that may require community engagement. Other coursework that may help students prepare for working in different cultures includes cultural anthropology, sociology, or geography. Finally, an understanding of the implementation and analysis of household surveys and interviews will allow students to make invaluable contributions to research focused on socioecological questions.

Study Abroad

Studying abroad is a great opportunity to create a foundation that will increase your ability to engage with diverse groups and enable success in the future. First, it prepares you for being an outsider and allows you to experience other cultures in a controlled setting. The experience of being an "other" is an important one and one that many wildlife students working internationally often experience. Being comfortable as an outsider and learning how to navigate those situations is something that is easily experienced but challenging to teach. Study abroad often gives students a taste of these experiences and helps prepare them for professional careers. Studying abroad also provides students with exposure to unique wildlife, biomes, and ecosystems, and the opportunity to understand them in situ. Exposure to new systems improves students' knowledge base and their ability to understand and navigate new ecosystems and wildlife challenges. Studying abroad also allows students to create contacts, boost their resumes, and narrow their particular interests. Perhaps most importantly, studying abroad broadens students' perspective of the world (Chieffo and Griffiths 2004, Anderson et al. 2006). It allows them to think more effectively about intercultural problems, create novel solutions, and communicate with diverse groups (Williams 2005).

When considering studying abroad, the first place to look for resources is at your local university. Most universities have study abroad programs or courses in a range of fields. These programs may be tailored to the field of study (history, ecology, etc.) or to the region itself (Rome, Cape Town, etc.). Similarly, language studies (potentially in a region you are interested in future conservation efforts) are also a great option for immersive study abroad programs. We recommend programs that allow students to work with or take courses with local students. We also recommend programs that allow students to visit small towns and large cities, wild areas and museums, and provide a breadth of experiences. Students should consider contacting older students in their own departments who have participated in study abroad programs. We suggest finding out about other students' experiences before deciding on a particular program. It is also important to note that there are not always as many opportunities to study wildlife abroad as there are for relevant study in other disciplines. This is one the reasons that international field courses, workshops, and volunteering opportunities focused on wildlife can be key to jumpstarting your international wildlife career.

Beyond your local university, many other colleges and universities offer study abroad programs that are open to other students. A great variety of programs are offered to students willing to explore opportunities outside their university. These include summer programs that allow you to visit a country to do research, volunteering, or service learning. Two highly reputable study abroad providers to consider are the Organization for Tropical Studies (OTS) and the School for Field Studies (SFS). These providers offer summer- and semester-abroad programs to study ecology and wildlife in the developing world that emphasize student-lead research. While there are many worthwhile study abroad programs, be sure to investigate them thoroughly to be sure they are accredited, affordable, and, if necessary, accepted for credit at your home institution.

Finally, many universities allow their undergraduate students to enroll directly in a foreign university. Directly enrolling allows students to take classes with local students and integrate into the country. Most universities with a department of ecology, zoology, wildlife conservation, or related field will offer classes on wildlife and ecology, often with a focus on local ecosystems. Some universities offer classes that include field trips or even field-based courses to local research stations that they operate. Even if a university does not have a relevant department, it may be worthwhile to investigate labs and projects at other local universities or nongovernmental organizations (NGOs) that accept volunteers. These might include conservation organizations, environmental education groups, or local parks.

Research Opportunities through Universities

Apart from study abroad, some universities offer funding to participate in or conduct independent research projects locally and abroad. Check your university's office of undergraduate research, professors in your department, and job boards (see below) for possible funding to conduct research projects. Some programs may ask you to develop a research proposal. Going through the process of writing a proposal as an undergraduate provides invaluable practice for grant-writing in the future. In many cases, undergraduate research projects may be part of larger projects and ongoing research. This allows students to be successful in a short period of time and to be a part of a larger team of researchers.

Another potential avenue for working on research abroad is as a field or lab technician for graduate students. Graduate students at many universities often conduct research projects in other countries and hire technicians. In addition, graduate students at universities based abroad may also be willing to hire American students or graduates, particularly in more developed countries. In some cases, technicians might have to pay all their own costs, and in others, some costs may be partially or fully covered (travel, room and board) or a small stipend may be offered. Some of these technician opportunities may be advertised or posted on job boards (see below). In other cases, one might consider researching professors and labs at foreign universities and writing to inquire whether there are opportunities to work or volunteer as a technician or assistant. Even professors that do not have opportunities available may forward your email to other colleagues who may be searching for assistants or technicians.

Job Boards

Job boards are a good place to start investigating potential careers and opportunities in international wildlife; they are a useful resource for finding opportunities both in the United States and abroad. Job boards serve as a clearinghouse for opportunities at universities, nonprofits and NGOs, private institutions, and government agencies. Job boards typically have listings for paid and unpaid work, temporary and permanent positions, and opportunities that may provide a good introduction to wildlife research. When looking at opportunities on job boards, be sure to check whether or not the posting has a project fee or any costs to pay up front in exchange for the opportunity to volunteer. Some opportunities with project fees are worthwhile, but others can be predatory and provide you with little relevant experience for a lot of money. Also, be sure to ask whether scholarship opportunities are available, or opportunities to volunteer additional work in exchange for help with travel costs and so on.

The following resources may be especially useful:

Texas A&M Department of Wildlife and Fisheries Sciences Job Board (http://wfscjobs.tamu.edu/job-board/) is an excellent resource. Most postings are based in the United States, but international jobs are sometimes advertised.

The Société Française d'Écologie (French Ecological Society (https://www.sfecologie.org/sfecodiff/consulter/) also runs a useful job board, which includes posts for jobs, graduate school positions, post-docs, funding, and conferences. The majority of positions are based in Europe, the United States, and Canada, but it does include some in Africa, Asia, and Latin America as well.

The Association for Tropical Biology and Conservation also runs a job board (http://tropicalbiology.org/blog/category/opportunities/) that includes a variety of opportunities, including jobs, internships, and graduate school.

The Ornithology Exchange (http://ornithology exchange.org/jobs/index.html) includes short-term positions, graduate school, post-docs, faculty, and nonacademic ornithological jobs.

Society for Conservation Biology (http://careers.conbio.org/jobs/) posts a variety of conservation-focused positions, including positions in academic, government, and nongovernmental organizations and institutions.

The Ecological Society of America, although US based, often has on their job board international opportunities and those that could lead to international positions (https://www.esa.org/esa/careers-and-certification/job-board/). The postings cover a wide range of opportunities across wildlife, ecology, geology, and other STEM fields.

Ecolog-L is a listserv run through the University of Maryland. You can set up a free subscription (https://listserv.umd.edu/archives/ecolog-l.html). The listserv has over 19,000 subscribers and provides information on grants, jobs, assistantships, and research opportunities.

The Wildlife Conservation Society also posts many jobs (in addition to other ways to get involved as an activist) in wildlife conservation internationally. Many of their positions are at zoos and aquariums, which offer a different route to get involved with education and wildlife conservation (https://sjobs.brassring.com/TGWebHost/searchopenings.aspx?partnerid=25965&siteid=5168).

The Wildlife Society, although focused on domestic wildlife conservation, occasionally has postings for jobs, internships, and volunteering opportunities that take place abroad (http://careers.wildlife.org/jobseeker/search/results/).

In addition, opportunities can sometimes be found posted on Facebook groups specific to country and/or taxa, such as Mastozoologia Brasileira.

Fulbright Program

The Fulbright Program is an educational exchange program run by the US State Department. It supports Americans for research abroad and citizens of other countries to study or research in the United States. The program operates in approximately 160 countries, but the list of eligible countries may change year by year and should be verified on the program website.

There are several types of grants under the Fulbright Program, and eligibility varies. The Fulbright Scholar Program is for professionals and academics with terminal degrees, while the Fulbright Student Program is for undergraduate seniors and early career and aspiring researchers with a BA, BSc, MA, or MSc and currently enrolled graduate students. Grant types include traditional research grants, which may be independent projects or part of a graduate degree, English-teaching grants (which may allow grantees to also pursue research projects), and the Fulbright-National Geographic Digital Storytelling Fellowship. Fulbright grants generally last one academic year (approximately 9–10 months).

A wide range of projects and topics can be proposed under the Fulbright programs, and scientific proposals are generally encouraged. Proposals are judged by both the US-based Fulbright Commission and the Fulbright Commission of the country being applied to. Preferences for subjects vary by country, and some may accept proposals for projects relating only to specific themes. The number of grants and percentage of proposals that are accepted vary by country. English-speaking countries generally receive the most applications and tend to be more competitive. Proficiency in the language of the country being applied to is generally an advantage.

Graduate School Abroad

One option for students wishing to conduct wildlife research abroad is to directly enroll in a program at a university abroad. Overseas graduate programs are a great way to become established as an international researcher. It is a chance to link yourself to an international institution, conduct research, and build a strong international network. Not only will you be working in a different country, you also have the potential to work with a different group of international students than you might experience at a graduate program in North America. Some excellent universities that offer wildlife and conservation graduate programs in English include James Cook University and Queensland University in Australia, Victoria University in New Zealand, University of Cape Town in South Africa, Hong Kong University, and the National University of Singapore.

Application requirements and funding may vary by country and university. In some cases, international, English-based curricula exist, while in others, students may be required to take classes or write in the national language. In many cases, dissertations may be written in English. The structure of a graduate degree can vary widely. In some cases, it may be similar to research-based graduate programs in American universities. However, doctoral degrees tend to be shorter (often only three years) outside the United States and often involve less coursework and teaching. Additionally, fairly or not, some of these shorter international doctoral degrees can be viewed as less rigorous.

In some countries, universities may charge tuition fees, while in others, university is free and even foreign students may be eligible for stipends. Funding and fees may also vary substantially between public and private universities. See the job boards section for resources on finding advertised positions. It is advisable to contact professors early in the application process and to communicate with department or university administrations for support, information on funding, visas, and so on. It is also important to note that the amount of available resources may vary widely by country and university. Universities, especially in less developed countries, may have limited resources for equipment and reagents or infrastructure for laboratory work. The European Union Erasmus Programme offers both master's (http://eacea.ec .europa.eu/erasmus_mundus/results_compendia /selected_projects_action_1_master_courses_en .php) and doctorate (https://www.em-a.eu/en/erasmus -mundus/erasmus-mundus-joint-doctorates.html) degree programs hosted by consortiums of universities, some of which are relevant to wildlife ecology and/or conservation. Some consortiums include universities outside the EU or may involve an internship outside the EU. Scholarship support is available for most degrees but is competitively awarded.

Teaching and Mentoring

Conducting wildlife research abroad can also present opportunities to give back and help build local capacity. An affiliation with a local university, as a part of graduate work, Fulbright grant, or other opportunity, can facilitate such a relationship. You may be able to teach or serve as a teaching assistant for a course or begin a journal club. There also may be a demand for editing and/or correcting by native English speakers.

International Research from US Institutions

Another option for conducting research in wildlife ecology and conservation as a part of a graduate degree is to join a lab that has a research program abroad. Many professors at universities in North America have ongoing collaborations and projects in foreign countries that provide opportunities for graduate students. You can get involved by contacting potential advisers and expressing your interest in international research. You can also check job boards for posted graduate degrees with an international research component. International research as a graduate student can be funded in a range of ways, including private grants (Idea Wild, National Geographic Explorer, the Explorers Club, Mohamed bin Zayed Species Conservation Fund, zoos and aquariums, taxa-specific grants, etc.) and government grants (NSF GRFP, NSF GROW, etc.).

National Science Foundation Graduate Research Opportunities Worldwide

Anyone considering graduate school in the United States should seriously consider and apply to the National Science Foundation Graduate Research Fellowship Program (NSF GRFP). Although fellows may enroll only in US-accredited institutions, they are allowed to conduct research and form collaborations abroad. The stipend and benefits are far greater than what most universities offer. Bringing outside funding gives graduate students a wide berth to pursue independent research, and the generous stipend also eases the financial burden of graduate school and research travel.

Further, NSF GRFP fellows (and non-fellows from a few other selected universities) are eligible for the National Science Foundation Graduate Research Opportunities Worldwide (NSF GROW) program. There are two main tracks: NSF GROW and NSF GROW with USAID. With NSF GROW, fellows establish a research collaboration with a university outside the United States. Currently, 17 countries in Asia, Europe, and South America are eligible. Each country has specific rules governing the length of the stay, host institution eligibility, and subject areas. GROW with USAID is focused primarily on a selected group of developing countries, which may vary by year. In some years, host institutions propose projects that may be searched in an online catalog, and applicants may consult with hosts to further develop or tailor the project to their interest and skills set. NSF GROW and GROW with USAID include a grant for travel and living expenses and financial support for the host institution. Previously awarded fellows may reapply as long as they remain eligible (NSF GRFP fellows in good standing).

Volunteering

Volunteering is ideal if you want to have a career in international wildlife ecology and conservation. Most people working in the field have spent time volunteering in many different regions and on different types of projects. Some volunteer positions can lead to permanent jobs and connections that will prove invaluable in the future. The field of wildlife conservation is quite small, and making connections within the network can give you a leg up in future job searches. Volunteering for any organization should be an opportunity to build skills, get exposure to jobs in the industry, and make connections. When looking for volunteer opportunities, you should be prepared to balance your specific career interests with a desire to learn and do new things and be exposed to novel career paths.

A great place to start volunteering is with local natural resource agencies. Most local, state, and federal agencies have internship and volunteering opportunities within their offices. These agencies (US Fish and Wildlife Service, US Geological Survey, state wildlife departments, and natural resource departments, etc.) tend to work locally and can provide invaluable experience. Opportunities at these agencies can range from hands-on work in the field with a range of taxa to working with the public to a variety of office jobs. And while you are unlikely to transition directly into international work from volunteering with a local agency, many have connections to international work and have employees that have worked or continue to work internationally.

In addition to government agencies, local private organizations have numerous opportunities to get involved as a volunteer. Often small local conservancies and private land management organizations need help with outreach and restoration and have volunteer and internship options. Larger international organizations such as The Nature Conservancy, Wildlife Conservation Society, Greenpeace, or Conservation International have wide-ranging projects both locally and internationally, and there are many ways to get involved. Additionally, taxa-specific organizations, such as Bat Conservation International, the Jane Goodall Institute, the Audubon Society, and others, sometimes have opportunities to volunteer locally and with international projects. Finally, the job boards discussed previously also regularly post volunteer and internship opportunities.

Nontraditional Wildlife Experiences

Typically, when people think about working with wildlife, they think about biologists, ecologists, and managers who work directly with wildlife and nature. People tend to forget that wildlife management and conservation is a complex field that requires dedicated people with an incredibly wide range of

skills. Whatever your interests are, there is a way to apply them to international conservation. If you are interested in working in international wildlife, consider some of the nontraditional routes that can lead to worthwhile careers.

Photography

Some of the people who have had the largest effects on international wildlife conservation have been photographers. The images of Ansel Adams, Art Wolfe, Joel Sartore, and countless other photographers have bolstered the conservation movement and maintained interest in wildlife across the world. Wildlife photographers and videographers have the ability to transport people around the world to ecosystems and landscapes they could never visit on their own. Photographers inspire people with their images and create an emotional connection to wildlife that enables long-term support for international wildlife conservation. If you are interested in photography, considering a career in wildlife photography could be very rewarding. Creating a career in wildlife photography requires an understanding of wildlife behavior and the fundamentals of photography, and a great deal of patience and persistence. Start by photographing wildlife locally, hone your skills, and take photography courses. Get your name and your portfolio out there. Enter your photos into contests, donate them to museums. Partner with researchers at local universities or agencies (such as the Audubon Society) to join them as a photographer. And while it is difficult to make a career as a wildlife photographer, it can serve as an excellent introduction to wildlife conservation.

Journalism, Science Communication

Another nontraditional route to international wildlife careers is via science communication and journalism. This includes writing, reporting, and the arts. There is a growing need to translate science into more digestible and entertaining formats for a broad audience. The skills necessary for both understanding and translating scientific writing are highly coveted and could lead to successful careers working internationally. This could include careers such as a scientific reporter in the media, creating publications or marketing for an international wildlife organization, working with museums or arts organizations to make science more accessible to visitors, or working with organizations to increase environmental behavior. The first step is to assess where your interests lie and where your skills are strongest. Is it writing, design, visual arts? You can then tailor your education, volunteering, and career trajectory to allow you to hone those skills and apply them to international wildlife work. Also worth mentioning is how wildlife photography can lead to or be combined with careers in ecotourism and leading tours for avid photographers and taxa enthusiasts.

Law and Policy

A final nontraditional route that can lead to much-needed change in international wildlife conservation is via law and policy. International wildlife conservation organizations are in desperate need of lawyers and policy makers with expertise and passion for wildlife conservation. International wildlife trade treaties, such as CITES (the Convention on International Trade in Endangered Species), and the bodies who govern them need skilled litigators and policy makers to ensure that wildlife are protected. National and international bodies such as the IUCN (International Union for the Conservation of Nature) and USAID (United States Agency for International Development) also need experts all over the world who are passionate about nature, sustainability, and development and the laws that govern them. If you have an interest in law, consider a career in environmental law. If you are interested in public policy, consider learning about international development and environmental management. Many government bodies and international organizations are regularly

looking for people to help navigate and litigate international wildlife issues.

Ecotourism

Ecotourism and guiding customers on tours is a bourgeoning field in wildlife conservation, and there is a need for naturalists with good communication skill who can connect with people. Students can get certified as Nature Guides and be hired by tourist lodges and research stations to take guests on guided nature walks, safaris, photography tours, hikes, treks, and the like. Here are a couple of programs that provide accreditation:

Texas A&M Wildlife Nature Guide certification: https://naturetourism.tamu.edu/online-courses/wildlife-guide-certification/
The Field Guide Association of Southern Africa: http://www.fgasa.co.za/Campfire Academy South Africa: http://campfireacademy.co.za/2017-nature-guiding-courses/
Ecoguide Certification: https://www.ecotourism.org.au/our-certification-programs/eco-certification-5/

Advice for Professionals

It is never too late to engage in a meaningful and rewarding international experience. In fact, early career professionals often bring cutting-edge tools and other demonstrable skills to the table, and these individuals can add substantial value to multi-investigator and interdisciplinary efforts. Seasoned professionals, of course, have the distinct advantage of bringing with them a proven track record of research productivity, management experience, and/or leadership acumen. With that said, engagement in the international wildlife and conservation arena will require some effort, regardless of years of experience or size and number of tools in the toolbox. We offer here some general advice and considerations for professionals wishing to venture for the first time into the international workspace.

Get connected and do not expect to get paid right away. An unsolicited invitation to work abroad on your dream project is unlikely; be proactive and assertive. Remember also that the relationships leading to the most significant opportunities are generally built over time, and immediate remuneration for services provided should not be a primary influence of one's interest in a particular activity. If you know of a person that you might like to work with or a project you would like to be involved in, express your interest. Ask a mutual friend or colleague to provide an introduction. Send emails and follow up with phone calls. Try to arrange a meeting time at a conference or workshop. There is no substitute for an in-person interaction, and, if practical, it might be worth your while to simply get on a plane to initiate a meaningful dialogue.

Be aware that many, if not most, international wildlife and conservation-related projects in the developing world are carried out through large networks of local and other governmental entities, NGOs, and international partners. Identify key players in these networks and seek ways to communicate with them. Go to international conferences (International Congress for Conservation Biology, ICCB; International Wildlife Management Congress; IUCN World Conservation Congress; and the International Mammalogical Congress), participate in international working groups (International Wildlife Management Working Group, IUCN Species Survival Commission Specialist Groups, and the International Affairs section of the Ecological Society of America), engage, and contribute. People will notice your level of interest and commitment, and your network will quickly expand.

Ways for Professionals to Get Involved

A way for professionals and organizations to begin building an international network of collaborators is to bring international scholars and wildlife managers to North America. In academic settings, we have found that international students and postdoctoral fellows can be exceptional resources to build and grow international research and teaching programs. Additionally, we recommend inviting international wildlife professionals to visit your department, institute, or agency. A week of formal and informal talks and socializing with potential collaborators from overseas can help catalyze relationships and projects that last throughout your career.

There are also a number of opportunities for professionals to live and work overseas. Fulbright, for example, offers a broad suite of opportunities (http://www.cies.org/), particularly for academics, which are likely to provide a foundation or springboard for sustained and meaningful work abroad. Every Fulbright experience is unique and influenced, of course, by the geographic location of the work detail and the language abilities of the individual scholar. We have yet to encounter a Fulbright scholar who did not speak highly of their experience in the host country.

Of course, a sabbatical can afford an accomplished faculty member many of the same rewards as a Fulbright with additional flexibility. We note that sabbaticals are not limited to those with a university affiliation. In fact, sabbaticals are increasingly commonplace in industry. The benefits of a sabbatical are many and include, for example, simply retooling, such as learning new skills and/or languages. A sabbatical also can and should provide an opportunity to interact with new colleagues, and in the context of this chapter, this would include colleagues in another country. Exposure to alternative perspectives from associates abroad can stimulate new ways of thinking that, in turn, pave the way for novel collaborations and innovation. If you are eligible for a sabbatical, we strongly encourage you to take advantage of the opportunity.

Creating study abroad programs can also help academics establish a long-term presence internationally. Increasingly, universities are encouraging faculty members to develop two-to-four-week study abroad programs for students. Even for faculty members who have not been abroad, it is often possible to find local contractors willing to help provide logistics and content for university courses. Before creating something from scratch, we suggest contacting faculty that have established successful programs for contacts and advice. Once the course is established, it can become a great platform from which to start research projects and build a network of collaborators.

Often, we fail to recognize the full range of employment opportunities in the wildlife and conservation arenas that exist outside of academia or government. Numerous consulting firms, for example, function on the international stage. A number of our colleagues travel across the globe to conduct ecological research and monitoring, as consultants or working for consulting firms. A host of institutions, such as the World Bank, United Nations Development Program, and the International Union for Conservation of Nature, also offer opportunities to engage in wildlife-related projects around the globe.

Summary

There is a world of opportunity in which to put one's skills to use and engage in meaningful activities that yield not only personal satisfaction, but also societal value. We reviewed a number of options for professionals and students interested in international wildlife management and conservation. The message of Chapter 1, on the opportunity for and necessity of international collaboration, is important, and the recommendations provided in this chapter are meant to assist in facilitating engagement. You are never too young or too far along in your career to start on the path to meaningful engagement as a wildlife professional overseas, but you have to have patience, determination, and willingness to take the first step.

LITERATURE CITED

Anderson, P.H., L. Lawton, R.J. Rexeisen, and A.C. Hubbard. 2006. Short-term study abroad and intercultural sensitivity: A pilot study. International Journal of Intercultural Relations 30:457–469.

Cheruvelil, K.S., P.A. Soranno, K.C. Weathers, P.C. Hanson, S.J. Goring, C.T. Filstrup, and E.K. Read. 2014. Creating and maintaining high-performing collaborative research teams: The importance of diversity and interpersonal skills. Frontiers in Ecology and the Environment 12:31–38.

Chieffo, L., and L. Griffiths. 2004. Large-scale assessment of student attitudes after a short-term study abroad program. Frontiers: The Interdisciplinary Journal of Study Abroad 10:165–177.

Cox, T.H., and S. Blake. 1991. Managing cultural diversity: Implications for organizational competitiveness. Academy of Management Perspectives 5:45–56.

Gibson, C.C., and S.A. Marks. 1995. Transforming rural hunters into conservationists: An assessment of community-based wildlife management programs in Africa. World Development 23:941–957.

Jehn, K.A., G.B. Northcraft, and M.A. Neale. 1999. Why differences make a difference: A field study of diversity, conflict, and performance in workgroups. Administrative Science Quarterly 44:741–763.

Muhumuza, M., and K. Balkwill. 2013. Factors affecting the success of conserving biodiversity in national parks: A review of case studies from Africa. International Journal of Biodiversity 10:1–20.

Nielsen, B.B., and S. Nielsen. 2013. Top management team nationality, diversity, and firm performance: A multilevel study. Strategic Management Journal 34:373–382.

Nilsson, D., G. Baxter, J.R.A. Butler, and C.A. McAlpine. 2016. How do community-based conservation programs in developing countries change human behaviour? A realist synthesis. Biological Conservation 200:93–103.

Williams, T.R. 2005. Exploring the impact of study abroad on students' intercultural communication skills: Adaptability and sensitivity. Journal of Studies in International Education 9:356–371.

Index

Page numbers followed by "t" denote tables; page numbers in italics denote figures.